THE ENTREPRENEURIAL ENGINEER

The authors – educators and successful entrepreneurs – wrote this textbook with the goal of maximizing your chance of entrepreneurial success. It is designed to encourage those who want to start a business and those who have already begun. It includes guidance, instruction, and practical lessons for the prospective entrepreneur. The book focuses on early-stage financing of a start-up company, beginning with an emphasis on constructing an effective business plan, including writing techniques to help convey your message, constructing an effective PowerPoint-type presentation, and preparing solid financial statements. This "why and how" of writing a business plan is followed by recommendations on raising outside capital. Important topics include developing a marketing strategy, recruiting and managing creative boards and managers, and retaining effective employees. Legal structures, negotiation strategies, and economic evaluation of opportunities are also discussed. The book concludes with a chapter on project management. The book includes many engineering economy topics, sufficient for those who will be taking the FE Exam.

Dr. Michael B. Timmons is a professor in the Department of Biological and Environmental Engineering at Cornell University. Dr. Timmons is a successful entrepreneur and has led successful fund-raising efforts from both angel investors and venture capitalists.

Dr. Rhett L. Weiss is the Executive Director of the Entrepreneurship and Innovation Institute at Cornell's Johnson Graduate School of Management, where he teaches graduate students in the entrepreneurship and venture capital areas. Before joining Cornell, Dr. Weiss had more than twenty-five years of experience in leadership and management roles.

Daniel P. Loucks is a Professor in the School of Civil and Environmental Engineering and the Institute of Public Affairs at Cornell. His research focuses on the application of systems analysis, economic theory, ecology, and environmental engineering to problems in regional development and environmental quality management.

John R. Callister is the Harvey Kinzelberg Director of Enterprise Engineering in the Mechanical and Aerospace Engineering Department at Cornell. He teaches Introduction to Entrepreneurship and Engineering Enterprise in Mechanical and Aerospace Engineering and the School of Operations Research and Industrial Engineering.

James E. Timmons has more than thirty years experience managing complex development programs with budgets up to $250 million, with more than 1,100 personnel under his management responsibility. He is experienced in both the public and the private sectors.

THE ENTREPRENEURIAL ENGINEER

How to Create Value from Ideas

Michael B. Timmons

Cornell University

Rhett L. Weiss

Cornell University

Daniel P. Loucks

Cornell University

John R. Callister

Cornell University

James E. Timmons

CAMBRIDGE
UNIVERSITY PRESS

CAMBRIDGE
UNIVERSITY PRESS

32 Avenue of the Americas, New York, NY 10013-2473, USA

Cambridge University Press is part of the University of Cambridge.

It furthers the University's mission by disseminating knowledge in the pursuit of
education, learning, and research at the highest international levels of excellence.

www.cambridge.org
Information on this title: www.cambridge.org/9781107607408

© Michael B. Timmons, Rhett L Weiss, Daniel P. Loucks, John R. Callister, and James E. Timmons 2014

First published 2014

Printed in the United States of America

A catalog record for this publication is available from the British Library.

Library of Congress Cataloging in Publication data
Timmons, Michael B. (Michael Ben), 1949–
The entrepreneurial engineer : how to create value from ideas / Michael B. Timmons, Cornell University,
Rhett L. Weiss, Cornell University, Daniel P. Loucks, Cornell University, John R. Callister, Cornell
University, James E. Timmons.
    pages   cm
Includes bibliographical references and index.
ISBN 978-1-107-02472-4 (hardback) – ISBN 978-1-107-60740-8 (pbk.)
1. New business enterprises.   2. New agricultural enterprises.   3. Research, Industrial.
4. Technological innovation.   5. New products – Marketing.   6. Entrepreneurship.   I. Title.
HD62.5.T57   2014
658.1′1–dc23   2013012170

ISBN 978-1-107-02472-4 Hardback
ISBN 978-1-107-60740-8 Paperback

Additional resources for this publication at www.cambridge.org/Timmons

This book effort is dedicated to all those who have tried … whether they succeeded or failed

[Photo courtesy NASA]

Contents

7 Financial Accounting

13 Project Scheduling: Critical Path Methods, Program Evaluation, and Review Techniques

Preface

Where Are We Going?

If you ask a group of engineers, most of them would probably tell you that they have thought about starting a business. Most did nothing. Some others did some research and preliminary planning. Some of these then invested some personal funds in creating the shell of a company in some legal format. A rare few took their companies forward to the point that they became operational businesses.

Most of us will see opportunities to start a business. This book is for those among us who are contemplating the start of a business and those that may have already taken that big first step. As teachers and successful entrepreneurs, our goal in writing this book is to help you the reader maximize your chances of entrepreneurial success. You will find guidance, instruction, and practical lessons that will assist you, the prospective entrepreneur, with your first steps toward realizing a dream.

The focus of this book is primarily on early-stage financing of a start-up company. Most texts focus on venture capital–type financing, but this is not an option at the beginning. Most companies will be started with personal funds or funds from family and friends. If you can advance to the stage of having a product and some form of sales, then you can attract an angel investor. At this point you'd be looking for hundreds of thousands of dollars, which is still way below the funding level at which a venture capitalist becomes interested. Although the book focuses on the early-stage financing, the lessons and principles presented will be generally applicable as your company advances in maturity and funding needs.

Basis for Content

The book is based in part on the senior author's real-world experience of having started a technology-driven indoor fish company in upstate New York, an international consulting company, two nonprofit organizations, an indoor shrimp farm in Kentucky in 2007, and three other current ventures involving high-tech fish farms. In writing this book, we have incorporated real examples from our own companies and from other entrepreneurs to motivate and teach.

Our own entrepreneurial experience began with the tilapia fish company, whose objective was to produce food fish for the mass market, using a combination of proprietary and patented technology that we considered disruptive technology at that time and still do. We thought we could revolutionize the aquaculture industry (the jury is still out on that) by providing the means to produce fish in a sustainable and profitable manner. The company was launched in 1997, and became one of the largest indoor fish production facilities in the United States, producing more than 1 million pounds (500 ton) of tilapia for several consecutive years in the mid-2000s.

We first raised capital to fund this venture by self-financing and personal debt, which was then followed by "angel investor" financing, and then venture capital financing. Along the way, we became quite familiar with terms such as preferred stock, cram-down, down round, perfected loan, and disruptive technology – to name just a few. So our writing reflects the belief that there is no substitute for one's own experience, but that there are large dividends to be had (saving money by reducing your mistakes) by learning what we can from the experiences of others. Hopefully this book will help you minimize costly mistakes.

Book Content

An important part of this book is the emphasis on constructing an effective business plan and preparing solid financial statements. This "why and how" of writing a business plan is followed by recommendations on how to raise outside capital. Then, we suggest some writing techniques that will be helpful in conveying your message of opportunity to others. After that, we examine methods of developing your marketing strategy, recruiting and managing competent individuals at the creative and managerial levels, and, as the business grows, recruiting and retaining effective employees. Other aspects of starting and running a new business are also addressed, such as selecting a legal structure for your business, negotiating strategies, and performing economic evaluations of opportunities. We conclude with a chapter on project management.

In order to help you get a feel for some of the entrepreneurial concepts, we have included problems that require economic analysis of different opportunities or challenges facing a start-up venture. For example, chapter problems include how to determine internal rates of return (IRR) for a new enterprise or its net present value (NPV), how to evaluate alternative capital investments, and how to calculate basic financial terms such as EBITDA that would include depreciation calculations. The Appendix includes the standard equations used in engineering economic analysis for ready reference plus a collection of financial tables related to depreciation and classifications of real property and the interest factors for discrete compounding calculations.

Engineering Students and the FE Exam

For the engineering students using this text, at the end of each chapter is a section on engineering economics. After you've gone through the first ten chapters, you will have covered all the engineering economics concepts necessary to pass this portion of the Fundamentals of Engineering (FE) test conducted by the NCEES (National Council of Examiners for Engineering and Surveying; see www.ncees.org). The engineering economics material makes up 8% to 10% of the entire FE test, so it will behoove you to master these concepts. This economics text material is not intended to replace a full academic textbook on engineering economics. Please consult other texts on this material if you require more depth, for example, *Fundamentals of Engineering Economics* by Chan S. Park or *Engineering Economic Analysis* by David Newnan, Ted Eschenbach, and Jerome Lavelle.[1]

Supplements to the Text

Some chapter topics have been abbreviated by moving this material to the Web support page for the book provided by Cambridge Press; see www.cambridge.org/Timmons. The Web supplementary materials include a primer on dinner etiquette so you won't embarrass yourself at your first angel investor meeting by using the wrong fork or not knowing what to do with your cloth napkin at the conclusion of a meal. These chapter supplementary materials are noted by chapter number. Several business plans are posted there, most of which were written by students who enrolled in the senior author's entrepreneurial class at Cornell. We have collected all the engineering economic material from the end of each chapter and placed this on the website as a convenient reference tool to the engineering economics materials. We will also use the book's website to provide updated information. Additional financial tables are also provided there.

Acknowledgments

This book was made possible by senior author Michael B. Timmons being awarded a Clark Professorship in 2001 from Cornell University (see http://entrepreneurship.cornell.edu/). The Clark Professorships were established in 1992 by J. Thomas Clark (1963, MBA 1964) and Nancy Williams Clark (1962, MEd 1964) to expand and/or enhance the educational opportunities for undergraduate, professional,

---

[1] Chan S. Park, *Fundamentals of Engineering Economics*, 2nd ed. (Upper Saddle River, NJ: Prentice Hall, 2008); and David Newnan, Ted Eschenbach, and Jerome Lavelle, *Engineering Economic Analysis*, 11th edition (New York: Oxford University Press, 2011).

and graduate students in the areas of entrepreneurship, small business, and personal enterprise. These professorships provide funding for a limited term to faculty members selected in a competitive process to develop new courses, integrate entrepreneurship into existing courses, or engage in research in the areas of new business creation, innovation, and/or development. Professors (nineteen total) from eight Cornell schools and colleges have held Clark Professorships. These funds were used to develop this textbook, and we are extremely grateful for this help and sponsorship. Additional thanks go to Mr. John Jaquette, Director of the Cornell University Entrepreneurship and Personal Enterprise (EPE) Program. In addition to financial support for writing the text, the EPE program provides constant moral support and encouragement to each of us to continue to provide and improve the educational experience for Cornell students with an interest in entrepreneurship.

We would also like to specifically thank Professor Deborah H. Streeter, Bruce F. Failing, Sr., Professor of Personal Enterprise in the Department of Applied Economics and Management at Cornell University. Professor Streeter is part of the university-wide Entrepreneurship and Personal Enterprise Program. Dr. Streeter provided valuable guidance and input when we were first constructing the text, and she provided much of the marketing research content in Chapter 3. Also at the end of each chapter are several video clip links that are from her collection of 15,000 in-depth interviews on all aspects of entrepreneurship. That collection was the basis for a start-up company called Prendismo, LLC (see http://www .prendismo.com).

As we developed the new course, BEE 4890 Entrepreneurial Management for Engineers (a four-credit one-semester course), we relied on our personal experiences, class lecture notes, and several texts that we had read at that time as we developed lecture materials and material for the course text. In fact, one of the fish farm's angel investors had asked us if any of us had read Clayton Christensen's book on disruptive technology, *The Innovator's Dilemma*.[2] We all said no, but we'd pick up a copy ASAP. The book by Christensen was all about disruptive technology, which we then realized was describing our fish-rearing technology. So, some of the content in Chapter 2 has to recognize that some of the content was based upon Christensen's writings. The other two books that we found very useful in developing our course were *Leading the Revolution* by Gary Hamel and *Smartups: Lessons from Rob Ryan's Entrepreneur America Boot Camp* by Rob Ryan.[3] Both books are very good reads because they discuss the challenges of a start-up venture. Again, we are certain that some of the writing in the text that we've written will capture some of the essence of these two books; footnotes where material has been borrowed are presented throughout the text.

---

[2] Clayton Christensen, *The Innovator's Dilemma* (Boston: Harvard Business School Press, 1997).

[3] Gary Hamel, *Leading the Revolution* (Boston: Harvard Business School Press, 2000); and Rob Ryan, *Smartups: Lessons from Rob Ryan's Entrepreneur America Boot Camp* (Ithaca, NY: Cornell University Press, 2002).

Final Advice

We hope you enjoy this text. A few of our business associates have read some of the chapters and they say it reads like a story. That was our intention. We hope you can benefit from some of our mistakes and our successes. For sure, applying the lessons discussed herein to your own start-up venture will save you time and money. Both are critical to the success we wish you as an entrepreneur.

About the Authors

Dr. Michael B. Timmons, PE. Dr. Timmons has been successful in several entrepreneurial efforts and has led successful fund-raising efforts from both angel investors and venture capitalists. He received his BS from The Ohio State University, his MS from the University of Hawai'i, and his PhD from Cornell University. In 1997, Dr. Timmons's first major entrepreneurial experience was launching an indoor fish farm in upstate New York, called Fingerlakes Aquaculture LLC. This venture was a 1-million-pound-per-year tilapia farm. Building this farm required Dr. Timmons to personally guarantee a substantial percentage of the debt  equity used to build the farm. It was the lessons learned from this experience that prompted Dr. Timmons to write this textbook. The project was sponsored by J. Thomas Clark, a Cornell alumnus, who endowed Cornell University with funds to support faculty efforts that contributed to the entrepreneurial education of its students. Much appreciation and gratitude is extended to Mr. and Mrs. Clark for their financial help. Dr. Timmons was a J. Thomas Clark Professorship of Entrepreneurship and Personal Enterprise from 2002 to 2008.

Dr. Timmons is one of the recognized authorities on recirculating aquaculture technologies (coauthor of the text *Recirculating Aquaculture*, with J. M. Ebeling; see www.c-a-v.net). Dr. Timmons has worked in aquacultural engineering for twenty-five years as a researcher and extension specialist. He was one of the founders of the Aquacultural Engineering Society and has served in several officer positions including president. In the spring of 2012 (once again with his wife's blessings), Dr. Timmons launched Kentucky Natural Organics LLC, a 300,000-pound-per-year tilapia farm and hydroponic producer of leafy greens and tomatoes. These experiences may prompt another textbook! Stay tuned.

Dr. Rhett L. Weiss. Dr. Weiss is the Executive Director of the Entrepreneurship and Innovation Institute at Cornell's Johnson Graduate School of Management. In addition, he serves on its faculty, teaching or advising graduate students in the entrepreneurship and venture capital areas. Before joining Cornell, Dr. Weiss had more than twenty-five years in successful leadership and management roles.

He has served as a bank COO, directed a consulting practice at a Big-4 accounting firm, practiced law at a major international law firm, and holds a software and business method patent. Throughout his career, he has been involved in transactions worth more than $30 billion and in dozens of entrepreneurial ventures and innovation initiatives. Among them, he is Chairman and CEO of DEALTEK, Ltd., which he founded in 1999. From 2005 to 2010, Dr. Weiss served as Senior Team Leader – Strategic Development for Google Inc. He was chief designer and negotiator of several large strategic acquisition and development projects for
Google's global infrastructure, typically involving its legendary data centers. He headed key initiatives, contracts, and relationships with businesses, utilities, governments, economic development agencies, landowners, and other stakeholders. Dr. Weiss also created and conducted negotiation training workshops in Google's offices globally. He is a frequent presenter and writer on negotiations, entrepreneurship, and economic development. The transactions that he has led often have received industry recognition, awards, and coverage in business and technology news articles. Dr. Weiss holds a BS in Management with Honors (finance major) from Tulane University, a Doctor of Jurisprudence from the College of William & Mary, and an MBA-level Executive Certificate in International Business from Georgetown University. He also has held board chairman and other leadership positions at professional, educational, and civic organizations.

Daniel P. Loucks. Professor Loucks teaches in the School of Civil and Environmental Engineering and in the Institute of Public Affairs at Cornell. His research focuses on the application of systems analysis, economic theory, ecology, and environmental engineering to problems in regional development and environmental quality management. On leaves from Cornell he has taught at a variety of universities in the United States, Australia, and Europe and has been a consultant to international, governmental, and private organizations dealing with regional development issues
here in the United States as well as abroad. He served as an economist at the World Bank and as a research scholar at the International Institute for Applied Systems Analysis in Austria. Loucks was an aviator in the navy and eventually commanded the largest naval air transport squadron in the country, having detachments at Naval Air Facility, Detroit, Michigan; Andrews Air Force Base, Maryland; and Naval Air Station, Willow Grove, Pennsylvania. He was elected to the National Academy of Engineering in 1989.

John R. Callister. Callister is the Harvey Kinzelberg Director of Enterprise Engineering in the Mechanical and Aerospace Engineering Department at Cornell and a lecturer in the Sibley School of Mechanical and Aerospace Engineering and the School of Operations Research and Industrial Engineering. Callister teaches Introduction to Entrepreneurship and Engineering Enterprise, which provides a solid introduction to the entrepreneurial process to freshman students. One objective of the course is to examine and develop skills in the engineering work that occurs in high-growth,
high-tech ventures. This course is the first in a program in entrepreneurship and personal enterprise for engineers, but enrollment is not limited to students who elect to enter this program, and students from throughout Cornell University are welcome to enroll, regardless of major. Callister also teaches Entrepreneurship for Engineers, which examines the issues and skills necessary to identify, evaluate, and start new business ventures. Topics include competition, strategy, intellectual property, technology forecasting, product design and development, sources of capital, and manufacturing. Callister also teaches other engineering courses in the M&AE and ORIE departments.

Callister is a cofounder and vice president of Foxdale, Inc., an equipment leasing company in the blown-film polymer industry. He has participated in several entrepreneurship conferences in the past year. He joined the Cornell faculty in 1999.

James E. Timmons. Timmons holds a master's degree from Central Michigan University in Management and Supervision. His career to date spans more than thirty years, during which time he has managed complex development programs with budgets up to $250 million and other activities with more than 1,100 personnel under his management responsibility. He has experience with both government and private-industry programs and more than twenty-five years of experience in providing and managing the operational and maintenance support
requirements and structure for complex mechanical systems. He has developed, produced, maintained, and implemented support plans and programs for a variety of Department of Defense weapons systems and vehicle systems. Most recently, he has been involved with the definition and documentation of the mechanical, electrical, and fluid interface requirements between several of the International Partner modules to the International Space Station.

1 Entrepreneur's Primer

Entrepreneurship is the recognition and pursuit of opportunity without regard to the resources you currently control, with confidence that you can succeed, with the flexibility to change course as necessary, and the will to rebound from setbacks.

Bob Reiss[1]

1.0 Entrepreneur's Diary

I remain grateful for some good early advice. I had several inventions that had been taken to financial success by others. The benefactors sometimes said thank you. This made me all the more eager to take one of my own ideas to the real world. I had convinced myself that financial success and glory eagerly awaited me.

About fifteen years ago, I launched my first major entrepreneurial effort that revolved around proprietary technology used to produce seafood indoors in an environmentally responsible manner at a competitive price. The market for such a technology had to be huge I kept reminding myself, because seafood was (and still is) a significant contributor to the U.S. trade deficit.

Fortunately for me, before I committed my home equity into this venture, I got some good early advice from Walter Haeussler, president of the Cornell Research Foundation. He told me that before I started my fish company, I needed to convince at least one other person to invest in the opportunity. Walter assured me that I made a very persuasive case and that he was so convinced by the oral presentation of my start-up business "idea" – that he was almost ready to be an investor himself! But almost does not provide any start-up capital!

So, I took Walter's advice and convinced one other person (actually two): my two brothers. A loving brother will tell you *no* if you do not have a viable business opportunity. This forced me to develop a solid business plan, to analyze the market, and to develop a cash flow statement, and so on – before investing our precious cash.

But before you run out and try to start a business, try assessing your personal suitability for entrepreneurship as you read this chapter. What does it take? What

---

[1] Bob Reiss, *Low Risk High Reward* (New York: The Free Press, 2000), p. 6.

1

common personal characteristics do many successful entrepreneurs share? This chapter is designed to help you answer some of these relevant questions before you take the plunge into entrepreneurship by asking the right questions and letting you reflect on your answers. We'll also present some of the fundamental rules for starting a successful venture and some other early advice for you if you are just starting out in one of these adventures, and will provide some background on raising capital, your need to stay focused, and some discussion on when you might quit your "day job."

1.1 So You Want to Be an Entrepreneur?

Some people seem to be made for entrepreneurial adventures whereas others prefer the security of a paycheck from somebody else. One thing is certain: if it weren't for entrepreneurs, then most of us would still be working on a farm without the many simple things that enhance our quality of life, such as light bulbs, flush toilets, cell phones, personal computers, and a host of other products demonstrating technological advancement.

Small businesses of fewer than twenty people account for 90% of all businesses operating in the United States. Small businesses employing fewer than 500 employees accounted for 50% of all wage workers and 45% of the payroll of all businesses.[2] We rely on entrepreneurs to start new businesses and create jobs and thereby keep our economy moving. On the down side, only about 50% of these new businesses are still open after four years.[3]

Why would anyone become an entrepreneur when failure seems so common? Ask yourself, "What's the difference between me and Bill Gates or Michael Dell (besides several billion dollars of personal wealth)?" For many people, their fear of failure or the consequences of failure is the biggest thing holding them back. For some, it is simply the economic risk of failure; for others, it may go beyond the money to include the loss of status. These reasonable concerns are why most people are unsuited to the entrepreneurial role. Recall the words from the definition of entrepreneurship at the beginning of the chapter: *the recognition and pursuit of opportunity without regard to ...* If you really can do these things, you might be the next great entrepreneur.

Most people will never know, nor will they care about your failures, only your successes. Failure can be the best way to learn, provided you don't repeat your mistakes. Ask yourself, "Who learns more, the person in a successful venture or a failed one?" Both learn but history is full of examples of entrepreneurs who

---

[2] The gross national product (GNP) in 2012 was $15.7 trillion; the 2005 GNP was approximately $11 trillion; and the 2002 was $10.5 trillion. Other portions of the GNP are composed of government spending, investments, trade deficits, and so on. U.S. Department of Commerce: Bureau of Economic Analysis, http://research.stlouisfed.org/fred2/data/GNPCA.txt.

[3] Brian Headd, "Redefining Business Success: Distinguishing between Closure and Failure," *Small Business Economics* 21 (2003): 51–61.

have failed at several attempts, only to later succeed. They were persistent but flexible in that they changed their approach or business focus in response to the lessons learned.

In some cases, it isn't so much the "fear of failure" holding us back as it is the "cost of trying" in time and dollars that prevents us from the pursuit of our entrepreneurial dream. As we move forward in our careers, we accumulate increasing numbers of commitments (family and community) and so have less and less time available to pursue alternative endeavors. So, don't hesitate, begin now! Your best time to launch your entrepreneurial adventure might be as you leave college or right now, before the trappings of life limit your flexibility.

Core Values of an Entrepreneur

Clearly, not everyone is suited to being an entrepreneur, just as not everyone is suited to being a professional athlete. Someone without a certain set of personality characteristics is not destined to become an entrepreneur. This section is designed to give you a brief glimpse at those distinguishing characteristics associated with becoming a successful entrepreneur

RISK TAKER

Most successful entrepreneurs are willing to take risks despite the chance of failure. As a result, the entrepreneur will do their best to ensure against failures and to consider and take steps to minimize the impacts of identified threats. The entrepreneur is self-confident, but not so confident that they assume that their talents will carry them to business success without a good product and planning. Risk taking is often necessary as the entrepreneur will never have all the capital necessary to minimize all threats or to capitalize on all opportunities. There will be times when a lack of capital will require a decision on which of many critical needs receives the necessary resources and which do not. So as an entrepreneur, you will need to be able and willing to make these hard decisions. Some people can't do this and are frozen by various aspects of the process, for example, need more data, what-ifs, and so on.

PERSISTENT WITH FLEXIBILITY

It is a truism that there is nothing more important for achieving success than persistence. The successful entrepreneur doesn't give up. The successful entrepreneur believes that his or her venture can succeed and has the will to face and overcome those inevitable bad days and temporary setbacks. Persistence is not, however, blind. The pursuit of a doomed strategy, without flexibility, is a recipe for disaster. For example, when your product or pitch is better suited to a different market, make the adjustment! U.S. car manufacturers continued for decades to ignore U.S. demand for highly efficient automobiles and continued to give up market share to the Japanese and other Asian car manufacturers. It took the gas price rising above $4 per gallon before the recognition took hold. So, be not only persistent but also flexible "enough" to recognize when a change of strategy is needed.

CREATIVE

Successful entrepreneurs are able to recognize an opportunity and then creatively transform this recognition into a business venture. You will constantly be faced with problems during the process of starting a new business. You won't find the answer in a textbook, so being able to creatively solve these challenges yourself (or team) is essential. Being "smart" also helps, but smart and creative are not synonyms. Employing smart people is a good idea, because they are adroit at helping to implement the good ideas that you develop to address your problems. Walt Disney was a creative genius, but his brother Roy was the implementer of many of Walt's ideas, and the Walt Disney Company was the result.

ENERGETIC

Starting a business requires enormous amounts of time over a sustained period. The successful entrepreneur is a high-energy person with a strong work ethic. Fourteen hour days, six days a week are common. But, if you love what you are doing, the days will go by faster than you can imagine. In fact, you will have to be careful to provide balance in your life as the new start-up venture can become all-consuming. You can't work seven days a week for any length of time and simultaneously maintain your health and/or a family life. Be sure to save some time for some balance in your life. If your start-up results in financial success, but you end up with having no one to share this with, how happy will you be? People rarely go to their grave wishing they had spent more time at the office and less time with their families.

COMMUNICATOR AND LEADER

Creating a successful company requires the efforts and contributions of others, as well as your own. You need to develop and inspire your team by communicating your vision to them. In the beginning of your start-up, investors must see your passion and your will to succeed. If you cannot inspire a potential investor with your vision, the result is predictable, they will not invest. Investors avoid what they perceive to be a boring opportunity, even if the financial returns look large and probable. After obtaining start-up funds and launching your company, you must be able to constantly update your vision of success to your employees so that they embrace the company's mission with as much passion as you do. Without dedicated passionate employees, you will not be successful. One of our greatest political leaders of the twentieth century was President Ronald Reagan, who was also known as the great communicator. These attributes go together, leading and communicating.

INTEGRITY (HONESTY)

A business must be based upon integrity. Your company must be one that others trust. You must be a person of your word. You must lead your company by your own personal example of how you conduct business both inside and outside the company. Everything will not go smoothly, but being truthful and honest in your dealings with your employees and your clients/customers will go a long way toward finding a solution to problems in times of stress.

NOT ON THE LIST (GREED)

Some books on entrepreneurship include "greed" on the preceding list, because greed is sometimes aligned with market domination. I do not agree with this assessment. Greed in my definition is obtaining profits without regard to the needs of your own employees. Market dominance is a business strategy whereas greed is associated with one's moral compass. Although I believe that no level of greed is appropriate, I also believe that being success driven is a good thing. Our tenet is that an entrepreneur should strive for the natural returns that accrue to the founders of a carefully considered, well managed, and successful enterprise. The successful entrepreneur must go after the identified market with an attitude that you want your company to become the dominant player and capture all of this potential market. This is an infrequent result, but it does happen. Microsoft is a good example of a single company dominating a market segment. Is Bill Gates's wealth a result of his greed or his successful business strategy or his driven nature? Did you know that Bill Gates for many years has been the first- or second-largest donor to charities in the world?

1.2 Fundamental Rules for Success

For your business to become successful, you must recognize and follow a few fundamental rules. Your company's ability to sustain growth and profitability will depend on them. Study these rules carefully, and consider why they are essential. Do this before you begin your quest as a checklist to measure your idea and plans before you start, and again, after you have started, as an assessment of where your business is actually positioned. If you do not generate sufficient sales and profits, you will not be in business very long. Therefore, reflect on these rules before you take that entrepreneurial risk of spending your own and other people's money. Then, once you've started, if the situation does not develop as you had expected, what do you do? Do you pursue the original business strategy to the bitter end, or do you adapt?

SOLVE A PROBLEM

The number one rule is that your product should solve a customer's problem. If you make life better by solving some specific problem, customers will buy your product if it is priced right. This premise should be the foundation of your company. You will need to market your product, and the simplest definition of marketing is "solving customers' problems profitably." If you can solve a problem for someone, you have a good chance of business success, particularly if there are several people with the same problem.

> The simplest definition of marketing is "solving customer's problems profitably."

PROVIDE CONVENIENCE

Closely related to solving a customer's problem is making life in some way more comfortable or convenient. Customers have a long history of paying price premiums for convenience. Consider bottled water. If fifteen years ago, you had told someone that you were going to start a company that sold bottled water at prices similar to carbonated bottled drinks, their reaction would most likely have been "I can get a drink of water for free. Why would I pay for it?" Well, they were clearly shortsighted and did not see the entrepreneurial opportunity. What about packaged salad mixes that cost four or five times the cost of the individual components? These products were nonexistent a few years ago, and now they occupy significant shelf space in grocery stores. Or, backing up the clock some more, an automatic transmission versus a standard shift (with clutch) provides convenience. Today, a standard shift transmission is a special order vehicle!

DIFFERENTIATE

Once you have something that you think solves a customer's problem or makes someone's life more convenient, you are home free, right? Wrong. You would be startled by the number of similar products already in the market place or the number of start-ups that are somewhere in the development process working on a similar product to your own. Simply put, you are either trying to fill a need that the general population does not know they have (market not developed yet), or you will be trying to take customers away from a current direction of behavior or buying patterns. Why, then, should they choose your solution over an alternative? *Differentiation* is the buzz word here, and it is an extremely important concept to grasp. How is your product different? You must be able to articulate this difference to potential investors to the point that they can see the same vision that you see, and then you must be able to deliver this differentiated product to the market once your company has launched.

There is a common trap here, however, particularly for engineers who continue to provide differentiation beyond what the customer understands, needs, or wants. Differentiation loses its value and meaning when the features and functionality of your product exceed the customer's practical demands. Offering the customers more than they ever need in a product is a fatal error; for example, take an automobile that is $5,000 more expensive than a competing model because it has several "built-ins" that not everyone would want (hence the concept of base price plus optional add-ons). Eli Lilly in partnership with Genentech produced a 100% pure insulin using recombinant DNA technology. This was a magnificent feat of pharmacological engineering, and Eli Lilly was proud to offer this as the branded product Humulin™. They priced their new product at a 25% premium over existing generic animal-based insulin, which routinely contained 10 ppm impurities.

However, because the vast majority of insulin users could tolerate these impurities, there was no compelling reason for them to switch to the admittedly more pure but also 25% more costly Humulin™. Eli Lilly's original marketing effort failed, even though it had clearly differentiated its new product from the current competition. The "cost-benefit" comparison did not work in their favor. Increased cost for no benefit is not a winning approach. Then, Eli Lilly repositioned its Humulin™ product by marketing it to the fraction of insulin users that *do* have resistance to the generic products, and it is still on the market today (but not at 25% price premiums; see www.humulin.com).

TARGET NEW MARKETS

Existing large markets are dominated by *Fortune* 500 companies. These companies have lots of capital that they can use to squash any new start-up that appears to be threatening their current hold on their market. Recognizing this reality, your product and company success will be much easier to achieve if you have a product that will benefit an identified a market segment that is currently being ignored either because it is too small for the large companies or the large companies have not yet discovered it (and most likely they will not). This identification of un-served market is sometimes called "greenfield" competition or markets.

Small markets do not satisfy the growth needs of large companies. Don't be so proud of your technological breakthrough product that you believe that you can compete in a mature, well-established market. This would make you a competitor of the large companies from the very start, and they will smother you if you start to show signs of breathing success. Ideally, the major players in the existing market place will ignore you and the market you are addressing.

Your technology might fall into a category that is complementary to an existing technology, which works pretty well sometimes, particularly if the large company decides to buy your start-up. Generally, however, it is better for the start-up to stay in the greenfield category or to develop strategies to avoid head-to-head competition with the established providers.

If you still think your product can compete in the established big market, for your product, you should have a cost of production approximately ten times lower in comparison to the competing product to have a realistic chance of success. The big competitor will do all it can to take your customers by lowering its margin (temporarily), producing a bottom end low cost product (similar in cost to your product), and offering other sales incentives that you cannot match.

However, success sometimes happens for the little guy. Case example: Kionix (Ithaca, NY) produced a new technology Micro Electro-Mechanical Systems (MEMS) based accelerometer (the triggering device) for auto air bags that sold for $2.50/unit versus the $25/unit that the market dominant company (75% saturation) Breed Inc. received for their old technology (mechanical) airbag accelerometers. In fact, Breed was first offered the MEMS accelerometer technology in 1998 by Greg Galvin (CEO of Kionix), but it turned down the offer. Kionix decided to attack the market by itself with their ten times lower production cost advantage. To cap the story, Breed Inc. filed for Chapter 11 bankruptcy protection in 2000, which was only four years after Galvin had offered it the MEMS air-bag-trigger device. Greg Galvin will tell you that his company's initial market focus was not on accelerometers for the automotive market and that he really wanted to subcontract the device to Breed Inc., but he demonstrated the attribute on our list of required characteristics for entrepreneurs – persistence with flexibility. (This example also provides a peek at a topic we address later: disruptive technology.)

AVOID THE 1% MARKET SHARE TRAP

A final caution on this attribute of targeting new markets, don't fall into the trap of seeking 1% of a large market with the logic that the market is so large, you certainly can capture a few million dollars of it. It won't happen, and no angel/venture capitalist (VC) investor will ever accept this flawed argument. Start-ups need to think about capturing 30% to 50% (or more) of a new market and about being the dominant player.

In general, customers are reluctant to do business with small-new companies trying to compete in large markets served by large, established companies because of their fear that the small company will disappear (and this fear is well justified in most cases), leaving them with none of the product that they need, and consequently in a weak bargaining position relative to the dominant supplier. Conversely, if you are the only company selling some new technology that makes a customer's life easier, that is more convenient, or that solves a problem for them, they will buy your product and tolerate these other issues.

DON'T RUN OUT OF MONEY

It is a very unusual start-up that "gets it right" the first time out with its product. Even though you spent three years refining your product and doing market surveys, don't be surprised when a few months after formal product launch, your sales are not nearly reaching their projected levels and feedback is coming in that the product is not meeting the customers' needs. What now? Quit? First, this situation is almost to be expected. What customers say they want and what they tell you they will pay while you are doing your initial market research are often far removed from actual practice. Your sales, marketing, or product form will go through some iterative process of trial, learning, adjusting, and trial again.

> Number one rule in a start-up: Don't run out of money!!

Guessing the right strategy at the outset isn't nearly as important to your ultimate success as conserving enough cash so that you get a second or third attempt at iterating to a successful strategy and/or product. This means that you need to raise sufficient capital before you launch your business. But what is sufficient? ***Be prepared to fail early and inexpensively*** in the first stages of your company so that you have the resources to make the necessary corrections and survive to success. That's the definition of enough start-up capital.

LUCK

Yes, I almost forgot, it really helps if you have some good old-fashioned luck along the way. Although good luck might best defined as opportunity meeting preparation, sometimes it is just being in the right place at the right time. You will be hard-pressed to find an entrepreneur whose business successes and the path toward these successes got there without any luck at all, and most successful entrepreneurs will readily admit to this. That luck can take many forms: from having just the right members on your board of directors, which resulted in your next round of funding; from landing a major "breakthrough" deal instead of losing it either to the competition or to the prospect's choosing to do nothing; or from simply being in the right place at the right time instead of totally misreading the market's interest in the venture's offerings. Let's just say entrepreneurship is not easy, and hard work alone will rarely be the only necessary ingredient to ensure success.

> Good Luck:
> When opportunity meets preparation.

1.3 Rules to Raise Capital

As one progresses along the entrepreneurial path of fund raising, the odds continue to mount against you as you try to secure the needed capital, and thus, your need for persistence. So, what is a reasonable approximation of your odds of obtaining the needed funding? Most VCs will tell you that they receive in excess of 5,000 business plans each year and that they funded only two or three of the projects. Angel investors don't get this many, but some still evaluate 100 or more potential opportunities each year.

Think about this. If the odds are this bad, there really is no statistical correlation between submissions and funded proposals. In practical fact, you will be frustrated and sadly disappointed if you expect to obtain financing from either of these potential sources based on simply submitting a well-written proposal for review.

To get a fair evaluation of your proposal, and thus a reasonable chance of securing funding, you must establish some personal connection to the potential investor. At each level of success, your most likely connection to the next level of investment funding is from one of your current investors who knows someone at the next level of funding, for example, a family member who knows an angel investor, and then an angel investor who knows a VC.

Investors are experts at what they do, which is investing. They are not experts in every technology that they try to evaluate. They know you are putting the best "spin" possible in your presentation, and they know that they do not know all of the soft spots. Investors generally apply the following principles in looking for investment opportunities:

- new activity in proven areas of rapid expansion (you will recall the "dot-com" explosion in the late 1990s; this investment approach is called the herd mentality of the VC),
- a proven track record of the entrepreneur and other principals in the proposed venture, and
- compliance with whatever other guidelines they personally may have adopted to help them contend with the load of proposals on their desk.

You will improve your odds of success by approaching investors with realistic expectations on what levels of funding they would typically make. Understand the differences between family and friends, angel investors, and VC types of investors. In this book, as opposed to most entrepreneurship books that focus on VCs as target investors, we focus mostly on the family/friends and angel-investor stages and provide more details about how to go about this in later chapters.

An Often-Missed Point

A point generally missed by most new entrepreneurs is that the same steps and processes are necessary to raise capital from any type of investor. For example, raising money generally starts from family and friends (or yourself) and requires a well-written and well-researched business plan, for example, identification of markets, a marketing strategy, cost analysis, and so on. Then, raising capital at the next level, typically angel investors, means you need the same preparation as before, but now you should have a real prototype and probably some customer feedback and a more refined marketing plan. And for the highest and last level of fund-raising from VCs, they will have expectations of you having demonstrated some initial market success.

Some start-up businesses may never need venture capital or capital beyond friends and family. My point here is that the same principles for raising capital apply whether your targeted investor is a VC, an angel investor, a rich uncle, or a local bank. Each deserves your best effort.

1.4 Keeping Your Focus

Lest you become distracted as you proceed, stick to the original plan (sometimes called a business plan, hence you need a plan if you are going to stick to it and we'll help you to write one in Chapter 8). You probably entered into your current entrepreneurial activity because you and your partner(s) understand it better than any other activity at this point because that has been your focus. That focus and the considerable investment that you have made in time and energy and money in your current activity have enabled you to perceive other opportunities. These other opportunities can be included in a revised business plan, as circumstances and abilities permit. Eventually, they may in fact overcome the original activity. Howard Hughes started off with an oil-drilling bit. Ray Kroc started off with a 15-cent walk-up hamburger, and so on. Build your plan and sell it, if you think there is more and or easier money in alternative opportunity, either incorporate it or leverage it, or perhaps strike up a strategic alliance.

Often, the perceived peripheral opportunity will not create capability (read upside opportunity), but it just services the core activity focus (what makes your plan unique). Much like the difference between Mighty Maids (national chain providing maid services and cleaning functions) and your team of entrepreneurs, the support service activity can be done by someone else who is willing to do it for less than it costs you. But the cleaning company cannot do what your company is trying to do.

Staying on Plan

It is very easy to become distracted by what appear to be better or more profitable alternatives as you implement your business plan. Almost every potential investor will tell you in no uncertain terms that the challenge of implementing your business must be a lot easier than you make it out to be, so why do you need that much money? Why is this? The investor is probably seeing the "city from a distance."

Let's use an analogy. Imagine that you are going to a distant city. You stop to admire it from the top of a distant hill, and you are pleased by its beautiful skyline and symmetry of its streets and boulevards. As you get closer, you begin to notice that the streets are crowded and dirty, there are one-way streets that cause you problems, the buildings show evidence of damage or decay, and your overall impression of the once-beautiful city begins to change. Then, you see a bridge you didn't notice from farther away. It seems to go to a different part of the city than you had originally intended to go to, and the boulevard you see across the bridge is wide and nicely landscaped, and full of promise, but you can't see exactly where it will lead you. Do you go there, just because you see a bridge and it looks better than what you are looking at now? Or, do you press on to your original destination (staying on plan).

In this analogy, you might be trading a set of details that you understand and don't like, for the comfort initially provided by being distant from an alternative destination where details are not yet apparent. It would be logical to assume that when you get closer to your revised goal, you'll again see details you don't like, but now you're in a different place, where maybe you didn't intend to be when you started. You'll probably be even more discouraged than you were before. This whole process (getting there) would be a lot easier if someone had gone before you and had kindly provided directions, but in this case, you are the first. That's because you are a start-up. There are no reliable directions; just whatever plans you made based on the best data available.

The same analogy applies here as you commence on your start-up, when opportunities to expand your business model appear such as some type of satellite program effort, a franchise type activity appears to increase your market share, or some co-lateral opportunity is seen that just seems to be a good idea. Before you do any of these, consider the following negative aspects (the potholes in the streets):

- diversion of your time and attention from your primary operation (to a related product or market opportunity),
- diversion of financial resources away from your primary goal,
- liability resulting from taking this added direction or dimension of helping or enabling others in the same industry, and
- eventual competition (potential) from the source you helped create (satellite program or franchise).

Each of these areas is a valid and important concern. For early start-ups, try to keep things as simple and manageable as you can, for now. Kids will play games wherein they will invent very complex processes and rules, simply to have something to do, and failure is part of the fun. Start-ups can't afford that, because real money is at stake. The "rules" can't be too complex, as that opens up avenues for mistakes. Nor can the rules be too simple, or something important will be overlooked. They can't be too bold, or the risk factor may become too great. Nor can they be too conservative, or opportunities will be lost, perhaps never to be recovered.

So, maybe it is time to remember what the original intention of your entrepreneurial effort was. Go back to your business plan, and review it, revise if necessary or warranted. All the other possibilities that you are considering are like that "city in the distance," as you approach that particular city, you will see the details that reveal themselves as you get closer. Maybe you will decide not to go there at all, but right now, the city is visible on the horizon, and it has bright lights and a beautiful skyline. You just can't see the trash in the gutters yet. Be careful where you go and where you step.

1.5 Advice to Students

Many of you reading this text will be undergraduate or graduate students at some university. Your interest in entrepreneurship continues to escalate and probably you have taken one or two introductory courses in entrepreneurship and/or business management. Somewhere along the way, you have developed or imagined a product or service around which you think you can launch a company. You are probably quite aware that your university has immense resources, but where do you start? How do you get involved? How do you make those critical connections?

I am fairly confident that your university will have networking clubs through which you can make critical contacts and begin to establish your business network. Make contact with your university's patents and licensing division; it is in the business of technology transfer and commercialization, which leads directly to start-up opportunities. Do a Web search for "local" entrepreneurial clubs and outreach groups. If there is not one or if none appeal to you, then start your own club. You'll be surprised how much interest there will be.

Recently a senior student, Yongjoon Choe, BS-ECE 2013, in my Cornell entrepreneurship class faced this same problem. Yongjoon and his friends created a new club called "Cornell Startup Spirits." Their club hosts a career fair on campus where local start-ups and the start-ups operated by Cornell alumni have an opportunity to interact and brainstorm with others that have the same core interests. Just being at one of these type events can lead to your finding the "missing link" that you need to launch your company.

1.6 When Do You Quit the Day Job?

As we end this chapter, you might be asking yourself the question, "So, when do I really cast caution to the wind and start my first entrepreneurial venture?" Good question. Engineers tend to be conservative, so it might even be harder you. Remember this, all successful entrepreneurs started with the same amount of experience you have at starting a business – none! Many entrepreneurs will start their enterprise while maintaining some other form of employment, aka, the proverbial "day job." As a result, you end up doing two full-time jobs. This is one good way to mitigate your risk because it will allow some time for the new venture to gain some momentum before you jump on with your own salary needs. Note, that investors typically do not like this approach as they want to see their money being invested with someone who is going to live and breathe the start-up business twenty-four hours a day. And they want your economic lifeline to be 100% critically tied to the success of the business they are investing in. I do not agree with this approach, but sometimes it may be your only choice.

At some point, you will probably become time and energy constrained and will not able to give the proper guidance, direction, and activity required to implement your plans in the way they should be being implemented. This will probably be tolerable for a while, but at some point, you will become consumed in a time-intensive effort in literally every area of product development that exists for your company as it tries to go to the next level. In most cases, you will be doing virtually simultaneous development of concept, procurement, design, marketing, system fabrication, and distribution, among other functions, of some enterprise.

At this point, you the entrepreneur must take serious personal reflection on the next phase of your company. You must understand that the next phase of your company's growth is a huge undertaking that will require the continual best efforts of several people. You must evaluate your personal availability to this project. How much of what needs to be done can you really do? How much of what needs to be done are you actually qualified to do? Your vision most likely is overcrowded with the possibilities for the future. This is a very good thing, particularly if you can secure funding to do what you want and decide to do. So when do you quit your "day" job? That's a tough question that can only be answered by you. It is hoped that this book will help you to make a sound decision and if you do decide to go forward, the book will contribute to your success. Happy entrepreneuring!

1.7 Additional Resources and References

Christensen, Clayton M. 1997. *The Innovator's Dilemma*. Boston, MA: Harvard Business School Press.

Headd, Brian. 2003. "Redefining Business Success: Distinguishing Between Closure and Failure." *Small Business Economics* 21 (2003): 51–61.

Rogers, Everett M., 1962. *Diffusion of Innovations*. New York: Free Press of Glencoe, 367.

Ryan, Rob. 2001. *Smartups*. Ithaca, NY: Cornell University Press.

1.8 Video Clips

One of the unique aspects of this textbook is access to the Cornell entrepreneurial collection of video interviews of successful business leaders that have been broken down into short one- to two-minute clips on specific topics. We select two or three clips and highlight them in each chapter. The e-Clips collection was created by Dr. Deborah Streeter and contains thousands of video clips that were created from in-depth video interviews or presentations by entrepreneurs and other experts involved with supporting entrepreneurship and small businesses. Interviewees include start-up and experienced entrepreneurs, venture capitalists, bankers, angel investors, and employees of start-up companies. The collection of clips was turned into a private company in 2011, Prendismo, LLC.

Directions for viewing clips on the Prendismo Collection (formerly from www.eclips.cornell.edu):

- Visit http://prendismo.com/collection/. (You must subscribe with the site for full access to all features.)
- Click on the **Subscribe** at the top of the site. Students receive reduced rates.
- Once you have subscribed, type in the name or title of the clip in the **Search** at the top right corner of the screen.
- Click on the clip you wish to view.

- Click on the play icon to view the clip. If you have not become a subscriber, you will only view the first 20 seconds of the clip before being prompted to log in.

1. Michael Crooke (Patagonia)

   - BS in forestry and an MBA from Humboldt
   - Worked in the outdoor world for Yakima, Moonstone Mountaineering, and Pearl Izumi
   - CEO of Patagonia
   - Creative leader working for environmentally sustainability
   - Title of Video Clip: "Michael Crooke Shares Thoughts on the Definition of Entrepreneur"

2. Sheila Johnson (Black Entertainment Television)

   - Cofounded BET, the first cable channel aimed at African Americans
   - Executive vice president overseeing community and philanthropic activities
   - BET was acquired by Viacom in 1997 for $3 billion
   - An accomplished violinist, musician, and educator
   - Title of Video Clip: "Thoughts on Entrepreneurship"

3. Laurie Linn (Communique Design & Marketing)

   - Fifteen years of experience in marketing consultation for major financial institutions
   - Worked with Chase, Bank One, and smaller financial institutions around the world
   - Discusses challenges of having her husband as her business partner
   - Title of Video Clip: "Laurie Linn Shares Thoughts on Entrepreneurship"

1.9 Classroom Exercises

SITUATIONAL ENTREPRENEURSHIP

1. To get you "thinking" like an entrepreneur, consider the following situation (specifically designed to help you focus on the thought process instead of distracting you with the facts of a specific business scenario):

You and a friend are in the middle of a weeklong camping trip. At this point, you are at least a day or two from civilization. Naturally, you both have your hiking boots and other outdoor gear on. Your sneakers, other clothes, and typical camping equipment are in or on your backpacks. Collectively, you two have enough food and other provisions to last the rest of the week, with an extra couple days' supply just in case you get lost. Presently, you both are hiking through the woods together, toward the end of the day, just chatting away. Basically, your attention is now turning to finding a good site to set up camp for the evening. Up ahead is a small but flat open meadow with a nice view. It is perfect for pitching your tents and for making a fire to cook dinner. You two start to scout out the meadow when you hear some crackling and rustling. Just then, from the tree line at the opposite side of the meadow, a bear emerges. It takes one look at you both, smells the food in your backpacks, and starts chasing after you both at full speed.

Consider these questions:

What do you personally do? Do you and your friend work apart or together as a team? What are your decision choices? How much time do you even have to think about them? What do you want to achieve? What kind of "SWOT" Analysis (Strengths, Weaknesses, Opportunities, and Threats) races through your head? What or who is your competition in this situation? How do you define survival, success, and failure in this situation? Are survival and success equal or different here? Are survival and failure equal or different here? Do your answers to the preceding questions vary at particular points in time and relative positions of the players during this rapidly unfolding saga?

Whatever your answers happen to be at this moment, if this textbook and the professor do their jobs, your answers to this same question likely will change as you learn more about entrepreneurship.

2. What is "survival" in business? What is "success" in business? What is "failure" in business? Are these terms a description of a fact or of an opinion? How does one recognize any one of these or know that either one has occurred? Is either one important to measure? Are survival, success, and failure static or dynamic concepts? Does business survival equal business success? Are these two terms at least synonymous? At any phase? At all phases? If not, then do these concepts diverge at some phase or threshold? If they diverge, when?

3. How far can the business *progress* without luck? In other words, even if you "do everything right," at what point, if any, does the business still need luck to *succeed*? At many points? At all points? Does it need luck just to *survive*?

4. The concepts of survival, success, and failure are so interrelated that often only a very fine line separates them. To see for yourself, take a quick walk through a slice of business history. Specifically, look at the current composition of the "*Fortune* 500," "Global 500," "S&P 500," or "Dow Jones Industrial Average." Then, go back in time in ten-year increments for fifty to seventy years and look at those lists. Notice that many once-prominent companies now no longer exist (at least on those lists). Notice, too, the companies that emerged on the newer lists, but were nonexistent on previous lists. Why do so many companies "come and go?" Bad leadership? Bad management? Bad timing? Bad markets? Bad access to capital (financial, human, and/or technological)? Bad luck? Some combination of these factors? Other factors?

ETHICS

5. What is the difference between greed and effective business strategy?

6. Write your own personal definition of integrity and how it should guide your business activities. (This is a take-home assignment. Save this for your own personal reference statement and review it periodically. Are you becoming a better person?)

1.10 Engineering Economics

The primary reader we are focusing this text on is the engineering student. Engineering is a licensed profession and as such, graduating engineers are encouraged by their engineering college to take the first step toward licensure, the Fundamentals of Engineering (FE) test. The FE test has a four-hour morning component called the general test and a four-hour afternoon component of either general or specific discipline tests (mechanical, civil, electrical, industrial, environmental, and chemical).

Engineering economics makes up 8% to 10% of these tests. So, over the course of reading the first ten chapters of this book on entrepreneurship, we will introduce all the concepts you will need to pass the FE exam. Besides being useful to pass the FE exam, engineering economic analysis should be useful for any aspiring entrepreneur. We are using the same notation that is used in the National Council of Examiners for Engineering and Surveying (NCEES)-supplied reference manual that you use when you take your Fundamentals of Engineering exam (first step to engineering licensure). We have also extracted all the engineering economics sections from the end of each chapter and have compiled them into a single file available on the website that supports the text; see www.Cambridge.org/Timmons.

Evaluating Alternatives

I've just given a general overview of entrepreneurship and how you identify a potential entrepreneurial opportunity. It is very easy to convince yourself of all the glorious upside potential and that you are going to dedicate the next several years working seventy-plus hours per week until you have created a success out of your "great idea." You are aware now that the general beginning point is that you will have to have a product, a service, or an invention to show to someone you want to invest in your company. You've probably exhausted your supply of friends and family financing, and this means you will be looking to someone outside your intimate circle to a broader universe of potential investors.

Investors will evaluate any business opportunity by how much it is worth in real dollars. What does "real" mean? Let's flip the tables for a moment. Let's assume you are the wealthy individual two young entrepreneurs approach you as an investor. One entrepreneur shows you that your $1 million investment will be worth $10 million in twenty years and the competing entrepreneur says your same investment will be worth $5 million but only in three years. Which one is the more attractive investment?

Engineering economics is one way to evaluate these opportunities. The major problem affecting the comparison is time and inflation. So, to properly evaluate investment alternatives, it is best to put all the investment opportunities back or forward to the same point in time. This is addressing a principle called the **_time value of money_** and it gets back to that earlier question of what are "real" dollars. Think of "real" here as being able to compare alternatives at the same

Table 1.1. Compound Interest Variable Definitions

i   = Interest rate per interest period (normally periods are one year)

i_n = nominal interest rate

i_c = compound or effective interest rate

f   = inflation rate

d   = rate of return earned above the rate of inflation

n   = number of interest periods

P   = a present sum of money

F   = a future sum of money. The future sum F is an amount, n interest periods from the present, that is equivalent to P with interest rate i.

A   = an end-of-period cash receipt of disbursement in a uniform series continuing for n periods, the entire series equivalent to P or F at interest rate i

G   = uniform period-by-period increase or decrease in cash receipts or disbursements; the arithmetic gradient

g   = uniform rate of cash flow increase or decrease from period to period; the geometric gradient

r   = nominal interest rate per interest period*

m   = number of compounding subperiods per period*

*Normally the interest period is one year.

point in time. To do this, we'll have to introduce several engineering economic principles and definitions to allow you to function. By the end of Chapter 8 of this book, we'll have introduced and given you a chance to work with all the important economic factors that affect your entrepreneurial venture, for example, taxation, depreciation, alternative choice selection, net present value, pro forma statements, internal rates of return, and equivalent annual costs, among others. First, let's start off with some of the basics by defining the variables that are generally used with compounding interest and some associated definitions. See Table 1.1. We will use these throughout the rest of the text.

The Compound Interest Tables are given in the Appendix for the following:

- Single Payment: F/P, P/F
- Uniform Payment Series: A/F, A/P, F/A, P/A
- Arithmetic Gradient: A/G, P/G

You should become familiar and proficient with using these tables because it will make doing the problems much easier for you.

1.11 Interest, Compounding, and Cash Flow Diagrams

Simple Interest and Compound Interest

The two most basic parameters affecting the real value of money are interest rate and time. You've heard the comment "All it takes is time and money ..." Well, they go together. The value of money is time dependent; value is always changing with time. The key is to determine the equivalent value of alternatives at a common point in time. Or, some present value is equal to some future sum of payments. This is called *equivalence*. Equivalence is the basis of the preceding examples in which investors were deciding whether to invest in your company or to allow an improvement to increase productivity in an existing company, by putting values to the same point in time to determine the impact of inflation, loan interest rates, or required rates of return.

First, what is ***interest***? Interest is defined as the return on an investment or the fee charged by a lender by a borrower for the use of borrowed money, and which definition you use depends on whether you are in the position of the lender or the borrower. Most commonly, the interest is expressed as a percentage of the borrowed money, and a compounding period is specified. An example of this would be 8% interest compounded monthly. The interest if there are no compound periods is called ***simple interest***, and interest over multiple compound periods is called ***compound interest***.

The interest rate applicable in an economic transaction is affected by the perceived risk or probability of nonpayment in the transaction. A bank may lend money to a low-risk customer at 7.5%, but an entrepreneur may have to borrow money at much higher rates from potential investors, because high risk and bank loans are mutually exclusive.

An interest rate has three components:

1. A risk-free component based on an economic concept called the marginal productivity of capital. Many economists believe this rate is about 3% or 4% in a risk-free and inflation-free environment.
2. A risk component to compensate for uncertainty or the possibility of nonpayment.
3. An inflation component applicable in periods of inflation. In periods of inflation, lenders demand higher rates of return to compensate for the decline in purchasing power between the time money is lent and the time it is paid.

SIMPLE INTEREST

Simple interest is interest earned only on the original principal. The formula for calculating simple interest or earnings is

$$E = P \cdot i \cdot n,$$

(1.1)

where

E = simple interest or earnings
P = principal (amount borrowed or lent)
i = interest rate per year
n = number of years or fraction thereof

Example 1.1 – Simple Interest

If you borrow $100 for one year at 10% per annum, the interest for the year is

$$E = (\$100)(0.10)(1) = \$10$$

If you borrow $100 for three months at 10%, the interest is (note the assumption is made of 10% per annum, so three months is 3/12, or 0.25).

$$E = (\$100)(0.10)(0.25) = \$2.50$$

Note how the interest percentage must be converted to a decimal, and how the time must be expressed in terms of years or fraction thereof. We can demonstrate these ideas in a cash-flow diagrams as shown in the following figure (note that an arrow pointed downward represents a cost or an investment and an arrow pointed upward indicates a receipt or positive income flux).

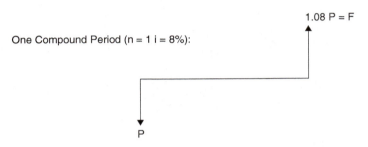

Multiple compound periods, still simple interest ($n = 4$, $i = 8\%$):

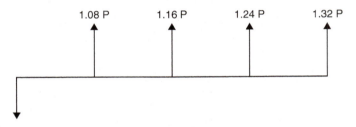

COMPOUND INTEREST

Compound interest is interest that is earned on both principal and interest. When interest is compounded, interest is earned on the original principal and on the interest accumulated for the preceding periods.

The *compound period* specifies the amount of time before the interest is added to the borrowed amount, and this new sum becomes the total principle owed. The alternative to this would be to have no compound periods, and this would mean that just before the debt is repaid, the interest fee would be added. If this were done, then there would only be 8% of the original added; in other words, you would be left at the end with 108% of the original. With compound periods, however, you add 8% to the investment every period and the amount increases. Therefore, shorter compound periods mean that the interest added to the borrowed amount will increase. The shortest compound period possible is continuous compounding, and we will present formulas for this later in the chapter. Most banks or investors use a daily, bimonthly, or monthly compound period.

Example 1.2 – Compound Interest

You borrow $500 on January 1, 2008, at 10% interest, compounded annually. You pay the loan on December 31, 2012. What will your total payment be? Answer: you must pay $805.26.

Year	Principal & Interest	Interest Added for Year	Total Payoff at End of Year
2008	$500.00	$50.00	$550.00
2009	$550.00	$55.00	$605.00
2010	$605.00	$60.50	$665.50
2011	$665.50	$66.55	$732.05
2012	$732.05	$73.21	$805.26

So how does this relate to our cash-flow diagram? Instead of displaying cash payments or receipts for when they are actually made, it is more convenient to sum all of the payments and receipts for each compound period because no interest would have been accrued during this time. In the diagram, each compound period is shown as a gap between payments, and shown at the end of each period is the total amount of money added or subtracted during that period. This money is added to or subtracted from the money from previous periods, which has already gained interest. Additionally, the payments, receipts, and interest over several periods can be summed to create a larger period in the diagram, which we cover later in this section.

So let's start with the easiest cash-flow diagram possible, in which there are no compound periods at all, or it may also be defined as a single compound period. Usually there is an initial payment, which we will call P, and then a future payment to repay the original payment plus the accrued interest, and we call this F. Interest is usually represented by a lowercase i, which in this case represents the effective interest rate per period (discussed later in this section). For one period, P is shown at the beginning of the period and F is at the end.

P and F are proportional to each other by the following relationship:

$$\frac{F}{P} = 1 + i \tag{1.2}$$

Or, in other terms, F is simply 100% of P plus i added as a fee for using the lender's money.

Therefore, if we only know P and i, we can solve for F with the equation:

$$F = P \ (1 + i). \tag{1.3}$$

Now let's add some interest periods to this equation. At the end of every interest period, the interest on the total amount up to that period is added to the sum. So if you had two interest periods, you could solve for F by the equation:

$$F = P \ (1 + i) \ (1 + i). \tag{1.4}$$

If you define the number of interest periods as n, then for n interest periods the equation for F would be

$$F = P \ (1 + i)^1 \ (1 + i)^2 \ldots (1 + i)^n. \tag{1.5}$$

Or more precisely,

$$F = P \ (1 + i)^n. \tag{1.6}$$

Similarly, if you rearrange this equation to solve for P you get

$$P = F \ (1 + i)^{-n}. \tag{1.7}$$

Symbolically, the equations for F and P are usually represented by the following functional notation:

$$F = P \ (F/P, \ i, \ n) \ and \ P = F \ (P/F, \ i, \ n). \tag{1.8}$$

The term (P/F, i, n) can be thought of as finding P given F (and vice versa), with interest i and number of interest periods n, and it symbolizes the equation $(1 + i)^{-n}$. The symbolic notation is usually used to avoid confusion and create more organized work in tables and equations. Notice that if you treated the P and F as units, then they would cancel each other out to make the equation true. Also notice that opposite equations will cancel each other out:

$$(P/F, \ i, \ n)(F/P, \ i, \ n) = 1. \tag{1.9}$$

The P and F equations are fairly simple, but you will see the need for notation when we present much more complex equations later in this section. Tables are given in the Appendix for various combinations of P, F, and n (periods, generally years) and different interest rates.

1.12 Nominal Interest and Effective Interest

Before we go further with this idea, we must clarify the different types of interest that you may encounter. Interest rates can be specified for years, months, weeks, days or any other length of time. The interest period n is usually one year and we will always assume the period is one year unless otherwise noted. Years ago, only nominal interest rates were used. Nowadays, the passage of truth-in-lending laws has resulted in the widespread publication of effective interest rates in advertising. This has resulted in a lot of confusion. It is hoped that the following discussion and examples will clarify this for you. **Compounding** is a process where accumulated interest from sub-periods is used to calculate an interest rate for an interest period. For example, if you have an interest period of n, but your interest is compounded daily at an interest rate of r, then each m, where m is the number of compounded subperiods within n, must sum to some higher interest rate for period n. We call the resulting interest rate a **compound interest** or **effective interest**. Effective interest rate is a single rate of interest that truly represents the interest earned over some period, typically one year, where interest is paid in each of some set of subperiods, for example, each month. Therefore, for compounding, we must specify a certain number of subperiods m, or **compound periods**, within the interest period since the interest rate given usually corresponds to a given compound rate. If, however, there are no compounded subperiods within an interest period, then we calculate the **nominal interest** for that period. Note that it is possible, however, not usual, for a compound period to be longer than an interest period; in this case, you would make the number of subperiods (m) a fraction equal to one divided by how many interest periods there are per compound period.

NOMINAL INTEREST RATES (APR)

Nominal interest rates are stated in terms of annual rates. Annual percentage rate (**APR**) is the annual rate that is charged for borrowing (or made by investing), expressed as a single percentage that represents the actual yearly cost of funds over the term of a loan. This includes any fees or additional costs associated with the transaction. Read more at

http://www.investopedia.com/terms/a/apr.asp#ixzz239ZzSVqa.

When interest is compounded for less than a year, the following steps are necessary before compound interest amounts can be calculated:

1. Divide the nominal interest rate by the number of compounding periods per year to determine the interest rate per compounding period.
2. Multiply the number of years involved by the number of compounding periods per year to obtain the total number of interest compounding periods.

Example 1.3 – Nominal Interest

Nominal annual interest rate:	12%
Interest compounded:	Annually for 3 years
Interest rate per compounding period:	0.12 ÷ 1
Number of compounding periods:	3 years × 1 compounding per year = 3 periods

Example 1.4 – Nominal Interest

Nominal annual interest rate:	10%
Interest compounded:	Monthly for 15 years
Interest rate per compounding period:	$0.10 \div 12 = 0.0083333$
Number of compounding periods:	15 years $\times$ 12 compounding periods per year = 180 periods

We distinguish nominal from effective interest by using i_n to symbolize nominal interest and i_c to symbolize effective or compound interest. These variables are defined in equation form by

$$i_n = r \cdot m \text{ (nominal interest)} \tag{1.10}$$

$$i_c = (1+r)^m - 1 = \left(1 + \frac{i_n}{m}\right)^m - 1, \text{ (compound interest)} \tag{1.11}$$

where

i_n = nominal interest rate,

i_c = compound or effective interest rate,

m = number of compound subperiods,

r = interest rate for each compound subperiod,

If the interest period (n) is one year, then

$i_a = i_c$ = effective annual interest rate (or APR).

Sometimes you may need to use both of these formulas to calculate an annual interest rate.

EFFECTIVE INTEREST RATES (EAR)

Many advertisements for automobile loans, home mortgages, and other financed consumer items, as well as financial investment products, include the effective annual percentage rate or "*EAR*." The EAR is merely the interest rate that would result in the same amount of interest paid per year if the interest were compounded annually (instead of monthly, quarterly, etc.).

Clearly the EAR is of very little value, but we must live with it, apparently. EARs are not useful in most calculations, so we must convert to nominal rates first (APRs).

The effective interest rate can be found from the nominal interest rate (APR) with the following equation:

$$i_{eff} = \left(1 + \frac{i_{nom}}{c}\right)^c - 1, \tag{1.12}$$

where

i_{eff} = effective annual interest rate or EAR

i_{nom} = nominal annual interest rate

c = number of compounding periods per year

Example 1.5 – Nominal-Effective Interest Conversion

For the previous example, we can find the effective annual interest rate (EAR):

$$i_{eff} = \left(1 + \frac{0.10}{12}\right)^{12} - 1 = 0.1047 = 10.47\%.$$

Note that if the interest is compounded monthly, the effective interest rate is greater than the nominal rate. This is merely the effect of the compounding of the interest over time. Hence, the two following statements are equivalent, and both are true:

1. The nominal interest rate is 10% per annum, compounded monthly.
2. The effective annual interest rate is 10.47%.

Unless it is explicitly stated, you should assume that interest rates are nominal, annual interest rates. This is also the case in the FE exam. Note that this nominal annual rate is only equivalent to the effective rate for annual compounding.

Example 1.6 – Nominal-Effective Interest Rate Conversion

Given a daily interest rate of 0.05% with monthly compounding, find the annual interest rate.

Solution:
First, convert the daily rate to a nominal monthly rate, and then this rate to an effective annual rate.

Assuming thirty days in a month,
$r = 0.0005$ $m = 30$
$i_n = r\,(m) = 0.0005(30) = 1.5\%$
With twelve months in a year,
$i_c = (1 + 0.015)^{12} - 1 = 0.1956.$

Therefore, the effective annual interest is 19.56% versus 10.25% (365 days * 0.05%/day). If we compounded 0.5% daily for 365 days, the effective annual interest rate would be 20.02%.

Example 1.7 – Loan Repayment

You need $20,000 to raise enough capital to start your own web-based consulting company. You plan to take on a four-year loan from the bank, to be compounded annually at 5%. If you have to return all the principal and interest in one lump sum at the end of the loan, how much would you have to pay the bank?

Approach A:

Amount owed:

At the end of the first year = 20,000 (1.05) = $21,000

At the end of the second year = 21,000 (1.05) = $22,050

At the end of the third year = 22,050 (1.05) = $23,152.50

At the end of the fourth year = 23,152.5 (1.05) = $24,310.13 (to 2dp)

Therefore, you have to pay back the bank $24,310.13.

Approach B:

Amount owed at the end of four years =

20,000 $(1.05)^4$ = $ 24,310.13

1.13 Problems

Note: the Compound Interest Tables are given in the Appendix:

F/P, P/F,A/F, A/P, F/A, A/G, and P/G.

1. Your mother has offered to lend you $20,000 at a simple interest of 5% (per year) to help you start your first business. If you plan on returning the principal and interest together at the end of four years, how much interest would you have to pay in addition to the principle of $20,000?

2. In this new business you are starting, you also decide that you need additional capital so you take a loan of $ 5,000 from the small business association (SBA). The loan has to be returned (paid back in full) exactly twelve months later, with interest. If the interest rate is 12% nominal, but compounded monthly, how much interest do you have to pay back?

3. Given a nominal interest rate of 10%, what are the equivalent effective interest rates given the following:
   a. annual compounding
   b. semiannual compounding
   c. quarterly compounding
   d. monthly compounding
   e. weekly compounding
   f. daily compounding

4. You are raising $10 million in equity capital from venture capitalists (this is probably the minimal amount of investment they would consider). Their company statistics are that it averages a 30% return on investment. However, the statistics also say that of their investments, only three of ten make money, with the rest either going bankrupt or limping along marginally with no significant returns back to them. So, for the VCs to be interested in your business, what will the value of their stock share in your company need to be at the end of four years?

An alternative way of phrasing the same issue is the following: If the terms of their investment were that you had to buy out their stock position at the end of 4 years (called redemption or redemptive rights), how much money would you have to return to them?

5. Question for the founders to consider, what happens if you cannot buy out the VCs in Question 4 according to the terms of their initial investment?

6. If a VC invested $1 million in your company in the form of a "WARRANT" that matured in four years at $10 million, what rate of return is the VC going to obtain? (Note: this rate of return is similar to what is occurring in problem 1.4)

7. (a) Most of the departments of the U.S. federal government have legislated that their budgets increase annually by a minimum of 8%. How many years will it take for the federal budget to double? (b) One can assume that tax revenues are proportional to GDP (gross domestic product), which generally increases around 3% per year. How many years does it take the GDP to double? Briefly discuss how this imbalance can be addressed (try to keep your answer to 200 words or fewer).

2 Recognizing Opportunity

*Men who are resolved to find a way for themselves will always find **opportunities** enough; and if they do not find them, they will make them.*

Samuel Smiles (Scottish author 1812–1904)

Entrepreneurs are simply those who understand that there is little difference between obstacle and opportunity and are able to turn both to their advantage.

Niccolo Machiavelli

2.0 Entrepreneur's Diary

I think that I have developed a disruptive technology for producing fish in an indoor water-recirculation system. My design dramatically reduces the costs and complexity associated with the water-reconditioning equipment used in indoor fish systems. My design uses less space, is more environmentally compatible, is more reliable, and uses much less energy than other systems that most people are using for this purpose. As far as I can tell, there is nothing that our system doesn't do better than any other system anybody is already using. What I don't have is a practical means of demonstrating this technology. It is tough to be the groundbreaker. How then can I proceed? The producers that are using the technology that I plan to replace aren't willing to convert their system to my system, because it takes them out of production for a considerable time, and to them, my system is unproven and therefore introduces a risk that they don't have now.

So how do I demonstrate my technology? I have to become my own customer. I am going to have to build a fish farm that features my technology and its benefits. Where do I begin? I have to design a large-scale facility capable of making a profit, then build it, and then operate it. Now, I have to ask myself, "What business am I in? The system technology business or the fish business?" Sometimes it is hard to tell the difference. The truth is that I was about to be in both businesses. My immediate focus, however, was on the fish business. That is where the long-term, continual profits lie and where I pitched my investment proposal to investors. Apparently, I had just recognized my opportunity. Recognizing opportunity, preferably centered around some type of disruptive technology, is what Chapter 2 is all about. But what's a disruptive technology? I'll try to answer that question for you in this chapter and how it applies to a start-up.

2.1 Disruptive Technology

How do we recognize opportunity? How do we measure progress? How do we deal with change? Most of the answers can be distilled into a single thought: it depends on how it affects us personally. Technical advancement is inevitable and will continue to profoundly change our world. "***Disruptive technologies***" are those technologies that have dramatic impact on the way we function and on how we conduct our ordinary daily affairs. Entrepreneurs should always be on the lookout for such opportunities because the upside economic returns can be very large. At the same time, missing an opportunity of which your competitors take advantage can marginalize your technology and eventually put you out of business. Successful entrepreneurship is about recognizing and acting on these opportunities. Remember that investors are always looking for big-win (opportunity) investments, too.

One aspect of the degree to which a particular advancement changes our world is the number of people affected (size of market) by it. As a basic example of this, consider the telephone. The telephone would be of absolutely no value whatsoever if there were only one person in the world. The concept of real-time communication with a distant person would simply not exist; that is, there would be no market. However, what is the value of the potential ability to communicate with others in real time across great distances in a world full of billions of people? We can't imagine what it would be like, living in our world today without the capability provided to us by the telephone.

Another aspect of the degree to which a particular technological advancement changes our world is its ease of use. That, too, is personal in nature. Some people can't type. Therefore, a word-processing computer program is not very important to them, because it does not directly improve their quality of life. Hence, the long-standing efforts on computer voice recognition, which is finally getting pretty good.

We measure progress by how technology affects us. Most societies regard technological advancement as being generally beneficial, and therefore, technological advancement equates to progress. Now, however, there is a caution. Is it the telephone that is so mighty, or is it the ability to communicate with others in real time across great distances that is the real progress? Today, we have many ways to accomplish this. Indeed, at the time the telephone was invented, a means to communicate with someone else in real time across great distances already existed. That capability, in the form of the electric telegraph, was invented by Samuel Morse. In his obituary published in the *New York Times*, April 3, 1872, the invention of the telegraph was acknowledged as a technological marvel and absolutely necessary for "the well-being of society." So much for the *New York Times* and its predictions.

The basic capability offered by the telephone was already available via the telegraph. Why, then, did the telephone quickly surpass the telegraph in popularity and usage? Why was the telephone a disruptive technology? Could it be because the telephone required no special skills to use and did not require subsequent processing on receipt of the signal? Yes. Did it cost more than the telegraph? Yes. No private home had a telegraph installed for ease of personal communication, even thirty years after it had been invented and was in widespread usage throughout the world. However, less than thirty years after the invention of the telephone, there were more than 3 million telephones in use in the United States. Furthermore, who was clever enough and quick enough to perceive this trend and to capitalize on it? Samuel Morse? No, he had passed on by then. It was Alexander Graham Bell and two investors who founded the dynasty of the Bell System. Was this disruptive technology? It certainly was to the telegraph people and to other businesses that soon became dependent on the new technology and learned how best to use it for their advantage. We'll discuss the basic tenets behind disruptive technology in the next section.

2.2 Application to Start-Ups

More than anything else, being an entrepreneur of a start-up requires the ability to find capital resources to implement your business plan. Capital resources are a synonym for investors. Knowing that most start-ups will fall short of expectations, investors are not looking for ways to make small returns but ways to make large returns. A catchy term that indicates potential for a large upside is using disruptive technology to describe ones technology. Everyone will have some type of technology, but "disruptive" means that large new markets are just around corner: you have invented the next "whatever," and everyone is going to buy one. I used disruptive technology in describing the fish-rearing technology around which I built my fish company. It probably helped me some in raising the capital for the fish company. In reality, my technology was more incremental than disruptive in its nature, but I still used the term.

Clayton Christensen has written a whole book, *The Innovator's Dilemma*, on the subject of disruptive technology.[1] In it, he goes into immense detail on all aspects of this subject. I will give you my abbreviated view on the subject here and what I feel is most pertinent to a small start-up company. I highly concur that having a disruptive technology is superior in concept to ***non-disruptive*** or incremental technology advances to form business plans around. However, there just aren't very many disruptive technology opportunities that come along every day! As mentioned before, this book tries to help you with start-up opportunities that are more realistic in terms of something that you might be given the opportunity to pursue; for example, $1 million companies are started a lot more frequently than are $100 million companies. And in the process of creating a modestly sized company, if you come across opportunities that are truly disruptive, you'll be prepared to take advantage of it when you see it.

[1] Clayton Christensen, *The Innovator's Dilemma* (Boston: Harvard Business School, 1997).

Examples of Disruptive Technology

Alexander Bell originally sought only to improve the telegraph, not to replace it. His experiments along those lines, however, revealed to him a whole new technology to accomplish the same basic purpose: real-time communications over long distances. Many advances in technology come about in this same serendipitous way.

My favorite example of disruptive technology is the steamship. When Robert Fulton sailed the first steamship, the *Clermont*, up the Hudson River in 1807, it underperformed transoceanic sailing ships in nearly every important aspect of performance:

- speed
- reliability
- operating cost

Imagine yourself as a salesperson for this technology. Why would anyone buy a steamship? Well, as we all know, it had one important advantage – it didn't depend on the wind to move! So, the marketing challenge became to identify where the steamship had an advantage over other boats. It wasn't in transoceanic voyages, but the steamship could be applied to inland waterways where the ability to go in different directions at command and against or in the absence of wind was critical. For example, the steamship was the best method of moving product up and down the Mississippi River. You can see how this market unfolded.

The take-away point here is that none of the makers of sailing ships who stayed with their aging technology until the bitter end in the early 1900s ever adapted to the new technology. Instead, they continued to produce and market their sailing ships. After all, it had served them and their predecessors quite well for more than 2,000 years. Consequently, not a single maker of sailing ships survived the industry's transition to steam power. You can probably think of other examples yourself. (See the Problems at the end of the chapter and the Classroom Exercises at the end of this section).

Your job as an entrepreneur is to recognize this pattern demonstrated in the steamboat example. It is natural that in a successful company, the sales force and management will pay most attention to their existing customers and continue to give them what they want. Remember that one of the required characteristics of the successful entrepreneur was imagination. Existing companies are generally not very imaginative, but often focus exclusively to make their product even better despite their efforts yielding diminishing returns as the product approaches perfection. The problem becomes that the product may already be as good as it needs to be, and no matter how clever the engineering and manufacturing team is, at some point, you have exhausted the needs of this particular type of client customer, that is, the market that was originally defined and has been served.

Large Company Weakness

Large billion-dollar companies have large growth demands. For example, the typical 20% targeted annual revenue growth needed for a $1 billion company implies that $200 million in additional yearly revenues are required. Obviously, a $5 million market (for example) is of almost no practical interest to the large company based on these numbers. Successful disruptive technologies almost always enter small, unserved markets that provide small revenue streams and, hence, are of little interest to the large company. A large company sees a switch to a less profitable, but "new," technology to increase its bottom line as being counterintuitive. As a result, large companies often ignore disruptive technology opportunities, even when they recognize them.

On the other hand, if you are in a start-up company, even a potential $1 million market generated by a disruptive technology at the beginning of its history would be attractive to you, particularly if you think your start-up can dominate this market and be recognized as the major player. Then, you build your start-up's cash flow and resources, improvements are made to the product, and, at some point, you are able to attack companion markets and perhaps displace the mature technology that has been in place for some number of years; for example, the steam-powered boats started to replace sailing vessels in more and more market niches. Now, the large company may decide to capture back market share by buying your start-up once you have captured a new market and demonstrated product success. For sure, this is how some large companies with sufficient capital from earlier successes are able to sustain themselves over many technology cycles and not just one.

Established companies will often overlook new technology opportunities simply because the market has not been proved, thus making the investment to develop the technology a significant risk. New markets are sometimes ignored because sales and marketing people from large companies are trained to listen to their existing customers in order to improve their products in order to capture more and more of their specific market. These methods do not work for disruptive technologies.

Large company executives do not like to make visible errors in judgment and hence are likely to avoid most new opportunities because they are steeped in risk and possible failure that could lead to personal humiliation or demotion. Large companies attempt to control and diminish this risk (or spread the blame in case of failure) by creating **intrapreneurship** avenues within the company. Here, smaller capital and personnel investments are made to see if new markets can be developed. Of course, these large companies are overloaded with committees, approval chains, and many experienced people who can see all the reasons why something won't work, and are much more challenged at identifying realistic opportunities.

Your key advantage as the start-up is that you have very little corporate history and momentum that keeps you going in the wrong direction, as in, making bigger sailing ships. You can identify and make strategic commitments to attack emerging market opportunities almost on the fly (remember an earlier comment in Chapter 1 about being able to fail early and inexpensively). Your advantage is your relative flexibility and ability to change your product quickly compared to established firms that will struggle to do so. You have the marketing flexibility to find new markets that value your product. Large companies often struggle with their own bureaucracy and are slow to make changes, all resulting in lost time in addressing a new market opportunity. This time lost by the large company is used by the start-up to gain a *first-mover* advantage.

Finally, if you work for an established large company, you will be in an ideal place to discover a disruptive technology opportunity, by watching for customer needs that are not being met by a current line of products. And if your company fails to divert resources in this direction (after you point them out), then there is clearly a market opportunity for a start-up.

2.3 Technology Cycle and the S-Curve

The general pattern of technology cycles can be described by an S-shaped curve, where the rate of technological progress or the relative performance of the new technology is plotted versus effort. Relative performance can be thought of as technological advancement or product sales, although the two are not necessarily directly related. Effort can be viewed as simply time in the market place or as the capital resources being dedicated to the product's development, refinement, and support.

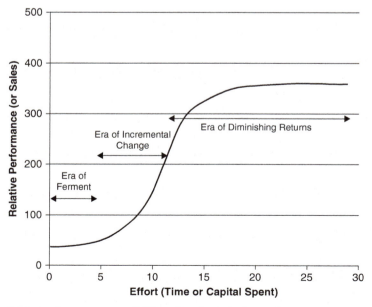

Figure 2.1   Relative performance versus effort for a disruptive technology cycle.

There are three distinct time eras for a product as described by the S-curve:

1. Era of Ferment
2. Era of Incremental Change
3. Era of Diminishing Returns

These eras are depicted in Figure 2.1.

Era of Ferment

Once a problem/opportunity is first recognized, research and development (R&D) on many different prototypes will be launched by various companies, generally resulting in some company bringing a product to market. This product will soon be joined by multiple competitors and copycats, competing with new and similar technologies and/or solutions. Competitors are responding to the "customer has spoken" and wants this "thing," so you better get one built and on the market quick. On the S-curve, this period is shown as the beginning of the S-curve with small incremental technological advances being attained as effort (resources) is committed (or time).

Companies commit effort to learn more about this particular new technology while searching for the best way to solve the customer need. Simultaneously, the new market must be expanded, and this will generally require a significant marketing effort. Key alliances should be made to secure a dominant market position for the new product, which then causes this particular design to become the accepted design technology (or product) standard. A good representation of this sequence in product domination is the VHS versus Betamax example. Betamax was introduced in 1975 by Sony and quickly captured 100% of the market; six years later, market share was 28%, and it has now long vanished from the market place. VHS, introduced in 1977 by RCA, became the standard videotape format despite being a weaker technology, with a recording capacity of four hours, and a cheaper price tag than the Betamax technology had. VHS won the marketing battle by securing contracts with companies that produced the videotape machines, and of course, RCA required them to use VHS tapes! As the *Era of Ferment* ends, signaled by a dominant technological solution having emerged, established companies using alternative technology face a critical decision as whether or not to make the transition. Typically, they don't (remember the sailing vessel industry). This then affords the start-up even more opportunity to capture a large market share by enjoying a superior competitive advantage with fewer competitors.

Era of Incremental Change

The *Era of Incremental Change* is the period of largest profits per unit sold, and the largest technological advances are made in the new product/technology per unit of effort applied. This era is the relatively linear portion of the S-curve.

Competitors have converged to a similar technology, and competition is largely based on product effectiveness and reliability. It is in this period that a start-up with a head start in developing a product must continue to improve through its R&D to stay ahead of secondary competitors. Sony Corporation took great advantage of this era with its Walkman radios then cassette-tape players. Once it had developed the technology to play the tapes, it developed a Walkman for every imaginable need. Sony was able to extract very large profits over a longer time simply by incrementally adapting its product to new markets. Then again, where are the Walkman products now? They now are obsolete and no longer available. They were replaced first with portable CD players and then with MP3 players. Sony failed to innovate in the face of these new technologies, and it lost a once-dominant position in the market for portable music devices.

The Era of Diminishing Returns

Perhaps the most difficult era to recognize as it starts is the ***Era of Diminishing Returns***. Mathematically, it is identified when you have passed the point of inflection (the second derivative is negative, so the technology is improving at a decreasing rate) on the S-curve. At this point, the product has become as perfected because it needs to be, and competition with other companies is based on price. The product is now in the commodity market category. A start-up ***should never try to enter a commodity market*** because it cannot compete. In this era, the advantage simply goes to the company that can deliver the product in the cheapest way to the market place, and rarely would a new player (start-up) be able compete in this era. Now is a good time for the owner of a start-up to either sell the company to one of the commodity players or jump to a new technology and opportunity and start all over again.

> A start-up *should never try to enter a commodity market* because it cannot compete.

Product Evolution

Product evolution, its key attributes, and how the product competes through these three eras can be described by

1. Functionality,
2. Reliability,
3. Convenience, and, finally,
4. Product price.

The steamboat first competed on function; for example, it could move in the absence of wind and could go forward and backward. Personal computers (PCs) first competed by providing the function of being "personal" as opposed to users

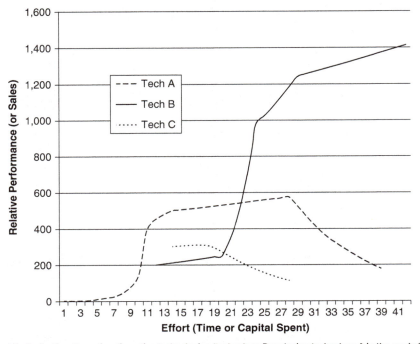

Figure 2.2 Life Cycles for alternative disruptive technologies (technology B replacing technology A in the market place; technology C being an unsuccessful product launch).

having to go to mainframe sites, and then PCs began to be more reliable as time went forward (many years). Then PCs became smaller (more convenient), and at last, price started to come down dramatically. Again, looking at PCs as an example (we should be a bit more familiar with this product than with steamboats), the pace of improvement in the disruptive technology is "high" relative to the established products. Thus, the disruptive technology will eventually compete and displace the entrenched technologies and will take over the mainstream markets because of the speed of its improvement. This is the key: the disruptive technology starts at the bottom of a neglected market and then takes over mainstream markets because of product evolution.

As disruptive technologies fully mature (the Era of Diminishing Returns), eventually other new technologies will replace them by being cheaper, simpler, smaller, or more convenient. Continuing with the earlier PC example, PCs continued to evolve and then started to get replaced with laptops, tablets, and similar devices, and even smart phones entered the market and became the preferred product. Other ready examples here would include the following well-known current technologies and the technologies they replaced in parentheses: matches (flint and steel), horse (walking), automobile (horse), airplane (ocean liner and passenger trains), telephone (telegraph), cell phone (analog phone tethered to a home/business wall jack), television (movies and radio), calculator (slide rule and pencil and paper), and computer word processor (typewriter), among others. Not all predicted disruptive technologies turn out to be disruptive; often they become simply product failures. Can you think of some (see Technology C in Figure 2.2)?

What is the moral behind these product introductions and subsequent evolutions? Established firms continue to try to push improved technology into their established markets, whereas successful start-ups find a new market that values the disruptive technology attributes, functionality being the first desired attribute. Think of more examples and go talk (or e-mail) your grandparents for a lively discussion on this subject.

2.4 Additional Resources and References

Christensen, Clayton M. 1997. *The Innovator's Dilemma*. Boston: Harvard Business School Press, p. 225.

Rogers, Everett M., 1962. *Diffusion of Innovations*. New York: Free Press of Glencoe, pp. 367.

2.5 Video Clips

Directions for viewing clips on the Prendismo Collection (formerly from www. eclips.cornell.edu):

- Visit http://prendismo.com/collection/. (You must subscribe with the site for full access to all features.)
- Click on the **Subscribe** at the top of the site. Students receive reduced rates.
- Once you have subscribed, type in the name or title of the clip in the **Search** at the top right corner of the screen.
- Click on the clip you wish to view.
- Click on the Play icon to view the clip. If you have not become a subscriber, you will only view the first 20 seconds of the clip before being prompted to log in.

1. Daniel Simpkins (Founder of Hillcrest Communications)

- This is one in a series of lectures from 2005 conference sponsored by Cornell University Engineering Alumni Association; the keynote lecture discusses "10 Laws of Entrepreneurship."
- Undergraduate and master's degrees in Engineering from Cornell University
- Title of Video Clip: "Daniel Simpkins States Importance of Thinking of What Is Not Possible"

2. Amy Millman (President of Springboard Enterprises)

- Amy Millman has established a successful twenty-five-year career working for and with business and government officials. As one of the founders and current president of Springboard Enterprises, she has built an organization that has assisted hundreds of women-led high-growth enterprises raise $1 billion in investment capital in less than three years. Prior to Springboard, Millman served as executive director of the National Women's Business Council, a federal statutory commission providing advice and counsel to the president and Congress on issues of importance to women business owners.

- Title of Video Clip: "Amy Millman Discusses Items that Are Stressed in Coaching Entrepreneurs on their Pitch Presentations"

3. Rob Ryan (Founder of Ascend Communications, sold to Lucent Technologies)

- Founder of Entrepreneur America
- Undergraduate degree from Cornell University
- Title of Video "Clip: Rob Ryan Shares Example of 3-D Technology Business" (First text of clip: "Greg Farvaloro. He had an idea for 3D, real 3D, spatial 3D display. He submitted the idea to the MIT business competition and was one of its winners, either the second or the first place **winner**. I met him at that point in time. Now, here's an example in his particular case ...)

2.6 Classroom Exercises

DISRUPTIVE TECHNOLOGY

1. What disruptive technologies can you name that started out as inferior or very limited for one market but then expanded and took over some other established market?
2. Can you think of companies that no longer exist or are much less dominant than they were ten years ago? (Interview some people who are fifty years old or older).
3. There are many examples of virtually simultaneous development of technologies to meet a perceived need in the market place. The Wright brothers were worried about several others who were almost ready to try heavier-than-air flight and displace their invention. Can you name other example of concurrently competing technologies where one triumphed over the others knocking them out of public awareness?

2.7 Engineering Economics

This section of the text is meant to guide you in learning engineering economics, which would be useful for any aspiring entrepreneur beyond the basic objective of passing the FE test.

Uniform Series Compound Formulas and Present and Future Value of Single Payments*

UNIFORM SERIES COMPOUND FORMULAS

Now that we have learned to calculate a cash flow and interest over one period, we can extend this knowledge to calculate a cash flow over several periods. To do this, we need to add one more term to describe the cash flow diagram. The payment or receipt at the end of each period we will define as "A" (see the following diagram), for example, $15 per period. You can think of this as a dividend you are going to pay your investors.

We will begin with the simplest form of A, which is when there is an equal disbursement at the end of each period. We call this a uniform series of payments. Depending on the purpose of the cash flow, the sum of the A's plus the interest added to them can equal an initial payment P (such as a loan or the investment in your company by the angels you have found), can equal some future receipt F (such as in a savings account or life insurance), or they can equal the sum of both an initial payment and a future receipt (possibly as in a successful return on an investment). This section will help you to determine the total value of a cash flow in terms of present worth, P; future worth, F; or annual worth, A.

The easiest case to understand is when there is an initial payment to you from a lender, P, and then you pay back this money in a uniform periodic sum of size A (your equal payments). This is also the case when you buy a house; the bank loans you the money needed to purchase the home, and you pay the bank back in a series of equal payments over some period such as twenty years.

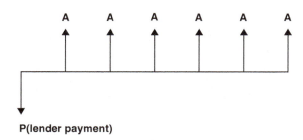

P(lender payment)

Usually you are calculating the value of A in order to pay off a debt, so we will try to find this value while knowing the initial debt, P. We start by looking at each A individually. Each A can be considered as a future payment F, which we already know how to calculate P from, and the sum of the P's from these A's equals the total P. For example, using equation 2 and knowing that all of the A's are equal (all of the F's are equal) we know that P must be the following:

$$P = A \left[(1 + i)^{-1} + (1 + i)^{-2} + (1 + i)^{-3} + \ldots + (1 + i)^{-n} \right]$$

Multiplying both sides by $(1 + i)$, we get

$$P (1 + i) = A \left[1 + (1 + i)^{-1} + (1 + i)^{-2} + \ldots + (1 + i)^{-n+1} \right].$$

Subtracting $P (1 + i) - P$, we get

$$P (1 + i) - P = iP = A \left[1 - (1 + i)^{-n} \right].$$

Solving for A, we get

$$A = P \left[\frac{i(1+i)^n}{(1+i)^n - 1} \right] = P(A/P, i, n).$$

$$(2.1)$$

You can rearrange this equation to solve for P:

$$P = A \left[\frac{(1+i)^n - 1}{i(1+i)^n} \right] = A(A/P, i, n).$$

$$(2.2)$$

Using the preceding equations, you can also solve the uniform series contribution to a future worth:

$$F = A \left[\frac{(1+i)^n - 1}{i} \right] = A(P/A, i, n)(F/P, i, n) = A(F/A, i, n)$$

$$(2.3)$$

$$A = F \left[\frac{i}{(1+i)^n - 1} \right] = F(A/P, i, n)(P/F, i, n) = F(A/F, i, n).$$

$$(2.4)$$

A diagram for a future worth with periodic payments is shown (six annual payments worth some future value F).

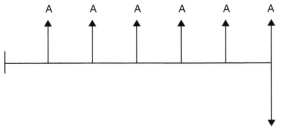

Future Value F,
cash received

Note that the (*A/F*) factor is also known as the ***sinking fund factor***.[2] The sinking fund is a fund in which annual payments worth A each are made to have a total of F in the account by the end of n years.

PRESENT AND FUTURE VALUE OF SINGLE AMOUNTS

Since cash flows have a time value, it is helpful to look at a time line to visualize these cash flows.

Future Value (F)

---

[4]The sinking fund is mentioned in the FE review manual.

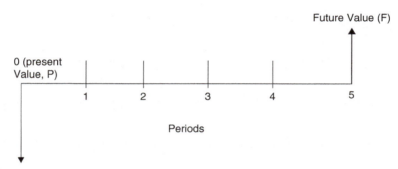

You deposit $100 in the bank at 5% per annum interest, compounded annually. What is the value of this deposit five years from now? The formula for future value of a single payment is

$$F = P(1+i)^n,$$

where

F = future value (the unknown in this case),
P = present value ($100 in this case),
i = the interest rate per period (note that this must be a nominal rate), and
n = the number of periods (5 in this case).

We get by substituting the above values into formula for F
$F = \$127.63$.

PRESENT VALUE

Given the future value, we can easily find the present value, using the same equation:

$$P = \frac{F}{(1+i)^n}.$$

[2] The sinking fund is mentioned in the FE review manual.

Example 2.1 – Single Payment Formula

How much do you need in your savings account today, which pays 4% interest compounded annually, if you wish to have a balance of $100,000 in five years?

Answer:

$$P = \frac{\$100,000}{(1+0.04)^5} = \$82,193$$

or in notation form and using the tables in the Appendix:

$$P = F\ (P/F,\ i,\ n)\ =\ 100,000\ (F/P,\ 4\%,\ 5yr)$$
$$=\ 100,000\ (0.8219)\ =\ \$82,190$$

Annuities

A *simple annuity* is a periodic payment of a fixed, constant amount (A). Despite the name, it is not necessarily an annual payment, it can be monthly, quarterly, semiannual, and so on. There are many examples of this: mortgage payments on a house, car payments, and bond interest payments.

An *ordinary annuity* is one where the payment occurs at the end of each period. The formulae for annuities use ordinary annuities, that is, payment assumed at end of a period.

An *annuity due* is the term used to describe an annuity that is paid at the beginning of each period. Rent is a common example. You will have to be a little clever to figure out the formulae for annuities due (hint: use the timeline; see the homework problems).

ORDINARY ANNUITIES:

$$F = A\left[\frac{(1+i)^n - 1}{i}\right] \tag{2.5}$$

$$P = A\left[\frac{1-(1+i)^{-n}}{i}\right], \tag{2.6}$$

where in both cases, A is the periodic payment and the parameters F and P are as previously defined.

END-OF-YEAR CONVENTION

There are some standard methods for use of the annual disbursement, A, in an analysis. One of these is the end-of-year convention, which just says that the disbursement for a period is always calculated for at the end of a period. The equations in this section, the economic tables found in the back of this book, and the analyses that is done in the following sections use the assumption that the disbursements are used at the end of the period. The alternative to an end of year convention might be a middle of year or beginning of year convention, which would yield the same results but would require slightly modified equations and tables. A diagram of three possible conventions follows. Notice that P still comes at the beginning of the period (and F would still come at the end), but only A that changes.

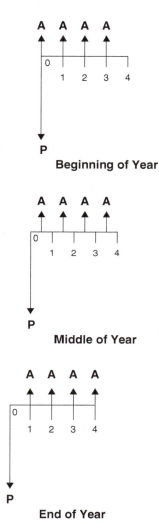

Three possible conventions:

Beginning of Year

Middle of Year

End of Year

The normally used convention is called end of year because the standard period of analysis in a cash flow is one year. Most business plans, cash flow projections, and payment plans for companies are on a yearly basis. The appropriate length of the period should be determined based on the circumstances. For example, an annual cash-flow review wouldn't tell you very much if it used a yearlong period rather than weekly, monthly, or bimonthly periods.

Another important assumption that the equations in this chapter make is that the periods are all of equal length. If you are given periods of unequal length, then it is possible to use the equations to convert all of the periods into the same length by combining several disbursements into one end-of-period disbursement or by breaking up one payment into several payments at the end of given periods. It is important to realize that if given a certain interest rate, compound period, and payment for a period, then it is possible to convert this payment to apply to any size of a period using the equations given in this section.

Finally, it is important to note that the interest rate used in all of the given equations is the effective interest rate for the interest period (n) and not a nominal interest rate or the subperiod interest rate. Along with this, these equations apply only to discrete compounding, or finite compound periods, and you must use separate equations for continuous compounding, which is described later in this book.

Example 2.2 – Uniform Series Formula

You buy a Honda CR-V EX sport utility vehicle, for $23,139. You must pay 7.5% APR (Annual Percentage Rate or i_{eff}), for a 5 year loan. Automobile payments use a monthly ($c = 12$), ordinary annuity payment scheme. Find the monthly payment.

Step 1. Find the periodic (monthly) interest rate.

$$i = \frac{i_{nom}}{c} = \left(i_{eff} + 1\right)^{1/c} - 1 = (0.075 + 1)^{1/12} - 1 = 0.006045 = 0.6045\%$$

Step 2. Find an annuity stream of sixty equal payments that has a present value of $23,139.

$$A = \frac{P}{\left[\frac{1 - (1+i)^{-n}}{i}\right]} = \$460.76$$

or

$$A = P\frac{i(1+i)^n}{(1+i)^n - 1} = \$460.76$$

Note that you will pay a total of $27,646 for the $23,139 Honda.

Example 2.3 – Uniform Series Formula

Imagine that you are transported back in time, to that era when steamships were just invented. As an intelligent budding entrepreneur, you recognize the advantage the steamships are going to have over sailing boats in transporting goods along the Mississippi River. You estimate that for the initial two years, you have a big enough market to sell ten steamboats a month, each at a fixed price of $250.

What is the present value of this yearly influx of income over the next two years? Use a nominal annual interest rate of 6%.

$A = 10 \times \$250 = \$ 2{,}500\ monthly$

$n = 2 \times 12 = 24\ months$

$monthly\ i = 0.06/12 = 0.005$

$P = A\ (P/A,\ i,\ n)$

$= 2{,}500\ (P/A,\ 0.005,\ 24)$

$= 2{,}500 \left[\dfrac{(1+0.005)^{24} - 1}{0.005(1+0.005)^{24}} \right]$

$= \$ 56{,}407.17$

2.8 Bonds

One way commercial companies and governments can obtain money needed for meeting expenses is by borrowing it. They obtain money from individuals and other organizations in return for written promises to pay a specified amount of money on specified future dates, up to an ending date, called the maturity date. These papers are called **bonds**. They are simply "IOUs", promising to that the issuer will pay their *face value* on the *maturity date* and interest payments periodically up to and including that maturity date.

Bonds are given a quality rating that reflects the financial strength and dependability of the bond issuer. The U.S. government bonds have always had (at least until August, 2011) the highest rating (AAA), as it is virtually certain that holders of U.S. government bonds will receive the money that is promised. Conversely, bonds issued by Ireland in 2011 were rated as junk bonds by Standard & Poors. Ford Motor Company is a good example of a corporate entity borrowing money from the public by issuing bonds. See http://www.bloomberg.com/news/2010–04–07/ford-sells-bonds-as-high-yield-debt-rallies-for-14th-month-credit-markets.html.

The risk one takes in purchasing a bond is that the bond issuer may not be able to pay what is promised, and this is called *credit risk*. The lower the credit rating given to a bond by companies who issue bond ratings such as Standard & Poors or Moody's, the higher the interest payments must be to attract investors. Investors will expect a higher rate of return on investments that are more risky. Hence, a lower credit rating forces the bond issuer to spend more money on interest payments.

Another risk associated with bonds is the market risk. The purchase price one pays for a bond at any time during its life, that is, while interest payments are being made, depends in part on what the going-market interest rate is for other investments that have similar credit risks. For example, a bond having a face value of $10,000 and annual interest payments of $1,000 up to and including the maturity date is obviously going to yield 10% to anyone who buys it for $10,000. But if the going interest rate at the time of purchase is only say 5%, then it seems reasonable that someone will be willing to pay more than

$10,000, and hence receive less than a yield of 10%. The purchase price is rarely the face value. The purchase price always depends on the going market interest rate.

Most individuals buy bonds hoping the market interest rate will drop while they own the bond so that they can sell it before it matures for more than what they paid for it. But the opposite may happen and the bond owner may lose money. This is the market risk every bond owner takes. Note that changes in the market interest rate do not impact the promised amount of interest that is to be paid on specified dates and the face value that is to be paid at the maturity date. It only has an impact on the purchase price one is willing to pay to buy the bond.

A *discounted bond* is one whose interest payments have been discounted to the present and this present value of all remaining interest payments is then subtracted from what would normally be the purchase price of the bond. Only the face value will be paid to the bond owner at the time the bond matures. Clearly this reduces the purchase price one is willing to pay for this bond that has no interest payments.

To compute a bond's purchase price value, (the maximum one would pay for the bond at any time at or after the time of its issue) the market interest rate is used to discount to the present all remaining future payments the bond promises to make. The interest rate is not used to compute the interest payments based on the face value of the bond. This is called the bond's interest rate, or the *bond rate*. It also used to be called the *coupon rate* when bonds had coupons on them that were exchanged at any bank for interest payments when they were due.

A bond *yield rate* is determined by dividing the total amount of interest paid in a year by the face value of the bond. Conventionally, a bond's yield rate is specified as a nominal rate (per annum), not its effective rate per year. If interest is paid more than once per year (typically twice or four times per year), the effective rate of interest will be higher than the nominal rate. The bond rate *has nothing to do with* the yield the bond owner receives, unless by chance the purchase price the owner paid was the same as the face value of the bond.

The Bond Equation

A bond consists of two payment streams:

1. The interest payments made every six months.
2. The payment of the face value at maturity.

As such, the price of the bond, *PV*, is equal to the sum of the present values of the annuity part (the interest payments) and the single payment part at the end of the term (FV, the payment of the face value).

The price (PV) can be found given the yield, and the yield can be found given the price, by using the following equation:

$$PV = A\left[\frac{1-(1+i)^{-n}}{i}\right] + \frac{FV}{(1+i)^n},$$ (2.7)

where

PV = price of the bond (when purchased at time zero),

A = period interest payment (calculated as detailed in the following),

i = periodic interest rate (calculated as detailed in the following),

n = number of six-month periods between issuance and maturity (adjust if payments are quarterly), and

FV = the face value of the bond.

Parameter Conversion

1. The term is given in years, such as n_y. We must convert this to the number of interest payments, n, by multiplying it by $c = 2$ (when payments are two times per year):

$$n = 2n_y.$$

2. The yield is given as a ***nominal***, annual rate. Thus, we must divide by $c = 2$ to get the periodic interest rate, i (i.e., the six-month interest rate):

$$i = \frac{yield}{2}.$$

3. The periodic interest payments, A, are *not* given explicitly, although the face interest rate is almost always stated, for example, 10% bonds. They are calculated as shown:

$$A = \left(\frac{face\ interest\ rate}{2}\right) \times (face\ value)$$

Example 2.4 Face Value

Simple case. A company issues (sells) bonds for \$100,000. The issuer of bond pays \$4,000 a year to holders of the bond, and will do so for ten years. At that time, the bond matures, and the bold holder is paid the face value of \$100,000. In the lingo of the fixed income world, we say that the bond has a face value of \$100,000, a coupon rate of 4%, and a maturity of ten years.

Example 2.5 Bond Value

A bond with a face value of \$5,000 pays interest of 8% per year in single annual payments. The bond is newly issued and matures in twenty years. The first interest payment is made a year from now. What is the most someone would pay for the bond to obtain a return of 10% per year, before taxes?

Solution:

$$Po = 5000(0.08)(P/A, 10, 20) + 5000(P/F, 10, 20) = \$4,148$$

2.9 Problems

1. You have done an analysis of the present value (P) of sales you can generate by selling hydroponic systems over a two-year period of $56,407. But should you go into this business venture? What will it cost you to generate these sales? The hydroponic systems sold for the next two years would cost about $125 per system to make. Assume that costs are incurred as sales are made (i.e., you incur a steady monthly cost corresponding to the number of systems sold (ten per month) as you only will make them on order). Also assume a yearly general cost of running the business of $12,000 a year. Again, using an annual nominal interest rate of 6%, what would be the present cost of selling hydroponic systems over the next two years? Would it be profitable for you to invest in this business venture using your personal funds?

Year	Cash Flow (in thousands)
0	−$450
1	$100
2	$300
3	$600
4	$800
5	$1,200

2. How much would an investment of $5,000 be worth in five years if the interest rate is 5%?

3. Consider $100 placed in a bank account that pays 9% effective annual interest at the end of each year. How much will the account have after one year? After ten years?

4. Consider a business that had a cash flow in the first five years as shown in the following table. Calculate the net present value (NPV) of this business using a discount rate of 10%. (Hint: do this the hard way; treat cash flow as unequal periodic payment A, and work out the present value of each cash flow in a year before summing it all to find the net present value)

5. You want to start a business selling tropical pet fish. You estimate that you would be able to capture 10% of a regional market that generates $2 million in yearly sales. Looking at a time span of five years, at a current market interest rate of 7%,

   a. What is the present value of your revenue over the next five years?

   b. What is the future value of your revenue over the next five years?

   c. If at the start of year 3, you plan to start selling pet turtles as well to earn an additional $50,000 yearly (you'll have three years of additional turtle sales), what would be the future value of your total revenue from both the fish and turtle after the five-year analysis period?

d. What would be the present value of your total revenue from selling both fish and turtle over the next five years span?

e. What is the uniform annual flow of revenue over the next five years that would be equivalent to receiving $200,000 from selling fish over five years and $50,000 yearly from selling turtle over three years? (Hint: find A using either the total F or P in part [d] or [e], respectively.)

6. In January of this year, Amy got a brilliant idea for a new chocolate flavor, but her chocolate company CEO decided not to implement her idea because the CEO estimated that it would only capture 1% of the chocolate industry market. Amy decides to strike out on her own. She estimates that chocolate lovers everywhere would pay $5 for her product and that she would be able to sell 15,000 pieces of her chocolate in December. The first eleven months of the year (this is your start-up phase where you need cash and you have no income, hence investors or self-financing), however, would be spent perfecting the product and would bring in no revenue. What would be her equivalent monthly income over the year? Use an interest rate of 4%.

7. What would the following notation get you: $A(F/A, i, n) \times (P/F, i, n) = ?$ Answer in terms of P, F, or A.

8. a. Mr. Johnson has invented a super vacuum cleaner. Cleaning company A offers him a contract to supply them with the vacuum cleaner for two years for a lump sum of $50,000 now. The rival company B offers to pay him a monthly sum of $2,500 over two years that you will immediately invest in an account paying a *nominal* interest rate (yearly) of 5% compounded monthly. What are the present values of each offer? If Mr. Johnson only wishes to maximize profits, with which company should he do business?

b. Are there other considerations you should make before making your final decision?

9. Henry wants to start up a business selling his original designed paper airplanes that can fly longer than any other (children's toy market). According to his estimation, the current U.S. market for paper planes is worth $500,000, but the market shows promise of increasing to $1 million after four years. What is the annual effective growth rate of the paper-airplane market?

10. One of Bob's dreams in life is to be a millionaire, and he decides to invest the $150,000 savings he currently has in his friend's new start-up business. If his friend promised an interest rate of 10%, how many years would Bob have to wait before his investment would be worth $1 million?

11. Derive the formulas (P value and F value) for annuity due from those of ordinary annuity. (Hint: consider the cash flow timings relative to the timeline.)

12. a. What annual cash flow (A) is equivalent to a series of salary bonuses your contract guarantees you that increase your base salary after the end of the first year by $10,000 per year for four years after you've taken your job. (Note that this means at end of year 1, you do not get a bonus, but in the

following four years, you receive $10,000, $20,000, $30,000, $40,000). Assume a market interest rate (you'll invest your bonus money) of 10%.

b. Demonstrate this calculation using the *P/G* and *A/P* functions.

Bond Questions

13. On March 1, 2014, Squeeks Corporation issued a $1 million face-value bond with 11% face interest rate and a maturity of fifteen years. The semi-annual interest payments are made on March 1 and September 1. The bond was issued at such a price as to yield 9.6%.
    a. What was the issue price of the bond?
    b. Prepare the journal entry to record the sale of the bond on March 1, 2014.
14. On March 1, 2014, Weeks Corporation issued a $600,000 face-value bond with 10% face interest rate and a maturity of five years. The semiannual interest payments are made on March 1 and September 1. The bond was issued at a discount for $554,400.
    a. What is the effective yield of the bond? (Use Excel)
    b. Prepare the journal entry to record the sale of the bond on March 1, 2014.
15. Trico Corporation bonds have a face interest rate of 8%, a face value of $1 million and will mature in twenty years. If you require a yield of 8.6%, what price would you be willing to pay for the bond?
16. On March 1, 2014, Peeks Corporation issued a $10 million face-value bond with 10% face interest rate and a maturity of fifteen years. The semiannual interest payments are made on March 1 and September 1. The bond was issued at a premium for $10,980,500.
    a. What is the effective yield of the bond? (Use Excel)
    b. Prepare the journal entry for the debtor to record the sale of the bond on March 1, 2014.
17. From the example problem, A bond with a face value of $5,000 pays interest of 8% per year in single annual payments. The bond is newly issued, and matures in twenty years. The first interest payment is made a year from now. What is the most someone would pay for the bond to obtain a return of 10% per year, before taxes?
18. Now, after owning the bond given in the preceding problem for ten years, the owner wants to sell it. The current market interest rate is now 5% per year for investments having equivalent levels of risk. What price should the current owner charge for the bond? If the bond is sold at that price, what would be the owner's annual return from owning the bond for ten years after taxes were paid on the interest payments and the capital gains obtained, assuming a tax rate of 30% on interest payments and of 15% on capital gains.

3 Defining Your Opportunity

Starting a new business happily taps into our spirit as well, because the process enlivens our heart and fuels our imagination – the linchpins of our existence. As an entrepreneur, your work is an expression of yourself. Without your ideas, beliefs, fortitude, skills and sense of adventure, your business cannot succeed.

R. Wolter[1]

3.0 Entrepreneur's Diary

Most games are not won because one of the players knows the rules better than the others. Most of us will soon know the rules of the entrepreneurship game (especially after you finish reading this text), but only a few of us who want to be successful entrepreneurs will become one. Why is it that some coaches seem to be consistent winners, even though the talent pool available is fairly similar? I think it is mostly doing your homework. That's what this chapter is about – doing the necessary preparation before launching your business. This is not fun, this is work.

Understanding your market, and therein recognizing whether you have a winning product, is probably the key aspect to creating a successful company. Your market analysis will be a key part of your business plan. But how do you analyze a market that does not exist? Markets always exist; they may be untapped, but they do exist or are ready to be created. Henry Ford could not analyze the market for automobiles when he was ready to start cranking out those Model Ts, but Ford knew quite well that there were markets for horses, carriages, trains, and other forms of transportation that he was going to try to capture. You will often have to use your imagination in trying to define and analyze your market.

Guest speakers are often invited to university classes to talk about how they successfully launched their business. Almost invariably, the speaker will start out by describing his or her previous involvement in some particular industry. And then, while involved in this previous business life and in the process becoming extremely knowledgeable about its market and associated dynamics, they

---

[1] Author, "Title," *Entrepreneur's Start-Ups* (December 2002): XX–XX. Wolter, Romanus. 2002. Stop dreaming-start your new business now. Entrepreneur December issue.

recognized an opportunity. The rest of the story goes on to describe how they achieved financial success. The stories may vary in particulars, but always seem to revolve around success being highly correlated to really knowing the market that one was about to compete in or against. There really is no substitute for knowing your market. Potential investors will often go to the market analysis of your business plan first to see if they think you have identified a significant market opportunity. You must convey complete knowledge of your market to be a successful fund-raiser and to execute a successful business strategy.

3.1 Mission Statements

You probably think you have a fairly clear idea of what your business is all about, right? OK then, describe your business to me in one sentence. This single sentence represents your mission statement. I think one of the most critical things you can do at the beginning of your entrepreneurial venture is to have a well written mission statement because "mission drives strategy," and if you don't have an effective business strategy, you will probably fail. We will soon get into doing an analysis of company's strengths, weaknesses, opportunities, and threats (SWOT analysis) – but before we can do that we need to be able to define what it is that we are all about. Analyzing something you cannot define is impossible. So, first things first, develop your mission statement.

How to Do It: Crafting a Mission Statement

A mission statement is a concise statement describing the general nature and direction of the company. A clear mission statement unites everyone in your organization and gives the team direction and helps increase customer confidence and employee morale. The mission statement should reflect your company's personality and its core values and should essentially be a statement of your company's reason for being.

There is no magic formula for writing a mission statement. It also means you have a lot of flexibility in writing one that fits you and your business. Here are some steps to get you started.

STEP 1: DETERMINE YOUR BASIC ATTRIBUTES

The mission statement can be one to three sentences, but it should be short and concise, for example, twenty to forty words. Your mission statement should answer the following questions:

- Who is your market?
- What is your product or service (and what benefits do they provide your market)?
- What is your special competence, strength or competitive advantage?

Mature companies who are no longer looking for investors will expand their mission statement to espouse general business philosophy (we will not harm the environment) and how the company implements their philosophy on a day to day basis (real-world action, e.g., we provide free fish every Friday to the homeless). Then, the preceding three-point list might be expanded to include the following questions:

- Why are you in business?
- What basic beliefs or values drive the business?
- How do you add value?
- What public image do you desire?

For this book, we focus on writing a mission statement for a start-up company (go back to the short list). In fact, potential investors would have a very negative reaction to a mission statement that was broadened in the form of the expanded list (philosophy) because they will think you lack focus.

STEP 2: CHECK OUT SOME EXAMPLES

If you are experiencing writer's block, consider some mission statements of other companies in your industry or other companies that you admire for some particular reason.

STEP 3: DEVELOP MULTIPLE MISSION STATEMENTS

Sit down with your team and start crafting mission statements, at least three to five. If possible, bring a couple that you have written to start the discussion and brainstorming. Starting with some examples will quickly generate opinions from your team and get the discussion moving. After writing several mission statements, your team can start editing towards a final (first) version.

STEP 4: LIVE WITH IT FOR A WHILE

Once you have your first version of your mission statement, before you print company letterhead and business cards (in quantity), live with your mission statement for a few weeks or months. Does it help you in making decisions between alternative choices or do potential investors seem to like its thrust? You might have to make a few adjustments or generate a completely new version as you move forward (or after doing your SWOT analysis). If you are interacting with potential investors, at all times you will need to have professional-looking business cards and so on – just don't print some enormous quantity that you have to throw away if you change your mission statement.

Mistakes to Avoid

While we just told you that there is no magic formula in constructing your Mission Statement, there are some common mistakes that you will want to avoid.

- **Too many buzz words.** Using buzz words is meant to impress the reader. However, buzz words can make your mission statement difficult to understand and create questions for the reader. At worst, it can convey a message you do not intend at all. Try to avoid using these types of words: quality, leverage, paradigm, synergy, empowerment, Web-based, proactive, empowerment, win-win.
- **Too long.** Keep it simple (twenty to forty words). Your business plan will present the rest of the details associated with your company. Think of the mission statement as a mini-abstract of your company's description.
- **Too focused on short term goals.** You are trying to state your reason for being. Aim for a mission that will last beyond your current product cycle.

Styles of Mission Statements

There are many styles of mission statements. Remember that for a start-up company, the mission statement is your chance to clearly state your company's focus. Thus, do not clutter this chance up with paragraph-style descriptions. Try to avoid a mission statement that is a series of bullet points:

Our company strives to:

- provide quality
- innovation
- Exceptional service
- innovation
- value

Another no-no for a start-up is to write a combined Vision and Mission Statement such as the following from McDonalds:

McDonald's vision is to be the world's best quick service restaurant experience. Being the best means providing outstanding quality, service, cleanliness and value, so that we make every customer in every restaurant smile. To achieve our vision, we are focused on three worldwide strategies:

1. Be The Best Employer. Be the best employer for our people in each community around the world. See the McDonald's People Promise.
2. Deliver Operational Excellence. Deliver operational excellence to our customers in each of our restaurants.
3. Achieve Enduring Profitable Growth. Achieve enduring profitable growth by expanding the brand and leveraging the strengths of the McDonald's system through innovation and technology.

The negative of a combined vision/mission statement is that it appears overreaching for a start-up. A start-up needs to singularly focus on their mission. Do not let your potential investor think that you are distracted. Most of what you would probably write in your vision description will be described in your business plan and through one-on-one conversations with your potential investors. So, back to examples of what we recommend: the Single-Sentence Approach.

Examples of Successful Mission Statements

Saturn (see http://humanresources.about.com/cs/strategicplanning1/a/strategic-plan_2.htm)

"Our mission is to earn the loyalty of Saturn owners and grow our family by developing and marketing U.S.-manufactured vehicles that are world leaders in quality, cost, and customer enthusiasm through the integration of people, technology, and business systems."

Denny's (www.dennyssupplierdiversity.com/site/pages/mission.html)

"Our Mission at Denny's is to establish beneficial business relationships with diverse suppliers who share our commitment to customer service, quality and competitive pricing."

Walmart (see http://cdn.walmartstores.com/sites/AnnualReport/2010/PDF/WMT_2010AR_FINAL.pdf

"Saving people money so they can live better."

Federal Express (see http://investors.fedex.com/phoenix.zhtml?c=73289&p=irol-govmission)

FedEx Corporation will produce superior financial returns for its shareowners by providing high value-added logistics, transportation and related information services through focused operating companies. Customer requirements will be met in the highest quality manner appropriate to each market segment served. FedEx Corporation will strive to develop mutually rewarding relationships with its employees, partners and suppliers. Safety will be the first consideration in all operations. Corporate activities will be conducted to the highest ethical and professional standards.

Fingerlakes Aquaculture, LLC[2]

The Company's mission is to become the dominant vertically-integrated supplier of the safest, freshest, highest quality seafood available through product branding and recirculating aquaculture technology.

Examples of Unsuccessful Mission Statements

At this point, you should be able to see what is wrong with the following two examples (example 1 is full of clichés, and example 2 is way too long):

1. "We will competently engineer fully tested imperatives so that we may pro-actively maintain world-class metrics as well as to credibly utilize orthogonal meta-services for 100% customer satisfaction."

2. "Acme Consulting offers high-tech manufacturers a reliable, high-quality alternative to in-house resources for business development, market development, and channel development on an international scale. A true

---

[2] Fingerlakes Aquaculture, LLC was started by the senior author of this text

alternative to in-house resources offers a very high level of practical experience, know-how, contacts, and confidentiality. Clients must know that working with Acme is a more professional, less risky way to develop new areas even than working completely in-house with their own people. Acme must also be able to maintain financial balance, charging a high value for its services, and delivering an even higher value to its clients. Initial focus will be development in the European and Latin American markets, or for European clients in the United States market."

Final Comments

Take your Mission Statement task seriously. The biggest joke about Mission Statements is that Enron's stated that the company prided itself on four key values: respect, integrity, communication and excellence. Among other things, all business dealings at Enron were supposed to be "open and fair." As Enron's story unfolded in Congress, it was obvious the former seventh-largest U.S. company wasn't living its own mission statement. Don't let this be an example for you and your company of having the Mission Statement simply being a creative-writing exercise and forget about it once you have your first round of significant financing. Practice what you preach.

3.2 SWOT Analysis

At some point along your entrepreneurial path, you are going to have to pull the trigger and jump into the water, but before you do, let's take a closer look at your potential business opportunity. You've just defined your company's focus through its mission statement, but now, the big question is whether or not your company can generate positive cash flow in some reasonable time frame and in an increasing, exponential fashion? Is there really a market out there that you can define and serve profitably? You better be certain of this before you go much further, because proceeding will entail spending increasing amounts of time and cash.

So take stock of the whole situation that surrounds this wonderful opportunity that you are envisioning. The commonly used analytical approach is to use a SWOT (strength, weaknesses, opportunities, and threats) analysis. Your investors will often ask you if you've done one. If you were to answer, "What is a SWOT analysis?" – you probably just lost that potential investor. So, we will cover this technique briefly, just to be sure you know how to do one. Sometimes, you may use some of these SWOT elements as section headings in your business plan, particularly the threats and weakness categories, and this might even prevent a lawsuit from a disgruntled investor if your business is unsuccessful.

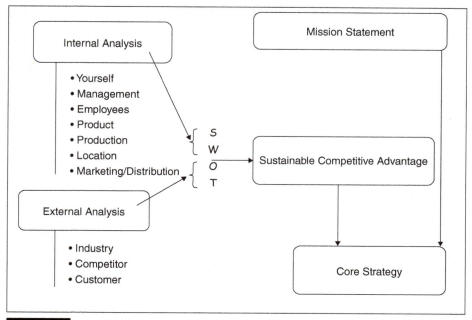

SWOT analysis.

Specific Steps to a SWOT Analysis

You have a basic idea of what your business is about and you know a lot about your product and its attributes, but just where should you position your business, what customer type will you target, what threats should you be worried about, and are you ignoring potential opportunities? Now is the time to analyze these issues before you try to raise additional money to launch your business.

The SWOT process can help you identify your core strategy by helping you to analyze the various aspects of your business and their relationships to external market factors; see Figure 3.1. The core strategy is related to the central mission of the company as described succinctly by your company's mission statement (and we just covered). The mission statement will help guide your core strategy. Your core strategy should result from a thorough SWOT analysis, and the mission statement will most likely be revised several times as you do your SWOT analysis and uncover new information about the market.

After writing your initial mission statement to give you some focus, the specific steps to your SWOT analysis are the following:

1. Do an internal analysis to identify strengths and weaknesses
2. Do an external analysis to identify threats and opportunities and to understand the driving forces in the marketplace
3. Identify the key success and risk factors from steps 1 and 2, which then allows you to define your sustainable competitive advantage (the whole point of the exercise!!).

After completing your SWOT analysis, you will have identified your *sustainable competitive advantage* and will be ready to develop your *core strategy*. Whether you are starting a small business that will likely stay small or you are headed for a high-growth opportunity, you have to do all these steps to formulate a sound strategy. Do not undervalue the importance of doing this analysis, and particularly the external analysis to understand the market. Rushing through your market analysis because it is time-consuming and tedious (I've never done this before, and I'm sure that I have a big market) is a foolish mistake. Even more foolish is to have a marketing plan that is not based on the findings of a detailed market analysis.

Strengths and Weaknesses

You must conduct a comprehensive inventory of the internal strengths and weaknesses (the "S" and "W" of the SWOT analysis of your business). This includes looking at all of the elements of the business, starting with you, the founder. Make a bulleted list of any particular strengths you bring to the business or weaknesses and vulnerabilities that you see.

S = STRENGTHS

Here are some examples of possible strengths for a start-up company:

- Strong competitive advantage of the company (you must describe very clearly what this is in believable and quantifiable terms)
- Management team in place
- Current status of the company (cash flow, money in the bank, existing customers)
- Potential of an uncaptured market (*requires* documentation)
- Possession of a revolutionary idea, novelty
- Legal protection of the idea (a favorite for inexperienced investor types, who feel that patent protection is a prerequisite to launch a technology-type company)
- Lack of competitors (be careful, see the section on threats)

Example. "Our license from the Cornell Research Foundation for a unique water treatment technology gives a competitive advantage in production costs of at least 25% over our closest competitors."

Consider such things as special technical or operational expertise, sales and/or marketing experience, network of contacts, or special ties to the marketplace. Next, examine the rest of the management team for its strengths and weaknesses. Do you have a good, balanced team? Or if you are a one- or two-person company at this point, what management skills are missing? Consider the strengths and weaknesses of your future employee base. For example, do you have access to a low-cost (or high-quality) pool of workers? Is there stability in your workforce?

Consider any special skills or training that your employees have or list those things that they lack.

Examine the features of your product as you build the S and W of the SWOT. List as strengths any proprietary technology, special features, or innovations you possess. You should also examine your manufacturing process and list any strengths. What is your input cost structure (e.g., high volume–low price) and are production costs vulnerable to change? Do you have flexibility in the production process; for example, can you scale up or down relatively easily? Your physical location may be a strength, for example, close to consumers, employees, input providers, strategic partners, or you may have a virtual presence that gives you added flexibility.

Your internal inventory should now be the basis of developing your business plan on your key strengths. Maintaining these strengths will be difficult as your company starts to succeed and competitors appear, so you should also be thinking of ways that your company can erect barriers to enter the market you are developing. That is, be thinking about what your sustainable competitive advantages might be now and how you will expect to maintain it in the future. You will need to articulate this business strategy clearly in your business plan and, in the process of doing so, clearly demonstrate that you understand the marketplace and where you intend to compete.

> The ultimate goal in formulating your business strategy is to discover your own sustainable competitive advantage.

W = WEAKNESSES

Weaknesses include things that you know may happen but that are somewhat beyond your control. This does not include competitors, because you can employ strategies to compete in your market against known competitors. Examples of weaknesses would be such things as

1. being dependent on a specific government regulation,
2. being dependent on an overseas supplier of a critical component,
3. being dependent on a nonexclusive license,
4. being dependent on weather events, and
5. being dependent on biological growth processes (if it does not grow or dies in the process, you have lost your product stream).

Example. "Our shrimp cocktail company depends on imported raw product from overseas. Future government regulations and import laws may result in duties that make our raw product cost much higher than the cost used in the current financial analysis."

Weaknesses can also be things you know about that could hurt your company. Just as location could have been a strength, your actual current location may be a weakness. If you are aware of any weaknesses, you are legally obligated to inform your potential investors. Your business plan should propose a method of addressing any identified weakness or should at least bracket the ramifications of such a weakness.

Opportunities and Threats

Your investors are going to want to know something about:

- the size of your market,
- the market history and associated trends,
- what portion of this market are you going after, and
- unique attributes of the market.

Remember that your potential investors are looking for a market opportunity with a large upside; they are not looking for 10% or 20% returns but for 100% or larger yearly returns on their investment. If they wanted a safe investment with correspondingly lower returns, they would invest in a mutual fund or a safe publicly traded company paying dividends.

O = OPPORTUNITIES

Opportunities are those future market opportunities for your company's product beyond your immediate market focus. This analysis requires documentation. Examples of opportunity might include the following:

1. Product can be used for other applications,
2. Product can serve multiple markets with minor variation,
3. Product manufacturing can be automated to reduce costs.

Example. "Our Kosher fresh fish fillets will command premium pricing and support our initial higher production costs compared to the costs of imported products. As our company becomes more efficient and our production costs become lower, we should be able and plan to compete in the non-Kosher markets."

T = THREATS

Threats are identified as something your company knows about that could ruin your company. Threats include other competitors (you must document these as well), and the lack of competitors would be a strength. Stating a lack of competitors in the Strengths section when there are competitors is a *fatal* flaw because the worst competitor is one you are not aware being there.

Thinking you have no competitors is a *fatal* flaw.

Investors will immediately lose any interest they have if they identify or are aware of competitors before you are. Today, most markets are global, so there are probably competitors well beyond your local area that you have not yet discovered. Claiming that there is no competition simply makes you appear naïve to potential investors. If you start generating a significant market, rest assured, your competitors will appear, even if they were not there when you started. Try to gauge the potential capability of any potential competitors to respond to your success.

A strategic analysis must take into account what is happening in the industry. Threats need to be identified and addressed. Whether it is supply-chain integration or environmental regulations, you must be aware of the larger environment in which your business exists. Information technology has made it much easier to understand what is happening in an industry as a whole.

You should be very cautious about being a single product company with a single function. Such products are often easily replaced by alternative choices in the market place or can be directly attacked by an existing larger competitor who makes it a point to put you out of business. Other examples of threats could be the following:

1. An alternative product that is better, cheaper, bigger, smaller, lighter, more efficient, etc.
2. Very few customers (only large ones that might make the product themselves or drive your margins down in order for you to keep them as customers)

Example. "We have an exclusivity contract to supply Walmart with its supply of organic whitefish fillets for two years; this will require 100% of our capacity to deliver the product. We have no guarantee after that point in time."

Investors typically are enthralled with growing markets for obvious reasons, your revenue grows even with a fixed market percentage (which is why you try to dominate an emerging market). However, a growing market can turn out to be a threat to your company. First, a rapidly growing market may attract large firms as competition. For you as a start-up, you will need to find additional capital for production and to cover cash-flow needs between the time a sale is made and when cash is received (sometimes this can be sixty days). Large companies will often have plenty of cash or lines of credit to cover this problem. Your pro forma analysis (see Chapter 7, "Financial Accounting") should clearly layout your future cash needs.

3.3 Intellectual Property

Intellectual property plays a very important part in the strengths component of the SWOT analysis. We provide further detail on this subject here so that you can understand more completely the characteristics of each form and possibly decide which might be most appropriate for your start-up. This is one area in which you should seek legal advice before proceeding, but it is good to know the basic rules for each type before you see an attorney.

Trade Secrets

Technology is one of the most valuable assets that can be protected by trade-secret law. Marketing plans, business plans, and customer lists may be protected under trade-secret law. Trade secrets are defined as information used in one's business that provides the business owner an economic advantage over competitors who do not have access to this information. Information is considered a trade secret when it is neither generally known nor readily ascertainable in the industry. A trade secret could be any of the following:

1. Process
2. Formula
3. Pattern
4. Program
5. Device
6. Method
7. Customer list

Trade secrets are not registered like patents. Trade secrets and patents are generally mutually exclusive, because a patent's purpose is to disclose information and a trade secret's is to conceal information. A trade secret must exhibit a substantial element of secrecy, so that except by the use of improper means, there would be difficulty in acquiring the information.

You must take reasonable steps to protect trade secrets to secure the protection provided by the trade-secret laws. The trade-secret policies must be in writing, available to all employees and contractors and should be discussed thoroughly with every employee who has access to trade secrets. Require your employees to sign the trade-secret policy for your company, typically appearing in the employee handbook. Trade secrets should be very limited in terms of who knows this information and should be limited to absolutely only those employees who "need to know" this information. Conduct exit interviews with departing employees (remind them of your trade secret policy and that they signed an agreement in this regard).

Clearly identify any written documents that contain confidential information. Require employees to mark all documents that contain trade secrets as "Confidential." Do not mark everything as confidential, because it won't be, and the courts would not regard this as a serious and valid trade-secret program.

Make sure you have confidentiality agreements with independent contractors (such as university consultants!), and you may need to use nondisclosure agreements with suppliers, customers, and potential business partners. It is important to have effective computer security systems. Also, you should review and update trade-secret protection measures on a regular basis.

OTHER MEASURES TO PROTECT TRADE SECRETS

Maintaining security and protection of proprietary and trade secret knowledge will require constant vigilance. You should follow these minimum steps:

- Be very careful when writing articles and giving conference paper presentations, the advantage being "free" advertising. Require that any publicly released written materials be reviewed and approved by company leadership before any dissemination or presentation, to ensure that no proprietary or trade secret information is disclosed.
- Control access to your facility. All visitors must be registered and be escorted at all times. Do not allow photographs of anything without permission on a picture-by-picture basis.
- Exit interviews should be conducted with departing employees to ensure that they are not taking any trade secrets from the company and require them to sign a document stating so. Consider having your attorney write a letter to the employee's new company informing the company that the employee had access to valuable trade secrets and warning the new employer against using any of these trade secrets.

MISAPPROPRIATION

If a product is substantially derived from a trade secret, the person who misappropriates the information cannot use it even if it is combined with independent improvements. Under New York law, to prove misappropriation, a company must establish the following:

- Your company possessed a trade secret.
- The other party used the trade secret in violation of an agreement with your company or as a result of improper discovery (they stole it).

Obtaining financial compensation for trade-secret violation from the court will require that you document your efforts of protection. In addition, your compensation will depend on

- how valuable is the secret information to your company and to your competitors and
- how easily could this information be properly acquired and duplicated by others.

Patents

The U.S. Patent and Trademark Office (PTO) is the government entity responsible for all issues dealing with patents (see www.uspto.gov). A patent application, once filed correctly, will usually require at least eighteen months (minimum) before any decision is made. Almost always, the PTO examiner will dispute some claim in the patent and return the application to you. Do not be discouraged at this point because often the process will go forward quickly from here on.

Sometimes, there may even be a telephone discussion between the examiner and the filer that speeds things up. The turnaround time after the first rejection will be much shorter, typically two or three months.

A noncomplicated patent filing will run roughly $10,000 to $15,000. Once a patent is awarded, you might find that some other company is using what you think is your patent-protected technology. This means that you will have to contest the patent against the abuser or someone may contest against you saying the patent was "obvious." Either way, this means you will have more legal fees. Some people feel that unless your patent has been contested and you have won the suit, then the current uncontested patent is of very little value. Before you go down this road, ask yourself if you have the resources to contest a patent? One of your investors might if you do not.

Any person may obtain a patent for a discovery or invention of any new and useful process, machine, manufacturing process, or composition of matter, or any new and useful improvement thereof. Laws of nature, physical phenomena, and abstract ideas are not patentable. Ideas or suggestions are not patentable. In order to be patentable, an invention must exhibit three characteristics:

1. It must be useful.
2. It must be novel.
3. It must not be an obvious extension of existing technology.

Patents provide powerful protection against competition, essentially giving you a legal monopoly for the period of the patent. A patent gives you a negative right, meaning you can prevent someone from using the features covered by your awarded patent. The patent application you file must also be enabling, meaning that it has been written so that people skilled in the art to which it pertains, or with which it is most nearly connected, can make and use the same (the invention) without having to perform "undue" experimentation. This means you have told them everything they need to know if they want to steal your idea! This public disclosure may give your competitors information you would rather they did not have. And after the patent coverage ends, the public is free to use the information. Do you really want to file a patent?

There are three types of patents: design, plant, and utility. ***Design patents*** cover the appearance (as opposed to the function) of an item, for example, furniture, golf clubs, running shoes (Reebok vs. Nike). These patents are valid for fourteen years from date of filing. ***Plant patents*** are for botanical life forms and are valid for twenty years from date of filing. ***Utility patents*** are the most common form of patents, which are for machines, processes, articles of manufacture, or compositions of matter and are valid for twenty years from date of filing.

If you think you have developed valuable patentable technology, be very careful about publicly disclosing the idea. There is a ***statutory requirement*** of nondisclosure before a patent will be awarded:

In order for an invention to be patentable it must be new as defined in the patent law, which provides that an invention cannot be patented if: "(a) the invention was known or used by others in this country, or patented or described in a printed publication in this or a foreign country, before the invention thereof by the applicant for patent," or "(b) the invention was patented or described in a printed publication in this or a foreign country or in public use or on sale in this country more than one year prior to the application for patent in the United States ..."

If the invention has been described in a printed publication anywhere in the world, or if it was known or used by others in this country before the date that the applicant made his/her invention, a patent cannot be obtained. If the invention has been described in a printed publication anywhere, or has been in public use or on sale in this country more than one year before the date on which an application for patent is filed in this country, a patent cannot be obtained. In this connection it is immaterial when the invention was made, or whether the printed publication or public use was by the inventor himself/herself or by someone else. If the inventor describes the invention in a printed publication or uses the invention publicly, or places it on sale, he/she must apply for a patent before one year has gone by, otherwise any right to a patent will be lost. The inventor must file on the date of public use or disclosure, however, in order to preserve patent rights in many foreign countries.

Even if the subject matter sought to be patented is not exactly shown by the prior art, and involves one or more differences over the most nearly similar thing already known, a patent may still be refused if the differences would be obvious. The subject matter sought to be patented must be sufficiently different from what has been used or described before that it may be said to be nonobvious to a person having ordinary skill in the area of technology related to the invention. For example, the substitution of one color for another, or changes in size, are ordinarily not patentable. (Source http://www.uspto.gov/patents/resources/general_info_concerning_patents.jsp#heading-4)

This requirement is to prevent unnecessary hardship on some unsuspecting company that has taken the public information and developed it into a product, only to have its usage of the product denied by a patent awarded later. This is why you often see "patent-pending" stickers on technology currently being used or seen in the public arena.

Trademarks and Service Marks

A *trademark* is a distinctive word, name, symbol or mark that is associated with a product and identifies the source of the product. Similarly, a *service mark* identifies the source of a service. To protect these forms of intellectual property, you must do the following three things:

1. Register the IP with the PTO
2. Be used commercially within four years of filing
3. Must be renewed every ten years

A key to establishing your rights to the trademark type intellectual property is to use a key word or symbol to distinguish the product or service. Perform a search before you start using the word or symbol.

A word or symbol can only be considered a trademark or a service mark if it is considered **_distinctive_**, meaning that is capable of distinguishing the goods or services on which it is used from those of competitors. Distinctiveness will occur in levels, and you should strive to be as distinctive as possible. The most distinctive trademarks will be names that have been invented for the sole purpose of functioning as a trademark and have no other meaning, for example, Exxon, Equifax, Kodak, or Xerox. Your next level down of distinctiveness will be a trademark having a common meaning that has no relation to the goods or services being sold, for example, Apple, Lotus, and Saturn (car). Finally, you can have a name that suggests something about your product but that requires some imagination to reach a conclusion as to what the product really is, for example, Microsoft or Netscape. Descriptive marks, phrases, words, surnames, or generic names are not granted trademark protection, for example, eloquent, powerful, Timmons, Rogers, modem, www, e-mail. The exception to this would be if you want to do a major advertising campaign to establish the trademark. If your surname trademark has been awarded, someone cannot use his or her own name to usurp the marketing value of the protected trademark; for example, Robert McDonald cannot open a restaurant called "McDonald's."

You can file for trademark protection online by going to the U.S. PTO Trademark Electronic Application System (TEAS) Home Page; see www.uspto.gov/teas. You will need to be able to provide a JPEG file of your actual trademark. The cost is minor, approximately $325.

Copyrights

Copyrights are our last form of intellectual property. You should probably copyright your business plan and anything else important that you write. Copyrights do the following:

- Provide an exclusive right to reproduce, distribute, and sell the protected material
- Provide protection for original works of authorship
- Protect material for the life of the creator plus seventy years
- Extends to motion pictures, sound recordings, books, articles and papers, photographs, and software

Copyrights are administered through the Library of Congress and can be obtained formally by you simply filling out a form, sending in $20, and in two months, you will receive a copyright notice. For the forms, go to http://www.copyright.gov/. You can place the © symbol on your piece of written work when you initially create it and file the formal application at a later date. If you want to bring a suit against someone you believe has violated your unfiled copyrighted piece of work, then you will have to formally file your copyright with the Library of Congress.

3.4 Conducting Market Research

All successful start-ups will have a deep understanding of the markets in which they are competing. There is absolutely no substitute for doing the research necessary to describe your intended market and where you will compete in this market. It is naïve to believe you as a start-up can compete in all segments of some established market, but rather, you typically are trying to identify some particular segment that is neglected or underserved in which to compete. Thus, you need to develop a clear understanding of and then be able to describe the larger market first, for example, trends in sales, customer profiles, competitors, and so on. Fortunately, these days, there are some very powerful databases that can be of use to you. You might start by visiting the local university's business librarian to get you started and perhaps even grant access to privileged databases.

Market research includes what is called ***primary research*** (data that you generate from your own efforts and firsthand sources) and ***secondary research*** (data gathered by others, such as the government). For example, it is quite common to use primary research to define specific customer needs and wants or for details on competitors – and secondary research to get the "big picture" about the customer base, the industry trends, and perhaps a list of competitors.

Gathering Primary Data

There are formal and informal ways to gather primary data. Informal techniques would include the following:

- Use the phone book to look at competitors listings and to call or e-mail potential customers.
- Test market your product with friendly potential customers. Use this method to see if the dogs will eat the dog food.
- Check out credit reports. You can search online for available reports at the Dun and Bradstreet site (http://www.dnb.com/) or purchase credit reports on your competitors.
- Visit your competitor. Whether it is exploring a competitor's website or sales literature or making a trip to the physical location, you can check out pricing, observe customers, and evaluate the niche and image of the company.
- Visit or participate in a trade show. Trade shows offer an opportunity to observe competitors and customers at the same time.

Formal techniques to gathering primary data include

- administering surveys and
- conducting focus groups.

In doing either, make sure you have identified the right questions to ask, so that when you are finished, you can identify questions needed to better define your market and sales capabilities or to answer some key question in your product development (do they really need that added feature) or distribution method (wow, they preferred the product in packages of ten to a box, not twenty-five!). When doing formal research, you can either obtain information about a broad range of topics without going into depth, or you can go into depth on just one or two topics; rarely can you achieve both depth and breadth in the same survey.

Gathering Secondary Data

Secondary sources of market information are good for getting the big picture and understanding the trends and technologies that might affect your product. Rarely will you find that secondary sources alone are adequate for making good decisions about a business strategy.

The process of obtaining secondary information has been changed profoundly by the Internet. It is now easier than ever to obtain government information, with many departments offering the option of downloading numerical data or written reports essentially for free. However, don't be fooled into thinking that everything is on the Internet. You should still visit a real library to explore everything that is available, particularly any fee-based commercial databases the library may be paying for that that you cannot access through the Internet. In looking for secondary data, consider some of the following sources.

INDUSTRY AND TRADE ASSOCIATIONS

There is a trade association for just about everything you can imagine. Many associations also publish newsletters and magazines with useful information for entrepreneurs. Many associations are also providing information online. As a starting place, look in the ***Encyclopedia of Associations***, a volume you should be able to find in the reference section of your library. There is an online source for finding trade associations: the Gateway to Associations (http://www.asaecenter. org/directories/associationsearch.cfm).

Reading association publications can provide some interesting background about key trends and concerns characterize the industry. It can also be an interesting source of information about the competition. Even the advertisements in trade magazines can provide useful information for the entrepreneur trying to get acquainted with a new industry.

GENERAL INTERNET SEARCHES

Using your own favorite Internet search engine, you may be able to turn up thousands of "hits" that match keywords such as your product category or industry designation. Searching through the list returned by the engine can be frustrating and time consuming. Be careful to verify the credibility of any source you pick up this way. One thing that is very interesting is to see if you can locate any competitors through general searching. The results can tell you such things as

- whether the competition has a strong internet presence,
- whether competitors are reaching out beyond local markets, and
- what pricing is standard among your competitors.

GUIDING YOUR SEARCH WITH COMMERCIAL DATABASES

There are an increasing number of subscription-based services that are intended to help guide your search of digital resources. Databases such as Lexis-Nexis American Factfinder allow you to search online archives of articles from popular press to scholarly journals. If you live near a university that has a business school, its library is likely to subscribe to such services. You may find access through your public library as well.

GOVERNMENT SOURCES

Most government agencies provide some data online. For market research purposes, the following are some key online sources:

- **FedStats** – (http://www.fedstats.gov/) – a guide to 100 federal agencies
- **Bureau of Labor Statistics** (http://www.bls.gov/home.htm) – wage data, demographics, industries-at-a-glance
- **U.S. Census** (http://www.census.gov/) – demographics, economic data and links to state-based statistics and international statistical entities
- **Bureau of Economic Analysis** (http://www.bea.gov/) – national and international economic data

INFORMATION ON INDUSTRIES AND PUBLIC COMPANIES

If your competitors are public companies, there are many ways to access relevant data. Most involve some charge, but some are free. For industry information, it is helpful to start by looking up what is called the Standard Industrial Classification (SIC) for your product. There are lookup services, such as the one found at http://www.express-advertising.com/sic_lookup.html. The SIC codes are being replaced over time by a new system of classification called NAIC (North American Industry Classification System), which can be determined by the lookup service at http://www.osha.gov/oshstats/sicser.html. Once you know the relevant code, here are just a few examples of places you can look for information on your industry:

- **PRARS** (http://www.prars.com/ibm/ibmframe.html), **Report Gallery** (http://www.reportgallery.com/), **Carol.com** (http://www.carol.co.uk/), and **Wall Street Journal** (http://wsj.ar.wilink.com/asp/WSJ1_search_eng.asp) – free sources for annual reports
- **SearchSystem** (http://www.searchsystems.net/index.php) – a site with free copies of public records

- **Hoovers Online** (http://www.hoovers.com/) – comprehensive list of industry studies available for sale
- **Bizminer.com** – (http://bizminer.com/) – market research reports, industry analyses, and local area vitality profiles for sale
- **Standard & Poor's Industry Surveys** – three volumes. New York: Standard & Poor's Corp., 1999. Print resource that includes an analysis of major industries, these quarterly reports give detailed information on fifty-two large industries, including trends and analyses of market share
- **Plunkett (** http://www.plunkettresearch.com/) – provider of business and industry information, market research, trends analysis, statistics, company profiles, and executive mailing lists
- **Business.com** (http://www.business.com/) – a business search engine and directory that has a section devoted to agriculture and another one to food
- **Zapdata.com** (http://www.zapdata.com/) – the Internet service of D&B® Sales & Marketing Solutions with a comprehensive virtual data superstore. Registration is required, but it requires no fee. Some reports are provided without charge; others are on a fee basis.

LOCAL INFORMATION

If you are looking for information about local markets, you might want to check the following sources (you can find the closest office of either in your phone book):

- **Cooperative Extension** (http://www.reeusda.gov/1700/statepartners/usa. htm) – this agency, associated with U.S. Department of Agriculture, may have reports, data, or bulletins that relate to your product.
- **Chamber of Commerce** – Often a local Chamber of Commerce will feature statistical data about the area and will have a list of businesses that may be helpful in checking out your local competitors.

3.5 Primary Market Research: Using Surveys and Focus Groups

Many businesses are focused on consumer products.[3] In such cases, the use of surveys or focus groups can be extremely useful. If the business involves direct marketing, you may have an easy forum for conducting customer surveys. If your customers are mainly located in a relatively small geographic region, it may be fairly easy to pull together a group for a survey. As previously mentioned, testing your product on real customers is what Rob Ryan (see "Additional Recommended Resources" at end of chapter) calls the "will the dogs eat the dog food" test, which can only be determined by putting the food (the product) in front of (in the hands of) the dog (customer).

---

[3] Material in this section was partially provided by Dr. D. Streeter, Applied Economics and Management, Cornell University.

Surveys versus Focus Groups

Conducting a survey involves talking to at least dozens and probably hundreds of your customers. A survey can help you gain perspective beyond the small inner group of people that were involved in your product development stage. If a survey is conducted properly (as detailed below), then you can generalize your findings to a larger population. Validating a market will require you to communicate with a relatively large group of randomly selected potential customers. Consult with a properly qualified statistician before going through the time and expense.

By contrast, a focus group involves talking to a small group (eight to twelve people) of highly qualified individuals. By qualified, we mean that they are a subset of your prospective customers who have special knowledge or characteristics. A good example of a focus group would be a group of people whom you view as potential ideal customers by virtue of age, family composition, gender, or other demographics. You may want to gather a group of industry experts, with special knowledge of your consumer base. Because focus groups involve talking intensively to a small group of people, they should be used to do activities such as

1. brainstorming product features,
2. reviewing and critiquing marketing materials, and
3. evaluating branding approaches.

You can never use results from focus groups to generalize on a statistical basis. They are used to connect with a small and specialized group and are not necessarily representative of the population as a whole.

Sometimes, if there is adequate time and money, surveys and focus groups are used in conjunction with each other. After you conduct focus group work, it may be clear what questions should go on a survey. Or surveys may raise issues that are best addressed in the in-depth format of a focus group.

Types of Surveys

The various options for administering surveys are shown in Table 3.1, which provides details on the pros and cons of each technique along with some tips. In all cases, it is important to remember that the quality of the survey depends on a list (called a "frame") from which you choose your sample. Choose a sample without a built in bias if at all possible. (You should consult a statistician for expert advice here; as a start, review Survival Statistics listed at the end of the chapter under "Additional Resources.") Consider hiring a market research company to design and administer your survey if financing permits.

DESIGNING THE SURVEY

Start with writing down the key strategic question(s) you are trying to answer. The number one mistake people make in writing a survey is that they do not

Table 3.1. Various Types of Surveys

Survey Type	Pros	Cons	Tips
Mail or written	Some see them as easy Don't need to administer one by one Can target a broad group Can make use of targeted mailing list	Can result in low response rate (5%–15%) No chance to clarify, follow-up No controls for accuracy Expensive	Lots of white space Short and clear Put easy questions first, delicate questions at end Test your instrument
E-mail	Easy to reach many customers at once	Hard to get a random sample Anonymity is an issue	User-friendliness is a crucial issue See "Additional Resources" for free software to design Web-based surveys
Phone	High response rate Can clarify, follow-up Can have some open-ended questions Can be used for other purposes (networking) Can reach a large geographic range	Time intensive for interviewer Must be short More difficult to avoid bias Requires good phone presence	Be sure you have a script Decide ahead of time how you will record Limit to 5–10 minutes Start with easy questions Identify who you are Test your instrument
In-depth, in-person, video conference, skype, etc.	Can take advantage of key contacts More opportunity for open-ended questions Can tailor the interview more to respondent Get both audio and visual clues about respondent	Very resource intensive Usually depends on preexisting network or contact with respondent	Use personal contacts and networks to find ideal respondents Schedule the time beforehand with respondent Make sure respondent is aware of the nature of the interview Choose a nonstressful time of day

focus carefully on crucial questions, but become lost in detailed questions that are not useful to the strategy of the business.

Write the survey questions to answer the key strategic question(s). Two decision rules in finding the right strategic questions and using them to guide your survey design are the following:

- Is the answer to the question crucial to deciding on the business strategy? If so, it qualifies as a key strategic question.
- Is the answer available elsewhere, that is, without asking survey respondents? If so, do not use the survey to answer the question.

You normally cannot develop a single key strategic question; instead, you should aim to break the strategic question into several smaller questions. The sum of these smaller questions will represent the underlying question you are trying to answer. Start with a list of things you want to know before finalizing your business strategy. To generate the list, ask yourself, "Why do I want to do this survey?"

Generating Survey Questions

Focusing on the key strategic questions, make a list of questions you would need to ask in order to answer the strategic question in a detailed fashion. Now write the actual survey questions, following these guidelines:

1. Keep each question clear and simple. Write for an eighth-grade reading level.
2. Check the question for potential ambiguities. You should test the survey with a small sample and then do a final review before sending it out or administering it in-person.
3. Be as direct as possible, without being offensive. Avoid questions relating to age, occupation, race, and gender unless they are crucial to your findings.
4. Put the question in a format that is easy to answer by using categories and check-off boxes or answers that can be circled.
5. Make sure that if you have categories, they are mutually exclusive and collectively exhaustive.
6. Do not ask a question that provides useless information.

Collectively, the answers should provide you enough information to make a strategic decision. For example, the survey could help you choose a promotion strategy, define required features of a product, or decide for or against targeting a certain market segment.

Reporting the Results in Your Business Plan

Many entrepreneurs fail to include the details of the market survey into their business plan. The reader will want to know how the survey was administered, how many respondents answered, and some of the demographics of the surveyed population. In your business plan, you should include the following information and qualifying statements about your survey results:

*   Put key insights and useful graphics in the text of the business plan to support your arguments that the market for your product is adequate. Consider visual ways (such as pie charts, graphs, and/or tables) to communicate the information to the reader quickly.
*   Provide the details of the survey and include a copy of the survey instrument if you think it necessary in the appendix of the business plan. Summarize survey results in spreadsheet-type formatting and include them in a well-labelled appendix.
*   If you do not have a random sample, you cannot test hypotheses about the larger population or make inferences in a quantitative way about the larger population. Instead, your survey provides some general feedback from a specific audience you have identified.

3.6 Additional Recommended Resources

General

Adams, Rob. *A Good Hard Kick in the Ass.* New York: Crown Business Publishing, 2002. Although this book is focused on high-growth, high-tech businesses, it is an excellent source of advice for most entrepreneurs. Adams focuses on the importance of knowing your customer, having execution intelligence, and properly staging your business.

Ryan, Rob. *Smartups: Lessons from Rob Ryan's Entrepreneur America Boot Camp.* Ithaca, NY: Cornell University Press, 2002. This volume is based on high-tech business but has a wealth of tips for anyone starting a business, especially in the area of knowing your customer base.

Small Business Town

http://smbtn.com/businessplanguides/bplan3_marketing.shtml

This is an excellent collection of online guides for market research, emphasizing a practical approach and identifying current sources of information.

Mission Statements

Grensing-Pophal, Lin. *Human Resource Essentials: Your Guide to Starting and Running the HR Function.* Alexandria, VA: Society for Human Resource Management, 2002.

Mission Statement and Consensus Exercise

http://www.squarewheels.com/scottswriting/mission.html

This is the very "homegrown" website of Dr. Scott Simmerman, an organizational consultant in South Carolina. The section connected to this link describes a simple but innovative means for getting people to talk about the mission statement.

BizPlanIt.com

http://bizplanit.com/vplan/mission/basics.htm

This site has good descriptions of mission versus vision statements. It also includes a section on mistakes to avoid.

Survey Work

CREATIVE RESEARCH SYSTEMS
http://www.surveysystem.com/sdesign.htm

This is a company specializing in survey design and implementation and its website includes a good basic discussion of survey design, including discussions of how samples can be biased and of the pros and cons of various survey techniques.

SURVIVAL STATISTICS

http://www.statpac.com/surveys/

This site is actually an advertisement for a book called *Survival Statistics*,[4] but it includes a wonderful tutorial on all aspects of survey design.

Here are several sites that allow you to create surveys online:

iNet Survey (http://www.inetsurvey.com/)

Surveykey.com (http://surveykey.com/)

Zoomerang (http://www.zoomerang.com)

SurveyMonkey.com (http://www.surveymonkey.com)

Market Research

SMALL BUSINESS ADMINISTRATION (SBA)

http://sba.gov/starting/indexresearch.html

This site lists a few key resources and also link to the online reading room maintained by the SBA.

Entreworld

http://entreworld.org

Follow the menu to Starting Your Business/Market Evaluation. You'll find a great list of links to all kinds of advice on market research.

Small Businesstown.com

http://smbtn.com/businessplanguides/bplan3_marketing.shtml

The free online guides at this site are excellent and provide practical advice along with step-by-step explanations.

Food Marketing Institute

http://www.fmi.org/

This site has a great list of institutes and associations related to food.

Researchinfo.com

http://www.researchinfo.com/

This site has a listing of resources for market research, including links to an archive of research software (including software for building questionnaires) and a directory of research companies.

Fuld & Company

http://www.fuld.com/

This is a private company that performs competitive intelligence. The website has excellent articles and suggestions about doing competitor research. Check out the list of sources for competitor information.

---

[4] David Walonick, *Survival Statistics* (Minneapolis, MN: StatPac, Inc., 2003).

Tools

OffStats

http://www3.auckland.ac.nz/lbr/stats/offstats/OFFSTATSmain.htm

A free service to help find statistics on the Web.

Here are several sites that allow you to create surveys online:

iNet Survey (http://www.inetsurvey.com/)

Surveykey.com (http://surveykey.com/)

Zoomerang (http://www.zoomerang.com)

SurveyMonkey.com (http://www.surveymonkey.com)

Farmer's Markets

http://www.cals.cornell.edu/agfoodcommunity/afs_temp3.cfm?topicID=272

For information on how to get started in farmer's markets, check out this website created by Cornell's Community, Food, and Agriculture Program.

Other

New York State Data Center (Online database) http://www.empire.state.ny.us/nysdc/

The resource is the product of a cooperative program between the U.S. Bureau of the Census and the New York State Department of Economic Development, Bureau of Economic and Demographic Information. It provides broad access to statistical series for and about New York's population and economy. It is constantly being updated to include new items. It also includes a directory of state data affiliates' resources as well as links to other data resources.

The population data include census information on town, cities, and villages and are, hence, extremely relevant to the local marketing research that the students need to do. Even more valuable are the economic portions, and especially the reports on personal income. All reports can be viewed or downloaded.

More Business

http://www.morebusiness.com/getting_started/primer/d992205228.brc

This site primarily features motivational pieces for the entrepreneur, but has an excellent short item with many suggestions for informal selling ideas for start-up businesses. It also has an essay on getting free publicity at the related URL:

http://www.morebusiness.com/getting_started/primer/v4n3.brc

Focus Group Work

Center for Urban Transportation Studies

http://www.uwm.edu/Dept/CUTS/focus.htm

This site includes step-by-step instructions, including how to create a discussion guide and tips on selecting participants.

Management Assistance Program for Nonprofits

http://www.mapnp.org/library/evaluatn/focusgrp.htm

This site includes step-by-step instructions for carrying out a focus group.

Bader, Gloria E., and Catherine A. Rossi. *Focus Groups: A Step-By-Step Guide* [Spiral-bound]. San Diego, CA: Bader Group, 1999.

Edmonds, Holly. *The Focus Group Research Handbook.* Lincoln, IL: NTC Business Books, 1999.

Kamberelis, G. and (Author), G. Dimitriadis, **2013** | *Focus Groups: From Structured Interviews to Collective Conversations*, Routledge, NY NY.

3.7 Video Clips

Videos have been selected from the Cornell e-clip collection. We will provide a description here of the video clips for this chapter.

Directions for viewing clips on the Prendismo Collection (formerly from www.eclips.cornell.edu):

- Visit http://prendismo.com/collection/. (You must subscribe with the site for full access to all features.)
- Click on the **Subscribe** at the top of the site. Students receive reduced rates.
- Once you have subscribed, type in the name or title of the clip in the **Search** at the top right corner of the screen.
- Click on the clip you wish to view.
- Click on the play icon to view the clip. If you have not become a subscriber, you will only view the first 20 seconds of the clip before being prompted to log in.

1. Mike Pratico (Founder of Davanita Design)

- Graduated from Cornell in 1998 with honors
- During school he formed, Davanita Design, an e-strategy and implementation consulting firm
- Responsible for most of the business operational needs and later, the sale of Davanita to Avatar Technology
- Owned and operated a women's fashion boutique
- Title of Video Clip: "Mike Pratico Discusses Importance of Mission Statements"

2. Steven Gal (ID Analytics)

- Undergraduate Cornell & Law Degree from Southern California
- Cofounded ID Analytics

- An attorney and former educator
- Experience in business, policy and legislative arenas
- Title of Video Clip: "Steven Gal Shares Thoughts on SWOT Analysis"

3. Elizabeth Ryan (CEO of Breezy Hill Orchard)

- CEO of Breezy Hill Orchard
- Breezy Hill is a fruit and value-added producer for New Yorkers
- Aggressively raised private equity to launch a national product
- Created a hard cider division of the company
- Title of Video Clip: "Elizabeth Ryan Shares Thoughts on Market Research"

4. Gabe Murphy (Founder of CommuniTech.Net)

- CommuniTech.net one of the world's most successful web hosting companies
- 24,000 clients in 130 countries
- Title of Video Clip: "Gabe Murphy Discusses Importance of Market Research Reports – Especially in Getting VC Funding"

3.8 Engineering Economics

This section of the text is meant to help you in mastering the principles behind engineering economics, with a focus on material covered in the Fundamentals of Engineering (FE) exam. In Chapters 1 and 2, we covered simple and compound interest, annual payments, and present and future value of single or uniform series of payments. Now we introduce the effects of **inflation** and constant value analysis.

One of the risks of investing in "cash," such as putting your money in an insured savings account, into a money market fund, or just locking it away in a bank box, is that over time, the purchasing power of that money may decrease faster than the after-tax interest earned by such investments. This is called **inflation**. What you could buy with $100 twenty years ago is more than what you could buy with it today. And what you can buy today with that $100 will be more than what you will be able to buy twenty years from now. The amount you are paying for tuition, room, board and books this year will not be enough for the same items next year. Again, this is due to inflation – the decreasing purchasing power of money.

The rate of inflation varies over time and differently for particular goods and services. The U.S. Bureau of Labor Statistics (and similar departments in other countries) keeps records on the increase in costs of various groups of goods and services in various geographic regions of the country. The overall indicator of inflation is called the consumer price index (CPI). The CPI is an index of the prices of a set of goods and services. From 1997 to 2007, the CPI yearly increase ranged from 1.6% to 3.4% with an average rate of 2.6%; the current 2007 CPI is 207.3, compared to 100 from 1982 to 1984. This means that in a little more than twenty years, we are now paying approximately twice what we used to pay for the same basic commodities before. Some countries experience inflation rates that exceed 100% per year. In the early 1980s, the United States experienced inflation rates of more than 13% per year; borrowing money to purchase a first home was almost impossible for most consumers. After the USSR was broken into individual states (~ mid-1980s), inflation rates in Russia were so high that the amount of rubles one could get for a dollar were changing even while waiting in line to exchange dollars for rubles at a bank!

Many factors contribute to inflation, but the main one is an increased money supply in the market and the willingness of people to spend it. If merchants can sell their goods at higher prices, they will charge those higher prices. If consumers have more money to spend, they will be willing to buy goods and services at higher unit prices. Increasing wage rates to account for increased costs of living puts more money into consumer's pockets. People with more money are willing to pay more for the same goods and services. This in turn provides incentives for producers to raise prices. If governments print more money to pay off debts, this also increases the supply of money in the market. The result: more inflation. Price inflation means it will take more money to buy a fixed amount of goods and services.

ACCOUNTING FOR INFLATION

Assume it takes P_0 dollars today to buy a bag of apples. If the future annual inflation rate is f, then the price, P_n, for that bag of apples n years from now is $P_0(1+f)^n$. The price P_0 is expressed in end-of-year 0 dollars, and P_n is expressed in end-of-year n dollars. Each price is expressed in the purchasing price of dollars at the time the purchase is taking place. We call it then-current dollars. Now suppose you invest P_0 dollars at the beginning of year 1 in an investment that pays an after-tax interest rate of i per year.[5] The amount of dollars you will have at the end of year n will be $F_n = P_0(1+i)^n$. Using these funds to purchase the cost-inflated apples, the number of apples you are able to purchase at the end of year n with F_n dollars is

$$\frac{F_n}{P_n} = \frac{P_0(1+i)^n}{P_0(1+f)^n} = \frac{(1+i)^n}{(1+f)^n}. \tag{3.1}$$

This increase in the number of apples you can buy at the end-of-year n reflects the real rate of interest earned over and above the rate of inflation. Letting this annual real (un-inflated) rate of return be designated by i',

$$\frac{F_n}{P_n} = \frac{(1+i)^n}{(1+f)^n} = (1+i')^n. \tag{3.2}$$

Hence, the overall market interest rate, i, is made up of two components, the inflation rate, f, and the uninflated real rate of return or growth, i':

$$(1+i)^n = (1+i')^n(1+f)^n \text{ or (taking the nth root of both sides),}$$

$$(1+i) = (1+i')(1+f). \tag{3.3}$$

Rearranging

$$i = (1+i')(1+f) - 1 \quad \text{or}$$

$$i = i' + f + i'f, \tag{3.4a}$$

or if equation (3.3) is solved for the rate of inflation, f, then

$$f = \frac{i - i'}{1 + i'}. \tag{3.4b}$$

From equation (3.4a), you can see that when there is no inflation, then market rate interest, i, will be equal to i'.

To find the past or future prices of things based on present prices, these present prices are either discounted into the past or compounded into the future, by using the inflation rate. These past or future prices will be expressed in then

---

[5] We add the qualifier "after tax" interest rate here, because the government would take some percentage of your yearly return; hence, your net accumulation is subject to your tax rate. For simplicity, you could ignore the qualifier, recognizing that there are possible tax implications to the money you accumulate.

current dollars. Clearly, the price of goods and services expressed in today's dollars are what they cost today, regardless of when they are purchased. So, if a bag of apples costs $10 today, that is the price one paid twenty years ago, and that will be the price one will pay twenty years from today, if those prices are *expressed in today's dollars*. Obviously they will be different if expressed in then-current dollars. If we express those prices in then-current dollars, we will need to, using the annual inflation rate, discount today's price back twenty years to get the then-current price twenty years ago, or compound today's price twenty years into the future to get the then-current price twenty years from now. Assuming the inflation rate is constant and equal to f, expressing prices in

- then-current dollars, the price twenty years ago was *$10(P/F, f, 20)*;
- then-current dollars, the price twenty years from now will be *$10(F/P, f, 20)*; and
- today's dollars, the price is $10 regardless when the bag is purchased.

To find what any then-current price is at the end of any year *n*, simply use the inflation rate, *f*, to convert a known then-current value at some known year, say, *m*, to a value at the end-of-year n.

Thus, for then-current dollars going from, say, 2000 to 2020:

The price of a bag of apples in 2020 = The price of the bag in *2000 (F/P, f, 20)*

Expressing the then-current 2020 price in today's dollars:

Price in *2020 (P/F, f, 20)* = Price in 2020 expressed in year 2000 dollars.

Hence, keeping everything expressed in year 2000 dollars:

Price in 2020 expressed in year 2000 dollars = Price in 2000 expressed in year 2000 dollars.

Expressing everything in year 2020 dollars:

Price in 2000 expressed in 2020 dollars = Price in 2000 in then-current dollars *(F/P, f, 20)* = Price in 2020 in then-current dollars.

CONSTANT-DOLLAR ANALYSES

Many analyses are made to show the rate of growth of investments based on a constant dollar value at some fixed or base year. Such analyses exclude the effects of inflation. After all, when you invest your money you want to be able to buy more goods and services after you have invested it over a period of years than you could before you invested it. If you simply have more money at the end of an investment period but can only buy the same or even less goods and services, your real rate of return, that is, your return after inflation is taken into account, is zero or negative. Such constant-dollar analyses are made using the real (uninflated) rate of return, i'.

Suppose you want to save money to pay for tuition in the year 2020. The amount you would need to put into a savings that had an overall interest rate of i per year (this is market rate interest that includes inflation), after taxes, would be

$$P_0 = Price\ of\ tuition\ in\ 2020\ (P/F,\ i,\ 20). \tag{3.5}$$

This would be the same as using the price in the year 2000, expressed in year 2000 dollars, and the after-tax real rate of interest, i':

$$P_0 = Price\ in\ 2000\ (P/F,\ i',\ 20). \tag{3.6}$$

Note that the price in 2020, expressed in 2020 dollars, is the price in 2000, expressed in 2000 dollars, multiplied by the factor $(F/P, f, 20)$. Discounting this back at the overall interest rate, i, gives us the amount we need to invest. This is the same as discounting the tuition expressed in year 2000 dollars at the real rate of interest, i'. Thus, the amount needed for investment expressed in year 2000 dollars is

$$P_0 = Price\ in\ 2000\ (F/P, f,\ 20)\ (P/F,\ i,\ 20) = Price\ in\ 2000\ (P/F,\ i',\ 20) \tag{3.7}$$

$$P_O = Price\ in\ 2000 \cdot \frac{(1+f)^{20}}{(1+i)^{20}} = Price\ in\ 2000 \cdot \frac{1}{(1+i')^{20}}. \tag{3.8}$$

Cancelling out the "Price in 2000," we can obtain equation (3.3).

EXAMPLE PROBLEM 1: INFLATION

Suppose you want to borrow money to buy a car. The loan company will allow you to pay off the debt in equal end-of-year payments for n years in constant dollar values, that is, today's dollars. Yet when you make each payment you must pay in then-current dollars. What are the actual payments?

Solution. To solve this problem first find the constant end-of-year payments expressed in today's dollars using the real rate of return, i'. If you know the overall rate of return (market rate including inflation), i, and the inflation rate, f, you can compute the real rate of return from equation (3.3).

$$A = Debt\ (A/P,\ i',\ n) \tag{3.9}$$

This A is expressed in beginning of year 1 dollars, that is, today's dollars. The actual payments A_y at the end of each year y will be increasing by the rate of inflation, f.

$$A_y = A(1 + f)^y \tag{3.10}$$

This is the same as applying the geometric gradient factor using an interest rate of i' and a gradient rate of f.

$$A_1 = Debt\ (A/P, f,\ i',\ n) \tag{3.11}$$

$$A_y = A_1\ (1 + f)^{y-1} \tag{3.12}$$

Alternatively, you could wait until you have enough money to buy the car. What equal end-of-year payments, expressed in then-current dollars, would you have to pay into a taxable savings account, offering an annual before-tax effective interest rate of i, in order to buy the car at the end of n years?

To answer this question, one has to determine the price of the car at the end of n years, and then multiply that by a factor to convert that future price to a series of uniform annual end of year payments, A, using the after tax rate of interest:

$$A = \text{(Price of car today)} \ (F/P, f, n) \ (A/F, i(1 - tax\ rate), n). \qquad (3.13)$$

EXAMPLE PROBLEM 2: REAL RATE OF RETURN

On your bank CD, the bank pays you a before tax annual interest rate of 3%, the inflation rate is 2%, and your tax rate is 28%. What is the real (uninflated) after tax rate of return on your CD, i'?

Solution. Using equation (3.2) (where we adjust the before tax annual interest rate by multiplying by 1 minus the tax rate), and remember that i' is the real rate of interest earned over and above the rate of inflation:

$$i' = \frac{1 + 0.03(1 - 0.28)}{1 + 0.02} - 1 = 0.0016. \qquad (3.14)$$

In this example, the adjusted for inflation interest rate is only "slightly" above zero (0.16%). If the inflation rate increased to 3% per year, the real rate of return would be −0.0082. You would be losing purchasing power, but less than you would if you put your cash under a mattress or in a bank lock box.

EXAMPLE PROBLEM 3: PRESENT WORTH

If the annual real rate (excluding effects of inflation) for interest and inflation have been 9% and 5%, respectively, what was the uninflated present worth of $2,500 three years ago?

Solution:

$$i = i' + f + i'f = 0.09 + 0.05 + (0.09)(0.05) = 0.1445$$

$$P = \frac{F}{(1+i)^n} = \frac{2500}{(1+0.1445)^3} = \$1,668$$

Summary on Inflation

Moving money over time at the real rate of interest keeps the dollar value (purchasing power) constant. Moving money over time at the inflation rate determines the price expressed in then-current dollars. Moving money first at the real rate of return to compute a constant dollar value at some year n and then moving that amount over the same number of years to determine the then-current dollar value is the same as moving the initial then-current amount over the same number of years at the overall interest rate. This overall interest rate, i, that is used in the other sections of this book includes both inflation, f, and the real rate of return, i'.

3.9 Problems

1. If inflation is 2% and annual real rate (excluding inflation) of interest is 10%, what is the uninflated present worth two years ago of an employee's wages that we pay $50,000 per year for today?

2. You have $75,000 in cash. You are debating between investing in a start-up business or buying a bank CD.
   a. Assuming zero inflation and an annual market rate (uninflated real interest rate) of 8%, what is the future value of your CD in eight years?
   b. If inflation is 3% instead, and annual real rate of interest is 8%, what is the future value of the CD in eight years?
   c. Compare the two values. Based on your observation, would a higher than expected inflation rate be a desirable thing for an entrepreneur (the borrower) or a venture capitalist (the lender)?

3. If the market interest rate is 14 % and the effective annual interest rate (real rate) is 9.5%, what is the inflation rate? Hint: solve for f using Equation (3.3) or use Equation (3.4b) directly; $f = (i-i')/(1+i')$.

4. You own a business in a university selling popcorn and lime juice, popular in-lecture snacks. You project that you would be able to sell 50,000 packets of popcorn and 75,000 bottles of lime juice in the first year. Sales are expected to increase 10% from the first to second year, but you expect sales increase to drop to 5% yearly from the second year onward because of competition. Popcorn sells for $3 each whereas lime juice sells for $1.20 each. You expect your selling price to increase each year the same as the change in the CPI (consumer price index), which is projected to increase 3% per year. What is your projected gross sales revenue in the third year?

5. Consider the three eras of product evolution. The era of ferment is when return on investment is slow. The era of incremental change yields the largest profits on investment, whereas the era of diminishing returns give decreasing profits for the same amount of investment.Now, inflation increases the price of goods and services. In which of these three eras, and under what conditions, would a high inflation rate make the entrepreneur richer by the largest gross amount (most increase in his or her purchasing power)? (Hint: inflation affects not only selling price but also production cost.)

6. When you borrow money from a commercial savings and loan at some interest rate (call it i, to be consistent with book terminology), what factors are built into this rate to protect the bank? Discuss in terms of i' and f.

7. You as an angel investor have been approached by a young group of engineers who present you with a start-up business plan. If you want your $1,000,000 investment to achieve a real rate of return of 10% over the next five years and you expect inflation to average 5% over this same period, then what would your investment need to be valued at in five years?

4 Developing Your Business Concept

The sign on the door of opportunity reads PUSH.

Unknown

There is nothing more difficult to take in hand, more perilous to conduct, or more uncertain in its success, than to take the lead in the introduction of a new order of things.

Niccolo Machiavelli

4.0 Entrepreneur's Diary

When I was trying to start my fish business, I had a very difficult time trying to raise the capital to launch the business; I needed about $500,000. The central idea of the business was to raise tilapia (no one even knew what a tilapia was back then) using indoor recirculating aquaculture system (RAS) technology, which was far removed from a tried-and-tested technology at that time. The problem was that just about every investment that had been made into starting fish farms had failed miserably. Investors often just laughed at you when you tried to initiate a conversation about investing in aquaculture. It even created a new term, "aqua-shyster." It really helps to convince a potential investor of your business's viability if you can take him or her to a working prototype of what you are trying to implement full scale. I could take investors to my research operation at Cornell University and show them a single tank setup, but I could not show them a working farm or even refer them to any existing farms in the United States that you could consider economically successful. My somewhat standard response became "If it were easy, everyone would be doing it already! " Being a first mover has the often-dreaded result of being a failure, but the excitement of a huge upside! Eventually, I found the necessary investors, but it was not easy. I guess I did convince some of them.

4.1 Overview

Your current employer has indicated that you can negotiate a license for a piece of intellectual property (IP) that you have developed in your own laboratory that

85

addresses a market need that you have identified. You understand how this technology works and have a pretty good idea of how much it will cost to manufacture. You think it is time to pull the trigger and launch your new business. You realize that this will take a lot more capital than you have, which you estimate to be around $300,000 to reach a breakeven basis. You do not have these funds, so you will have to raise the money from outside investors. How do you do this?

All successful fund-raising is centered around your being able to convince people that you will make them money on their investment. Being able to explain to others your company's business model or concept is a fundamental requirement and is what we focus on in this chapter. You can think of your business concept as being an expansion on your mission statement.

Part of the "secret handshake" rules in the entrepreneurial world is to use the distinctive jargon of this community. Engineers (the primary target audience for this text) have one distinct disadvantage in the entrepreneurial world: they are engineers and aren't familiar with the jargon. As undergraduate college students, engineers have had mostly four years of science, physics, and other technical courses that do not expose them to business-world jargon. One of the major objectives of this text is to familiarize you with these terms and phrases that the business and investment community throw around like engineers use mathematical terms. In this chapter, any time I use one of these terms, I highlight it in ***bold italics***. Try to use them yourself when talking or writing about your business and its opportunities.

Countless texts and journal articles seem to have been written on business models and concepts. Some writers even go to great lengths to expand on how a ***business concept*** evolves into a ***business model***. Conversely, some authors do not even differentiate between the two terms (*concept* vs. *model*) and will use them interchangeably. For the record, however, a general definition is that a business model is simply a business concept that has been implemented; that is, first you have to have a concept before you can implement it. I use the term *business concept* when we are developing the various parts of our model and *business model* to refer to our completed analysis and formulated strategy. Your business concept will rely heavily on first having done a thorough market analysis. We'll do that first in this chapter and then with this information discuss how you develop your business concept.

4.2 Market Analysis

All too often, inventors and engineers equate invention advantage with near-guaranteed commercial success. This can be a fatal trap in that even a competitively advantaged product will fail if it cannot be profitably delivered to an identified market segment. As an entrepreneur, you must seek to comprehensively understand and assess the nature of your targeted market. A respected marketing business manager[1] for a *Fortune* 500 company suggests the following questions to ask in this regard (and as a follow up to your SWOT analysis):

1. Is the target market definable?
2. Is the target market reachable?
3. Can the target market be profitably served?

Let's put this into really simple language: you have invented a "round peg" and now you have to determine if you can identify and quantify a round hole for your round peg. Unfortunately, many times, too many start-up companies try to force a square peg into a round hole; do not fall into this trap.

Is the Target Market Definable?

Your first step in market analysis is being able to define your market. Do you really know who will buy your product? This may not be the end customer, but a wholesaler or a retailer. In my fish company, my target customer group was not the people who eat fish in a 500-mile radius of Ithaca, New York, but the wholesalers buying our live fish at the farm who then distributed the live fish to small Asian retail shops in New York City and Boston.

> An investor must see a clearly defined market that your product will serve.

You will need to spend considerable time on defining your marketplace and become an authority in all its aspects (customers and their demographics, niches, suppliers, alternative products, trends and dynamics, etc.). You need to demonstrate that your start-up company is not going to be surprised by an unknown market niche, a competitor, or unseen trends in your marketplace.

The market usually can be defined in three dimensions:

- What (customer needs what, it solves what particular problem?)
- Who (customer descriptions by particular groupings or segments)
- How (how are customers' needs currently being satisfied)

You will need to gather information on each of these areas and then make judgments about the attractiveness of a particular part of some broader market for your start-up. You will also be documenting the dynamics of different market segments and how each dimension is changing with time, for example, declining demand, a segment that is increasing with time, and so on.

DIMENSION 1: WHAT

The first step before starting your business is to make sure that you are solving a customer's problem. This, of course, means you have to clearly define just

---

[1] In this case, this person is my younger brother, Matthew S. Timmons, who is a business sector manager for Lubrizol Corporation went private in 2012.

what problem you are solving, and this may require several written pages to fully document the size and scope of the problem. Once you have defined the problem, you need to determine what will motivate a customer to choose your product as means to solve the problem? Why would a customer choose to purchase your product over a competitor's? You will need to have some *value-added* advantage, which is defined as the enhancement added to a product by a company before the product is offered to customers. If you can couple solving a customer's problem by providing a solution at an affordable price, then you will likely sell your product.

Solving problems consist of saving someone time or money, providing for a special need, reducing boredom and tedium, or eliminating anything else that is unpleasant and someone wishes it would simply go away. Customers with disposable income have a long history of paying for convenience. Accordingly, it is important to know and document the following:

- Are the customers dissatisfied with current products?
- Do they have a positive reaction to your product idea?
- Do they have a need or want that you can fill?
- What is their willingness to pay?

In my opinion, this is probably the most important part of your entire market analysis. Describing the problem that your company will solve is the most important first step in this analysis. To have any chance of being successful or of raising investment capital, you must clearly define a current problem or a need that is being neglected or unrecognized. Your business plan might label this part of the market analysis as "the problem," and it will be the first section of your business plan's market analysis. The executive summary also will highlight a description of the problem and your hook to catch the customer. Giving your potential investor a clear idea and description of the problem you are solving is absolutely critical.

Go back to the horse-and-carriage days. Henry Ford recognized that people really were not that fond of feeding their transportation vehicle (hay and grain), cleaning up after the horse (waste management), providing a stable (not optional while a garage for a car is), and wanted to get from A to B much faster. Ford believed his automobile invention would solve a lot of problems associated with horses as a means of primary transportation. Another, more current example might be the need of young working parents to monitor their toddlers at their day-care center. Would a day-care center that provided remote monitoring for the parents have a competitive advantage?

DIMENSION 2: WHO AND ITS SEGMENTATION

Having clearly defined a problem that you believe your company can solve, you need to quantify the extent of the problem. Put simply, how many potential customers might want to buy your product? Investors think in terms of *market segments*, which are the geographic, demographic, or other customer classifications that can be used to break down a much larger market into smaller components. Once you have defined the various segments of a potential market, then your initial strategy should be to identify which segment(s) you will focus on. Dominating a small segment and then expanding is often a good strategy for a start-up.

Some of these segments might be much more reachable for your start-up, often by using geography as an initial definer of reachable (a later section expands on market reach-ability). Start-ups will typically need to provide a dis-proportional amount of support to their customers (compared to later years when your product is more mature and needs less customer support). Being able to hop in a car and be at the customer's location in a couple hours or less to quickly deal with a problem is a good way to create a base of satisfied customers and to minimize overall company cost in doing so.

Making strategic choices as how to deliver your product or where to locate the business will depend obviously on knowing the location of your customer base. Creating maps that show potential customers can be very convincing to potential investors, for example, we are going to start here in central New York (1,000 potential customers) and then expand from there into the regions shown here on our map as Phase II (10,000 customers), phase III (1 million customers), and then phrase IV (20 million customers). Knowing the regional characteristics of your target market will help you to implement an effective promotion strategy and estimate how much it costs to acquire and service customers. These same principles apply if your product can be delivered electronically. However, it is still a good idea to do your initial product launch in a fairly tight geographic area that you can easily service with real people showing face time. It is always a fundamental principle to build on a base of satisfied customers. Get it right with ten customers before you expand and have to deal with the problems of a thousand customers.

This entire information gathering is related to quantifying your customer base and understanding their purchasing behavior in order to identify your target niche, clarify your product concept, and fashion an effective marketing strategy. From these data, you can estimate market demand for your product, which is critical for building your financial model of the business and constructing your pro forma analysis.

Be careful in estimating market demand (expressed in dollars) based on some sales price you "think" you can receive for your product and the number of units that might be sold. The willingness and the desire of the consumer to purchase a product will be affected by its price, which then affects market demand in dollars. Apply your pricing strategies cautiously as you may eliminate the desire of the consumer to purchase your product. Remember that the product of price and quantity determines market demand.

DIMENSION 3: HOW

Closely related to the "Who" dimension is the "How" dimension, which is how the current market is being served; that is, who are your competitors and what are their characteristics. This will require determining how many dominant participants are in the current marketplace and defining the current total available market for all players including niche competitors. You should provide a concise description of the dominant players in terms of their gross sales, office(s) locations, numbers of employees, marketing strategies, and any other defining characteristics to their approach. You can obtain considerable information about your competitors by obtaining their annual reports, if they are a public company. You can determine a lot about a private company by reviewing their website or even buying products from them; for example, it is amazing how much information a competing company will give you if you want to buy something from it. In addition, studying your competitors can lead to the discovery of successful strategies you may wish to mimic, or unsuccessful approaches to avoid.

A dominant player is a company that has 30% or more of market share, and niche players will have 3% to 5%. There is considerable debate on what is an appropriate target for market share. The more your product is a disruptive technology, the higher you want to target your market share, for example, 50% or higher. Remember that it is much cheaper to maintain market share than it is to capture market share, which is usually done by using a much lower price than the current products serving the market. So for new markets, go for domination and let your competitors try to take market away from you.

> It is much cheaper to maintain market share than it is to capture market share.

Targeting (market strategy) a niche market share may be appropriate when you are only trying to capture some specific segment of some broader market, for example, the 3% to 5% mentioned earlier. The key to making a small market niche work for you is that it should be truly a neglected market, one that the big players in the market know about but have chosen to neglect. Neglected markets that are also declining (somewhat the opposite of a disruptive technology market, one that you expect to expand) will be a very tough sale to investors and ones you should try to avoid. Just remember, small players are easily marginalized by bigger companies. If your market share starts to grow, you will become a target for elimination or marginalization. So unless you truly have a disruptive technology, be careful here.

AFFECTING ALL DIMENSIONS: TRENDS

You will need to clearly establish the current trends in the broad market you are addressing. If you can, it is always ideal to start a business in an industry that is growing, but there can also be market opportunities when industries go through change. Either way, you need to be aware of the industry trajectory and of how it might have an impact on your business over time. The key here is to understand the current dynamics of your target industry and then to provide a clear description in your business plan.

If your product is new to the marketplace (Henry Ford and the automobile), you will be identifying current products (horses, trains, etc.) that meet some similar need or function. Of course, there is a chance that your new market will never successfully emerge, for example, Webvan – a $375 million IPO launched in 1999 that failed trying to deliver groceries to the home via Web orders and Coca-Cola's introduction in 1985 of New Coke. These failure-to-emerge markets are called "ghost" markets and are one of your worst nightmares if this should happen to you.

Your goal should be to position your company for future growth. You need to identify the key trends and technologies that are influencing your industry. Are there trends in the industry that point to increased and or stable demand for your product? What competing technologies are having an impact on the industry? Are there or will there be technologies that lower the barriers to entry, which will increase your competition? Will you need to reposition yourself strategically to use new technologies to keep people out of the market niche you plan to inhabit? These questions are not all-inclusive but should stimulate your thinking relative to your ability to compete effectively.

The important point in the preceding discussion is that your research documents the marketplace trends that are occurring and the type of market you are addressing. Investors do not like surprises. Be sure to properly understand the dynamics of the market as much as possible in your intended marketplace and develop your marketing strategy accordingly.

Is the Target Market Reachable?

A reachable market means that your company can effectively compete for customers in some particular area, usually defined by geography. In other words, do you have the physical capacity to deliver your product to the customer? If I had identified the Asian retail stores for my live tilapia product, I would not have been able to reach them. I did not have a live hauling truck to move the live fish product to their stores, nor did I or any of my staff speak Chinese.

Once you have defined your various customer segments, you will have to determine if you can deliver your product cost effectively to any of them. Having a product manufactured in the United States that is ideally suited for a Japanese customer is not viable to you if it costs you too much to deliver the product to this market. This may mean establishing local manufacturing or establishing relationships with strategic partners that are located in the immediate geographic area to reduce delivery costs and supporting costs. If your product is virtual with no real shipping costs, this will create additional markets that are reachable for your company.

Legal Issues

Make sure you have found any legal issues surrounding the market and your ability to compete in it. Do you need a government license or permit to sell your product? How long do these permits/licenses take to obtain? Are there any new laws or regulations that are bringing about changes that might affect your potential business? Be particularly careful here when your production process requires sourcing components or delivering products through international channels. Generally, use local legal assistance because it will be more familiar with local and state laws.

Can the Target Market Be Profitably Served?

If you cannot deliver the product to the customer and make a profit doing so, you do not really have a market. This may mean that your customers are too far from your product and transportation costs destroy your capability of making a profit. Several factors besides transport costs may determine whether you can deliver your product to some target market in a profitable manner.

You must be sure you have defined all the cost aspects of producing, delivering, and servicing your product and supporting your customers. Defining your company's overhead cost structure (fixed costs) will be critical in this regard. It is not sufficient to estimate manufacturing costs only. What other services and supporting personnel are needed to effectively compete in the marketplace. Some of these cost factors may further define your reachable market (just discussed).

4.3 Formulating Your Business Concept

Your business concept will generally have the following parts:

1. Core strategy, what you bring uniquely to the market
2. Marketing strategy, based upon a market analysis (how you address the 4P's of marketing)
3. Strategic resources, things you have that other people don't or things you have that are better than anyone else's (intellectual property)
4. Value network, contacts, and special capabilities of your group

Thinking ahead a little bit, raising capital to start your business will require that you write a business plan that has to make coherent sense and should embody the preceding components. However, you will not write a business plan that is a linear creation of the business concept components listed earlier. More typically, you will develop your business model by first doing a complete market analysis that allows you to develop your market strategy. Then, you will implement the market strategy in some manner, which we refer to as your ***business strategy***.

Got it? To keep the difference between business model and strategy separate in my mind, I like to use the automobile as an analogy. A particular automotive line is a model, for example, a passenger car, a light truck, a bus, and so on. Based on what you think you will use the vehicle to primarily accomplish (after doing a market analysis), you choose some particular vehicle as your business model. How you choose to use the features on this particular vehicle defines your strategy.

Now, it is pretty easy to extrapolate that from this same automobile (business model) you could choose to operate it in a variety of manners, speeds, or to follow different paths to go from A to B or even change the path to C to D. Making these changes in execution does not require you to make changes to the model. Conversely, you may find out that you really would be a lot better off if you could make a few changes to the basic model so that you could be more effective in reaching your current goals. Again, for example, say that your original model was to make short trips picking up middle-aged adults and taking them fewer than five miles. After doing this for a few months, you find out that you are typically being requested to pick up senior citizens, often with limited mobility, and take them to an airports that are 50 to 100 miles from your home base and that they have considerable luggage with them. So, if your original vehicle choice had been a small compact car, you will probably have to buy a different (change your model) vehicle to address what you found out to be your market niche or segment. Now, the better your initial market analysis is, the smaller chance there will be in having to change your model or in the preceding example: your vehicle.

Business Concept: Core Strategy

To me, core strategy and describing to an investor how your company is different from all the others is the critical first step to securing their investment. "Core strategy" means simply the way you intend to achieve your company's objectives. Let's put it in even simpler language: How will your business make money? The investor needs to be told your strategy in a plain and simple way.

Your business opportunity must be unique in some way if you are to be successful. The operative phrase here is how is your business differentiated from the others in the marketplace? To a large degree, ***differentiation*** is your core strategy. Why will a customer buy your product as opposed to existing proven products already in the marketplace? Forms of differentiation will typically occur in one of three primary manners:

- lowest cost
- unmatchable service
- unique product features

> Differentiation is your core strategy.

Let it be clearly stated that you cannot combine lowest price with either of the other two attributes. Service might be combined with unique product features, but typically you will focus on only one of the three preceding product attributes. We expand on these in our "Market Strategy" section. Your differentiation and your core strategy are most effective when you can clearly identify what your ***sustainable competitive advantage*** is within the marketplace with respect to your other competitors. If you cannot maintain this advantage or create new ones, then your business will not survive.

Closely related to differentiation is your *value proposition* to the customer. A value proposition is the unique added value that your company is bringing to a customer through your operation. Your management team creates this value, through either their product-to-market excellence (no one else can offer what you provide) or your providing a unique service to your customer. Not only must you provide value to the customer; you must do so in a way that the customer understands and in a way that you can communicate easily in your sales pitch and advertising. Remember what problem are you solving for the customer?

Business Concept: Marketing Strategy

Marketing your ideas, products, or services will be one of the keys to making your business successful. If you do not generate sales, no matter how good your product is, your company will not be successful. In most cases, your marketing and sales teams are the most critical part of your management team. As such, developing a marketing strategy should command your highest attention. Describing this strategy in your business plan will be central to obtaining investors.

Marketing strategy will need to identify your value proposition, which is just another way of saying that you are solving a customer's problem. Remember the simple definition of marketing: marketing means solving customers' problems profitably.[2] Using this definition of marketing, develop your marketing strategy built around the 4P's of marketing,[3] each based on a thorough marketing analysis (discussed in previous section). Your value proposition will describe the following (couched in the 4P's):

- Product: what are the key benefits of your product and what problem does it solve?
- Price: you have a price that customers are willing to pay.
- Place: how do you deliver the product to your customers and who are they?
- Promotion: how do you tell your customer about your product?

When developing marketing strategy, marketing managers must create a strong interdependence of all 4P's to generate a positive response from the target market. The rapid transmittal of information and consumer instant awareness brought about by the Internet makes it imperative that a start-up business keeps itself constantly aware of changes in the marketplace and how the market is responding to your marketing strategy. Quick changes in strategy are sometimes mandated. Also remember that each of these aspects of marketing will incur costs that should be addressed in your operations plan and in your cost analysis. The more subtle costs that you might overlook include ownership costs of maintaining inventory, negotiating contracts for purchase or related events

---

[2] Randal Chapman, "A Marketing Definition in Six Words," *MarketingProfs.com*, August 19, 2003, accessed DATE, http://www.marketingprofs.com/3/chapman1.asp.

[3] E. Jerome McCarthy. *Basic Marketing: A Global Management Approach*. 14th ed. (City: McGraw-Hill Education Europe, 1994).

such as advertising, order processing and bill collection, and providing credit terms to customers.

The last thing to keep in mind as you develop your overall market strategy is to try to create some flexibility in your plans. By flexibility I am stressing the ability to alter or change your strategy or product format. It is not really until you enter the marketplace that you will understand how well your product is being received. At this point, you need to be able, or should be able, to make a change in your product, to get it "right" this time. Not surprisingly, it generally takes about three major alterations to your product before you can settle in on the final product format. This also goes along with conserving your investors' money until you get it right. And once again, remember that if you run out of capital before you do get it right, you just lost your company. So as you develop your initial plans for your 4P's, stay agile and keep your fixed costs and breakeven prices as low as possible to prolong the time you can stay in the game.

THE 4 P'S: PRODUCT

Tangible, physical products as well as services are included in the definition of the term *product* as it pertains to marketing strategy. The product is the "what" being sold and needs to be clearly defined for the customer. Make sure your potential investor has a clear description of what you in fact are selling. It may be useful to define the product or service in terms of its features and benefits. The features are the product characteristics, such as size, color, design, horsepower, hours of operation, and so on. Benefits are the positive results derived from the product features. For example, a feature may be elevated omega-3 fatty acids in my tilapia fillets and the correlating benefit is improved cardiovascular health for those who consume my tilapia fillets. Benefits can be physical (improved health), emotional (buyers feel better after using) or financial (improved economic standing) or saving the user time, that is, provide convenience.

Your product must have some ***marketing advantage*** and you will need to identify and stress this advantage when you write your business plan. Hopefully, you can even demonstrate that you have some type of ***unfair advantage***. Unfair advantages can be leveraged into creating a successful company, hopefully for the long term. John Nesheim devotes an entire book, *The Power of the Unfair Advantage*, to describing this "(see Additional Resources at end of chapter).

Will anyone buy your product? Good question! Take some advice from Rob Ryan, a pioneer in the high-tech industry who founded Ascend Communications in 1989 that provided firms with the infrastructure they needed to keep up with the rapid growth of the Internet in the 1990s. Mr. Ryan (a Cornell University graduate) is a favorite guest lecturer at Cornell University for entrepreneurship classes, and he emphasizes the need to talk to customers. In fact, one of the first questions he will ask a new entrepreneur with a proposed business start-up is, "How many customers have you talked with?" Without firsthand knowledge of your customers, he argues, your business planning process will lack credibility. Rob Ryan's mantra in this regard is "Will the dogs eat the dog food?" We can talk all we want about what a great nutritious dog food we have, but if the dog won't eat it, then it is useless! As you go forward, you will need to document if customers will, in essence, eat your dog food.

> Remember Rob Ryan's mantra: "Will the dogs eat the dog food?"

THE 4 P'S: PRICE

First, let's get this basic premise established: cost of production has nothing to do with product price, other than to define when a loss occurs. This is particularly troublesome for engineers who almost always think in the simple approach: that product price is simply some fractional increase over costs. Price should be based on the value received for your product. It probably costs a less than a dollar to produce a CD that I pay $15.95 for, but I am happy to pay it because I think I receive $15.95 worth of value in the CD. Value received is the key.

Pricing in general will depend on the following:

- Just how valuable is your product/service to the consumer? Does it save time, money, provide satisfaction, or solve a problem?
- Is there a directly competing product, and how much do they charge?
- Is there a competitor with an alternative solution, and how much do they charge?

Built into the preceding considerations is your understanding what type of market you will try to compete in (niche/specialty markets, commodity, or somewhere in between). Niche markets have products with high perceived value and have minimal competitors, but will generate low product volume. Commodity markets are the exact opposite. Commodity products are basically generic (no differentiation) and companies compete on price and generate profits through high volume sales.

From your market analysis, you may have identified multiple segments in your marketplace. Each segment may permit a different pricing strategy because it is seeking different attributes and should be willing to pay different prices. This is why companies such as Toyota Motors have several different models and even different lines, for example, Toyota and Lexus. This provides Toyota Motors with different products and different marketing channels.

Now that you have identified your target market, do a careful analysis on how much it will cost to produce your product or service. This determines the minimum price needed to recoup your production expenses and avoid negative cash flow. It is often a fatal mistake to start your business by selling at a loss and say that you will worry about the sustainable pricing level once the company has been established. Accordingly, you should reject start-up strategies for which sales cannot cover the costs early. We call this the bleed rate: how much money are you burning per month, and how long can you last before you run out of money (generally not long enough)? Running out of money and company death or failure are one in the same. It is very painful. Often these situations go right back to having unrealistic predictions of costs of goods and/or anticipated number of product units being sold.

Now, after knowing your minimum price, you should ask two questions before you set pricing:

- "How much sales volume can I gain from a lower price, and can we earn extra profit doing so?" and
- "How much sales volume can we lose and still make more profit from a higher price?"

In other words, the preceding questions will help you try to establish a price elasticity curve. And again, the answers to these questions will be largely determined by the type of market you are addressing and the perceived value of your product.

This whole issue becomes problematic when related to your pro forma statement (discussed in a later chapter in detail). The foundation of a pro forma statement is the cash flow statement and, in particular, net revenues from sales. This assumes that you have a clear handle on predicting your costs and revenues, which is the product of sales price and the number of product units you sell. But the quantity of products produced at any one time, that is, economies of scale, can have a great impact on the cost of sales. There is a fine balance between obtaining lower production costs associated with larger production orders (lower cost per unit and higher margin) and tying up precious capital and keeping capital available by having smaller production runs (and having smaller margins on product sales because of higher unit costs). As you can see, with most things that initially appeared fairly simple, pricing is a lot more complicated when you get into it.

VALUE-BASED PRICING

I'll start with value-based pricing, because I think that in a start-up, you really need to focus here. If you are a start-up, then you certainly are unlikely to be a low-cost producer. More typically, your product will have features that are unique in the marketplace and as such should command a high price based on value, that is, value-based pricing is what the customer is willing to pay for perceived value. Value-based pricing is difficult because it depends on customers recognizing and wanting the added value of your product versus products the competition is offering. Determining how much a customer will pay will require you to do some careful market research to assess how the target market values the product and how much pricing flexibility you might have in order to respond to competitors' prices.

Of course, getting the customer to use your product for the first time so they experience these added-value features can be a problem. This goes back to Rob Ryan's mantra "Will the dogs eat the dog food?" In this approach, the answer had better be yes, and not only just yes; they also love the dog food and will pay a lot for it. As you raise capital for an expansion of your business, you will need to have clearly documented that customers are in fact willing to pay the prices you are using in your pro forma analysis.

The main problem with this approach is that customers sometimes are not willing to share the price they are willing to pay with the seller so your marketing surveys and testing may give you bogus information.

And, finally, a lesson to be learned (hopefully, without too much pain): if you are a totally new start-up and have no data on a customer's willingness to pay, be careful, cautious and dubious of what your potential customers who have never used your product say they are willing to pay; in actuality, they often do something very different.

COST-PLUS PRICING

Cost-plus pricing is probably the simplest but *worst* method of setting a product price. There are three primary reasons for this:

- Hard to determine your product's costs
- Costs vary with sales
- Costs can quickly change

Cost-plus pricing is a weakness in strong markets. Even when cost-plus pricing is used as a reference, it tends to act as a cap. You may fail to see the logic of increasing prices even though your company is being overloaded with orders (the marketplace is seeing value in your product). Cost-plus pricing precludes you setting prices based on perceived value, which is where much higher profits are found.

Here is an example of how a pricing method change (value added to cost plus) led to the ouster of the founder of Compaq computer (be careful). Compaq was started in the summer of 1981, when Joseph R. "Rod" Canion, James M. Harris, and William H. Murto, three senior managers from Texas Instruments (TI), decided to start their own company, each investing only $1,000! These three former TI employees didn't even know what their company would produce and market; managing a Mexican restaurant or manufacturing storage devices for minicomputers or beeping devices for finding misplaced items were among their original ideas. They settled in on building IBM portable PC clones and went out and obtained $2.5 million of investment capital from Ben Rosen, president of Sevin-Rosen Partners, a high-technology venture capital firm in Houston. Rosen became Compaq's chairman and offered an initial investment of $2.5 million. After making profits for ten straight years, Compaq had its first negative quarter in spring of 1991, and CEO and founder Rod Canion was dismissed by the Rosen-led board of directors (see how quickly things can change). Here is how the evolution in pricing strategy changed:

- **Late 1980s:** high price, high cost computers, depended on 'added value' (very select market)

- **1992:** Rod Canion, founder and CEO, is ousted (remember this when you are a company founder leading a group); new CEO Eckhard Pfeiffer (COO at the time for Compaq). The new strategy imposed by Pfeiffer was to make Compaq a low-cost manufacturer and to compete in the commodity market ('cost plus' pricing). Canion, in fact, wanted to reduce costs and compete but not dramatically enough to satisfy the board.
- **Old model:** PCs are like fine jewellery (limited market, high margins).
- **New model:** PCs are like milk (not much difference in the product and just about everyone needs one; let's go for volume sales, which is cost plus, and compete in a commodity market). Pfeiffer moved the company rapidly forward in a profitable manner, slashing costs of production by 50% and reducing at the time the cost of manufacturing to 13% of total gross sales (down from 30% earlier). Amazing story. Surprisingly, Compaq still exists (see www.compaq.com) ! Look it up their story on the Internet: http://www.fundinguniverse.com/company-histories/Compaq-Computer-Corporation-Company-History.html.

DISCOUNTING

Offering discounts is a slippery slope to start down. Be careful, and if you do discounting, be sure there is a clear logic behind it so that customers can easily understand the basis for obtaining a lower price. Avoid negotiating a price with every customer. There is a difference between a promotional campaign that introduces your product and discounting to obtain market share. On a promotional basis, you might offer your product at its listed price, but give a reduced price for the first month or for so many initial units. This allows the retailer to introduce the product more effectively with the goal of generating future purchases while clearly knowing the future purchase price will be at the nondiscounted price. I prefer to call lower prices for quantity orders simply my pricing schedule and to not use the term *discount*. Finally, sometimes companies apply a discount on payments when a customer pays by a certain time of the month; this is to reduce the accounts outstanding balance.

THE 4 P'S: PLACE (DISTRIBUTION)

This "P" is not where your business is located! Here, *place* is a verb and describes how you distribute your product or service to the target market in the most efficient manner. Distribution implies there is a channel for achieving delivery. So, this P is really a "D" for distribution, but "3P's and a D" doesn't sound as catchy or academic as the "4P's" of marketing.

Assuming you have customers who want to purchase your product, the basic question becomes how to move your product from its point of manufacture to the end user. Your product can go either directly from the producer to the end user (direct sales), or you can move the product indirectly (indirect sales) through as many as three intermediaries:

1. Producer → *Retailer* → Consumer

2. Producer → *Wholesaler* → Retailer → Consumer
3. Producer → *Manufacturer's Rep* →
   Wholesaler → Retailer → Consumer

For a start-up in which you probably have some value-added-type product and are focusing on some particular niche of a broader market, you will most likely focus on direct sales to the end customer. You might consider multiple channels and use of an intermediary as you gain some footing in the marketplace and start to understand your customers' needs more clearly and how an intermediary can expand your net sales.

Marketing channels make the flow of goods from a producer through intermediaries to a buyer possible. As a start-up, you need to determine which intermediary or combination of intermediaries is best suited to take your product to market. Whatever your channel design becomes, your distribution system will have to address the following customer needs (or at least know that you are ignoring them):

- A location to purchase the product (where)
- Information about the product (how does it work?)
- Customization
- Quality assurance (warranty, etc.)
- Lot size
- Assortment (how many models)
- Availability (how long does it take to receive the product?)
- Before and after sales service

As you go through the preceding list, it is a very good idea to listen closely to some of your first customers (test markets) as to what their needs are in order to obtain the product easily. Until you receive some of this customer feedback, don't try to do too much too soon. It is much less expensive to add capability to address customer needs than to have customer support features that are not needed or whose cost will not be borne by the customer. As you analyze just what to provide the customer, you should also review what your competitors are doing in this regard. It might be an opportunity for doing something different and, in doing so, for providing added differentiation between your product and your competitor's, allowing you to capture some particular niche of the broader market.

I consistently have had bad experiences using intermediaries in my start-up businesses. Remember this clearly: selling to an intermediary means that you have distanced yourself from the customer. It also means that you are relying on the intermediary to represent your product's value to the end customer. Your self-interest and the intermediary's self-interest may not be and generally are not the same. Use of the intermediary means you can lose control of your overall product quality and, in the end, lose customers.

When the intermediary represents more than one product (almost always the case), this means that your product is competing with the other products the intermediary is representing. Now depending on where your product ranks in terms of generating value to the intermediary, your product may be less vigorously represented than if you were doing this task. For example, if your product has features that the other products in the intermediary's bag of products do not have, then it may be against the interests of the intermediary to differentiate your product, particularly if it points out a negative aspect of the competing product. Think about this. The intermediary promotes your product and then you revert to direct sales, leaving the intermediary with only their other inferior products and you take their former customers leaving them with fewer customers for their remaining products!

Using intermediaries does have an advantage of potentially reducing your fixed overhead, which is a good thing. Another reason to use an intermediary might be that the intermediary could provide servicing support for the product once the customer has purchased it, for example, installation, maintenance, and/or repairs. These services become the responsibility of the retailer and would require only one level between you the producer and the end customer. By carefully selecting the retailer and doing some negotiating, you may be able to avoid some of the obvious conflicts of interest; for example, require that the retailer represent only your product and not your competitors' to meet some targeted customer need.

If you do choose to use a retailer as an intermediary, then you will need to consider if you will use any particular type of criteria in selecting your retailers. For example, you might establish certain criteria and then only use retailers that meet these, such as selling exclusively only through upscale retail stores or only through catalogue stores. Maximizing availability through inclusive distribution may work against you, because it may not allow the most profitable venue to compete.

Given the potential problems with intermediaries I have mentioned, you still might want to consider two primary options in your marketing channel. For example, you might use direct sales for large customers and indirect sales for small customers. In my tilapia production company, we established two channels. We sold directly to the wholesalers who then delivered to the small Asian groceries where customers wanted to buy live product, and we developed a website for selling to individual customers and restaurants from a fixed set of products. Although our price to the restaurants was twice the price we sold to our wholesaler, we still made much less money selling directly to the restaurants. Before you decide which channel to emphasize, make sure you estimate the costs associated with servicing each channel and the profits that will be generated from each. You may want to equally promote both channels. Remember that direct sales typically will require a sales staff and their associated expenses, or you will need an effective Internet sales site for customer servicing.

POTENTIAL PROBLEM WITH DIRECT CUSTOMER REQUEST

A potential customer calls you about your product. After discussing the needs of the customer, you offer to introduce this customer to your distributor. Your customer cheerfully replies that he or she would prefer to deal direct. Do you accept this offer? With a commodity product, you may not be able to support direct sales, as in this last example. In direct sales, you will supply value in the form of product information, warranty, and so on. But this added service and value does not come without a cost to you, so you will need to set a higher price than that for the indirect sales. Be careful you do not let your sales channel drive you into a nonprofitable situation or that you disenfranchise your retailer distributors by taking their customers. The key here would be for you not to duplicate capabilities between what your company does and the type of direct customer you sell to (minimal to no service or support after the sale) and what your primary distributors would be providing (support after the sale).

THE INTERNET

It is hard to imagine having a business these days without having a company website. If you have a website and if your product allows direct sales via the Internet, you will have some definite advantages in streamlining your distribution channels to only one (direct sales, producer to end user). The following are examples:

- Your store is always open.
- Your manufacturing location is irrelevant (can locate for low costs of production and/or overhead).
- Your store is virtually everywhere (customer enters store via the Internet).

Direct sales over the Internet do not allow the customer to touch and feel the product before purchase, functions provided by a retailer. Make allowances for this handicap in your sales and marketing strategies, for example, free trial period or free returns.

INVENTORY

Closely related to your pricing issues will be deciding the amount of inventory you are going to maintain and where you are going to maintain it. Larger inventories mean you can produce more of a product at one time and gain some economies of scale; for example, the first unit costs $10,000, and each unit afterward costs $500. Once you have an inventory, it will require a place to be stored, and this ties up your capital until you can sell and collect revenue from the buyer. The larger production runs may mean that you need to rent warehouse space or have your distributor network to maintain your inventory; all this comes with a cost.

Inventory control also has an impact on your shipping costs. Typically, small companies will contract their production to an outside manufacturer or fabricator. You might then ship directly from your manufacturer to select distribution points (distributors) to eliminate the double shipping cost. Sometimes these first receivers are called *master distributors*.

THE 4 P'S: PROMOTION

Promotion is communicating the relevant information about the product to the target market. Because promotion is the link between the market and the other three P's as well as generally requiring most of the marketing budget, novices often make the mistake of only considering promotion when developing their marketing strategy. Avoid this mistake by remembering that promotion is only one aspect of a good marketing plan.

Research is important for the development of effective promotional tactics. Study the target market and competitors to identify useful strategies. Keeping the audience in mind will prevent your wasting money and time on useless tactics, such as advertising in magazines that a target market doesn't read. Visiting competitors' websites or reading trade journals are useful ways to identify what competitors see as important and are emphasizing in their advertising.

There are a large variety of tools and activities that can be used to develop a promotional plan that is tailored to your company's needs. These include, but are not limited to, the following:

- Marketing collateral – the production and distribution of brochures, newsletters, flyers, and posters
- Promotional activities – event sponsorship, attending trade shows and fairs (job fairs, health fairs, etc.), coupons, free samples, and contests
- Public speaking and attending conferences
- Media relations campaigns – development of press releases, press kits, and public service announcements
- Advertising – print ads in trade journals, magazines, and newspapers; direct mail and e-mail campaigns; outdoor ads (billboards and bus boards); and broadcast advertising (radio, television, or Internet sites)

Business Concept: *Strategic Resources* and *Intellectual Property*

In your discussions with potential investors, your goal will be to convince them that you have capabilities and strategic resources that no one else has, sometimes called **profit boosters**. This is not particularly easy if you are trying to convince them that your team's skill sets (experience) are a unique strength, because they will think that there are always other people who can do what you do and that, in their opinion, maybe can do it even better. You may also try to convince the investor that you have unique products or services, something in your input or output function that you are perhaps uniquely and really good at, and ideally you may have some unique intellectual property.

I can just about guarantee you that one of the first questions a potential investor will ask is "Do you have any patents?" In my opinion, investors overemphasize the value of a patent, but nevertheless, be assured they will ask you if you have one. A patent is only one form of **intellectual property** (IP) protection, but it is the most easily identified form of a strategic resource. IP is defined as tangible or intangible results of research, development, teaching, or other intellectual activity. However, many people, particularly engineers, will tend to grossly overestimate the value of their IP and, in particular, the value of a patent. A patent is a negative right, meaning it prevents people from doing a specific activity. Change the activity a bit, and people are no longer violating your patent. Maybe you should have kept your brilliant idea a trade secret?

If you are really fortunate (strategic) and you can force other companies to go through you by means of your patent (remember negative right: others can't use it unless you approve), you can create what is called a ***choke point.*** This will allow you to generate revenues through licensing or royalty fees even though you may not be directly participating in providing any product to the customer.

You can almost be certain that your potential investor is going to be very interested in your access to or ownership of IP. Your job will also be to convince the investor of its value and if you don't have patent protection on your product or service, then you will need to have a pretty good story for why you don't need a patent; for example, you might have a ***first mover's advantage***.

A first mover's advantage can be tricky, however, because first movers often fail. However, if you do have a viable product and you are first in the marketplace, you should be able to capture nearly 100% of the market and easily more than 50% (because others have not recognized the market, yet, or are ignoring it). Once you have the dominant share by being the first mover, then you can build in methods to rapidly incorporate improvements based on customer feedback, which should allow you to be able to maintain your dominant position. Market dominance can lead to other positive factors for your business by allowing you to build customer loyalty even in the face of competitors' products. This can be converted into what is called ***lock-in***, when you provide some type of discount or benefit to returning customers. The airlines try to do this by providing frequent-flier miles.

Now, if you are so fortunate to end up with these kinds of wealth generators, try to be careful about being greedy (we talked about this earlier in Chapter 1). Johnson & Johnson had the first mover's advantage with its coronary "stent" and had 90% of this market with gross margins of 80%. It priced the product too high, however (why?), and this allowed Guidant and Boston Scientific Corp. to take over 70% of the stent market in a forty-five-day period. So, if I could give you one of the funny little phrases I was taught by a successful entrepreneur, "pigs get fat and hogs get slaughtered!" So, being a happy, fat little pig means you can go on to eat fabulously the next day and the next day after that instead of going to the slaughterhouse after a few gluttonous meals.

Although a patent can be powerful, it is time-consuming and expensive to obtain. Some investors will not even place any value on a patent until it has been contested in a court of law, and this means you will have gone through the legal expense of prosecuting your negative right against some abusing company. You may want to keep your IP as a trade secret because your invention via the patent becomes public knowledge. In fact, the patent requires that you teach the skills of using your patent in its disclosure. Your trade secret will not be disclosed to the public, and in fact, you can use this as a very defensible reason of requiring a nondisclosure agreement from a potential investor. A trade secret doesn't cost you anything to maintain either. A more complete discussion of patents, trade-marks, and copyrights was provided in Chapter 3. An excellent resource site on IP is the U.S. Patent and Trademark website: http://www.uspto.gov/index.html.

Business Concept: Value Network

What does your company bring to the table that others likely do not? Potential investors will be positively impressed when you can demonstrate the following:

- special relationships with key suppliers (as in some favored status compared to your potential or current competitors)
- ownership partners that supply critical components to your company
- coalitions with other businesses that provide you an advantage over your competitors

If you are currently working in the industry that you intend to bring a product, then this component of the business concept will probably be one of your major strengths. It is hard to develop key relationships when you are not a part of the industry. If you do not personally have this experience, give serious consideration to adding someone who does to your team.

4.4 Names, Slogans, and Logos

We were all given a name by our parents. Names are helpful in communication, in recognition, and in identifying a recipient for praise and attention. And if you haven't yet, when you name your first child, you'll see that picking a name is not a simple thing. There will be a whole bunch of constraints imposed on your choice, mostly to properly honor such and such a relative, to maintain a tradition in the family, and so on. It is no different when naming your start-up. You will face a whole bunch of issues and selecting the right name is fairly important. Similarly, if you have named – or do name – your first child some really weird name, it can affect his or her success in life! As a sidebar, the old country-and-western singer Johnny Cash wrote a song about being named "Sue" by his father as his father took off for other parts shortly after birth. His father said later in the song that he knew this name would force his son to get tough or die! Well, I think you see my point. So, here we go with a short discussion on names, logos, and slogans, because these three things pretty much go together. It can be fun, so try to have some fun with it. And you do need to have these items for your start-up's letterhead, website, and, of course, business plan.

Names

First, do not think you have to hire a public relations firm and spend lots of money to come up with a name. PricewaterhouseCoopers (PwC) did this to name its consulting division that was being split off in 2002,[4] and they came up with "MONDAY"! Wow, but as you can see, choosing a name is a very subjective decision; for example, a lot of smart people thought MONDAY was perfect to describe the consulting division of PwC. IBM abandoned this name shortly after purchasing the asset.

Given there is no absolute right way to go about choosing a name, here are my suggested criteria to guide you in the process as you select one:

- It should be short and easy to remember.
- It should conjure up the "right" images.
- It must not conjure up the "wrong" images.
- It should translate well to foreign languages.

The last criterion (in case you do intend to go international) is to forgo the embarrassment that GM endured with the Chevrolet Nova, which in Spanish means "won't go." Great name for a car, isn't it? "I've got a brand-new Won't Go. Wahoo." A funny example is from Japan, where one of their favorite drinks is Calpis (which, phonetically in English, means urine from a bovine animal). Yum, yum. My Japanese friends didn't understand my reluctance to try this drink during my first trip to Japan.

It will take years to build brand recognition for the brand name you chose, and this will generally require a heavy advertising budget. The ad budget for a national retail brand is just astounding. One of the major pharmaceutical companies actually spends more on advertising than they do on research and development. You won't start with a large advertising budget, but you should start promoting your name using inexpensive ways, for example, names on your shipping boxes, press events with local, regional, or national news agencies. When you have a chance to use your brand name, use it. Think resourcefully in your promotion marketing effort (remember the 4P's) and the use of your name.

In naming your company, it is a good idea to take guidance from the names already being used for various brands in the arena of your competition. If you are in the food or consumer goods industries, then that would suggest you might choose a proper name (names of people or places) of some sort, one of the best examples being McDonald's. That is a name with no original connection to fast-food hamburgers, but Ray Kroc chose it and it worked out pretty well for him. Food brands seem to be proper names like Famous Amos, Mrs. Pauls, Budweiser, and Coors, among others. The notable exception to this is Coke/Coca-Cola, which was (originally) a product-descriptive name.

Your product name should conjure up a positive or intended image. We all recognize why Mr. Kroc didn't name his hamburgers "Kroc's." "McDonald's" generates images of frugality, sturdy folk, and value. Can you imagine an advertising slogan such as "That's a Kroc"?

---

[4] The consulting division that PwC named MONDAY was sold to IBM in 2002 for $4.5 billion. IBM did not keep the name.

The soap makers have made an art form of naming their products: Tide, Surf, Joy, Zest, and so on. What would be a similarly heroic name for say a branded fresh fish product such as tilapia? It should be associated with the sea, with good taste, with value, with ease, and with elegance (if possible). Let's think about this for a while. It might be Anglo-Saxon (depending on where you are focusing your market), or not overly ethnic in any way, unless you are specifically focusing on an ethnic product. Ask people who in the world is famous for fishing the ocean, and you will probably hear "English people or Vikings." So perhaps a name sounding as if it is from the United Kingdom? My first list of names for my tilapia farm tried to capture the suggested criteria, for example, Pegg's, Michael's, Dewey's, Erik's, and so on.

None of these names knocked our socks off either, and I eventually settled in on Fingerlakes Aquaculture to conjure up the image of my fish products coming from a pristine lake setting in upstate New York where my farm was located. Of course, this name doesn't mean much to folks outside of the Northeast. There is no perfect name. Coming up with the name for your company may take some time to trigger an inspiration that works for you.

Slogans

Slogans are phrases to express the essence of your company. Your slogan is actually part of your advertising program. As always in communication, keep your slogan to as few words as possible, so it is easier to remember and can fit nicely on your company letterhead, branding symbols, and labels. A good slogan typically has one or more of the following characteristics:

- Creates vivid pictures in the reader's mind
- Appeals to people's emotions or sense of pride
- Motivates customers to buy

A good slogan as defined by these points should accomplish one or more of the following:

- Directs people to act: "Only you can prevent forest fires." (Smoky the Bear)
- Invents a new concept: "Welcome to Miller time." (Miller Brewing Company)
- Provides focus on target markets: "First in Chicago." (*Daily News*)
- Mentions your services: "The greatest show on earth." (Barnum & Bailey Circus)
- Provides detail about your product: "99 & 44/100% pure." (Ivory Soap)

Great slogans are difficult to forget after being heard only one time. Sometimes, you might even use rhymes to make your slogan even harder to forget, for example, "Takes a licking but keeps on ticking" (Timex Corp). You will probably recognize most of the following examples that are some famous slogans from years past:

"Don't leave home without it." (American Express)
"Good to the last drop." (Maxwell House)
"Built for the human race." (Nissan Motor Corp.)
"We are driven." (Nissan Motor Corp.)

"It's the real thing." (Coca-Cola)

"Just slightly ahead of our time." (Panasonic)

"Let Hertz put you in the driver's seat." (Hertz Corp.)

"Breakfast of champions." (Wheaties breakfast cereal)

"For relief you can trust." (Tylenol pain reliever)

"Is it live? Or is it Memorex?" (Memorex Corp.)

"Let your fingers do the walking." (Yellow Pages, AT&T)

"We bring good things to life." (General Electric)

"That's Italian." (Ragu spaghetti sauce, Ragu Foods)

"We bring the world closer." (AT&T)

As you review the preceding slogans, ask yourself what characterized their keys to success and then try to mimic those attributes as you develop your own slogans. Whatever you do, try to avoid bad slogans, which usually have one or more of the following characteristics:

- Have confusing puns
- Are longer than they need to be
- Feed the company ego and not their customer
- Fail to create a strong identity
- Create images that are unclear

Logos

Once you have selected your company name, you should also create a logo to visually identify your business. Think of your logo as a sign. Your logo should have two key characteristics:

- simple
- unique

The first characteristic (simple) is the key: don't get carried away with too much detail or with having unnecessary elements to your logo. A logo can capture a whole idea in a single image, with the "swoosh" symbol of Nike®

being one of the most famous. Developing a logo to put on your products gives a sign to customers that whenever they see that symbol, they know what they are going to get. The McDonald's® golden arches is a good example here.

For a start-up, use your company's name in your logo. The companies who do not use their name in their logo, for example, IBM or Texaco (the red star with a big T), have spent years creating the association between their logo and their company. As a start-up, you don't have the time or the advertising budget needed to do this. So, your first logo should include your company name because it will go on all your packaging, letterhead, and business cards. You don't want to lose these advertising opportunities by having a logo that no one will understand or have trouble making the association back to your company.

Logos should have at most three colors, one or two is probably better. Can you recall any logos? Probably most have only one color (IBM, McDonalds, Nike). People tend to remember logos with fewer colors. Your logo should copy well into black and white because a lot of your materials will end up being copied in black, white, and gray. Using more than one color can also result in higher printing costs for your stationery or boxing materials.

Before settling in on your final logo design, PDF/FAX a copy to yourself and see how well it transmits. Try to avoid fancy features such as three-dimensional images and lots of crossing lines that appear impressive on high-definition screens, because they will not show up well on copies or black and white versions. Just keep it simple (I said that before, right). Stick to basic shapes and line strokes.

Create a high-definition file of your logo to use as an electronic paste object. You will need this file when you have your business cards created, and it will be useful when sending e-mail, creating your own internal version of a company letterhead, or at strategic places in your business plan. Remember to put your logo on everything. Any company gifts that you give to customers should always have the company logo on it. This is not only good advertising, but it also gives you a certain amount of legal protection in terms of what the law may consider a bribe; that is, something with a logo on it will be more likely to be classified as advertising or marketing materials instead of as a bribe.

4.5 Additional Resources

Business Concept and Core Strategy

MeansBusiness.com
http://www.strategyandcompetitionbooks.com/Strategy-and-Competition-Books/All-the-Right-Moves.htm

The preceding link is to a resource that is part of the website created by a company called MeansBusiness. The site allows you to purchase in an ala carte mode, portions of material from the popular business press. The resource associated with the link above references sections of a book called All the Right Moves, A Guide to Crafting Breakthrough Strategy, by Constantinos Markides.

Business Law

Bagley, C. E., and C. E. Dauchy. *The Entrepreneurs Guide to Business Law*. City, ST: St. Paul, MN, West Educational Publishing, 1998.

4.6 Video Clips

Directions for viewing clips on the Prendismo Collection (formerly from www. eclips.cornell.edu):

- Visit http://prendismo.com/collection/. (You must subscribe with the site for full access to all features.)
- Click on the **Subscribe** at the top of the site. Students receive reduced rates.
- Once you have subscribed, type in the name or title of the clip in the **Search** at the top right corner of the screen.
- Click on the clip you wish to view.
- Click on the play icon to view the clip. If you have not become a subscriber, you will only view the first 20 seconds of the clip before being prompted to log in.

1. Leonard Bisk (Founder of Ultra Fine Technologies Ltd.)

   - Entrepreneur who has started six successful companies
   - Founder and president of a start-up, Modigliani Imports
   - Graduate of Cornell University
   - Title of Video Clip: "Leonard Bisk Shares Thoughts on Patents"

2. Walter Wilhelm (Walter Wilhelm Associates)

   - Owned and operated a women's fashion boutique
   - Executive at the largest U.S. menswear manufacturer
   - GM of CAD/CAM group of Hughes Aircraft Company
   - Cofounder of Microdynamics, an innovative tech supplier
   - Founded Walter Wilhem Associates as a product development and supply-chain consultant entrepreneur who has started six successful companies
   - Title of Video Clip: "Walter Wilhelm Discusses Intellectual Property Issues"

3. Gabe Murphy (Founder of CommuniTech.Net)

   - Founder of CommuniTech.net, one of the world's most successful Web-hosting companies
   - 24,000 clients in 130 countries
   - Title of Video Clip: "Gabe Murphy Discusses Sustainable Competitive Advantage"
   - Title of Video Clip: "Gabe Murphy Discusses Challenges of Pricing Pressure"

4. Rachelle Cracchiolo (Cofounder of Teacher Created Materials)

   - BA and MA from California State University, Fullerton
   - Taught kindergarten through sixth grade in the Fountain Valley School District (California)
   - Cofounder of Teacher Created Materials, a company providing teaching materials to teachers

- Growth includes development of decorative and technology products
- Title of Video Clip: "Rachelle Cracchiolo States Importance of Understanding the Market"

4.7 References and Bibliography

Chapman, Randal. "A Marketing Definition in Six Words." *MarketingProfs.com*, August 23, 2003. http://www.marketingprofs.com/3/chapman1.asp, site for MarketingProfs.com

Hambrick, Donald C. and James W. Fredrickson. "Are You Sure You Have a Strategy?" *The Academy of Management Executive* 15, no. 4 (2001): 48–59.

Hamel, Gary, *Leading the Revolution*. Boston: Harvard Business School Press, 2000.

Kaplan, Robert S. and David P. Norton. *The Strategy-Focused Organization*. Boston: Harvard Business School Press, 2001.

McCarthy, E. Jerome. *Basic Marketing: A Global Management Approach*. 14th ed. City: McGraw-Hill – Education – Europe, 1994.

Nesheim, John L. *The Power of Unfair Advantage*. New York: Free Press, 2005.

4.8 Classroom Exercises

1. Can you think of examples of small market share products? Why did they fail, or why are they succeeding?
2. With some products, it may be difficult to tell where the "end-user" is. For example, the entire supply chain for a manufacturer of plastic milk containers is

   Plastic resin supplier → Blow molder → Dairy processor → Grocery store → Consumer.

   From the plastic resin supplier's point of view, who is the customer for the plastic resin pellets that he or she supplies? Can you think of other examples?

4.9 Engineering Economics

This section of the text is meant to guide you in learning engineering economics, which should be useful for any aspiring entrepreneur beyond the basic objective of passing the FE test.

Arithmetic and *Geometric Gradients*

Uniform series formulas cover the instances when the change in a cash flow over time is flat, or doesn't increase or decrease. But cash flows can often increase or decrease over time because of many factors such as pay raises, money inflation, or step payments on a debt. The slope of this cash flow change can be either linear (increasing by a fixed value each period) or parabolic (increasing by a fixed rate each period). If there is a linear change in cash flow then it is said to have an *arithmetic gradient* (also called a *linear gradient*), and if there is a parabolic or geometric change in the cash flow it is said to have a **geometric gradient**. Diagrams of the two kinds of gradients, applied to present worth, are shown in the following figure.

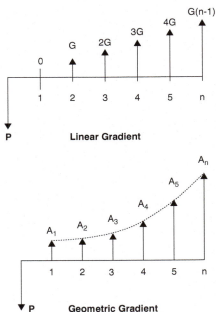

Note that the arithmetic gradient always starts with zero for the first year and that the geometric gradient does not. We use the term **G** to symbolize the amount that is added or subtracted each period for an arithmetic series and the term **g** to symbolize the uniform (constant) rate of increase or decrease of the cash flow for each year or period to period.

The equation for an arithmetic gradient can be deduced from knowledge of the equations for a uniform series and for a simple interest calculation. The gradient can be most easily visualized with the process of cash flow summation. Using this process, multiple cash flows can be added together to create a hybrid cash flow. For example, if a uniform cash flow of X dollars per year is added to a cash flow with Y dollars per year, then the result is a cash flow with $X + Y$ dollars per year. Similarly, an arithmetic or geometric gradient can be added to a uniform cash flow, as shown in the following figure:

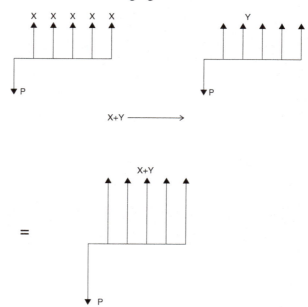

In the same way, a cash flow can be broken down into smaller components. In the example above, the uniform cash flow of $X + Y$ dollars per year can be broken down into a cash flow of X dollars per year plus a cash flow of Y dollars per year.

ARITHMETIC GRADIENT

The same process can be performed on an arithmetic gradient. Because the arithmetic gradient changes by the value G every period, then this is the same as a new uniform series of G being started in each new period and continuing to the end of the cash flow, as shown in the following figure:

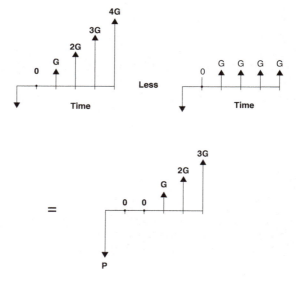

Knowing the equation for $(F/A, i, n)$, we can calculate what $(F/G, i, n)$ should be. Through a series of mathematical steps, the final equations to calculate the required annual payment (A) or the present value (P) for an arithmetic gradient of a uniform series of payments are[5]:

$$A = G\left[\frac{1}{i} - \frac{n}{(1+i)^n - 1}\right] \tag{4.1}$$

$$\text{or} \quad P = G\frac{(1+i)^n - i \cdot n - 1}{i^2(1+i)^n} \tag{4.2a}$$

or

$$F = \frac{G}{i}\left[\frac{(1+i)^n - 1}{i} - n\right]. \tag{4.2b}$$

(Notation for equation 4.2b[6]: calculates F given G or $F = G\,[F/G, i\%, n]$.)

Note that if cash flow for $t = 1$ is not zero, then we have to combine the uniform series formula to cover the basic cash flow, and the arithmetic gradient formula to account for the increasing portion is as follows:

$$P = A(P/A, i\%, n) + G(P/G, i\%, n)$$

or

$$F = A(F/A, i\%, n) + G(F/G, i\%, n),$$

where cash flow at $t = 1$ equals to A $(A = A_1)$ and increases with arithmetic gradient G over a period of n.

EXAMPLE – ARITHMETIC GRADIENT

Nora just opened her very own flower shop. Business was bare in the first month due to a pollen allergy scare however, and she earns nothing in the first month. Subsequently, however, she earns $500 in the second month, and her income increases by $500 per month during her first year of business as her fame grew. Using an interest rate of 12%, determine the following.

a. What is the present value of her first-year income?
b. If Nora had earned $1,500 in the first month as opposed to zero, what would be the present value of her first-year income.

Step 1: Find the periodic (monthly interest rate):

$$i = \frac{i_{nom}}{c} = \frac{0.12}{12} = 0.01\left(c = 12 \; for \, 12 \; months \; in \; a \; year\right)$$

---

[5] For details of a derivation, see Newnan et al., 2000.
  The reference for this should be: D.G. Newman, J.P Lavelle, T.G. Eschenbach, 2000, 11th edition, Engineering Economic Analysis, pp. 111–113; Engineering Press, Austin, TX.
[6] The formula for F/G is tested for in the FE examination is often not included in many engineering economics texts because it can be derived from the other formula; it is inserted here to make it easier for the reader.

Step 2: For (a), find the present value of a stream of payments with arithmetic gradient $G = 500$ over $n = 12$ months, where cash flow $= 0$ at $t = 1$ ($A_1 = 0$):

$$P = G(P/G, 1\%, 12)$$

$$= \frac{500\left[\left(1+0.01\right)^{12} - 1 - \left(12\right)\left(0.01\right)\right]}{\left(0.01\right)^2 \left(1+0.01\right)^{12}}$$

$$= \$30,284$$

Or use the tables in the Appendix for $i = 1\%$, $n = 12$:

$$P = 500\ (P/G,\ 1\%,\ 12)$$
$$= 500(60.5687)$$
$$= \$30,284$$

Step 3: For (b), find the present value of a stream of payments with arithmetic gradient $G = 500$ over $n = 12$ months, where cash flow is not zero at $t = 1$ ($A_1 = 1,500$):

$$P = A(P/A,\ 1\%,\ 12) + G(P/G,\ 1\%,\ 12)$$
$$= 1,500\ [(1 + 0.01)^{12} - 1]\ /\ [0.01(1+0.01)^{12}]$$
$$\quad + 500\ [(1+ 0.01)^{12} - 1 - (12)(0.01)]/(0.01)^2(1 + 0.01)^{12}$$
$$= \$47,166.96$$

GEOMETRIC GRADIENT

The analysis for a geometric gradient, as opposed to the preceding arithmetic gradient case, can be done in a similar fashion,[7] and you will eventually arrive with the following equations to predict the present worth (P) of a gradient set of payments:

$$P = \frac{A_1}{i-g}\left[1 - \frac{(1+g)^n}{(1+i)^n}\right] \quad \text{for } i \neq g \qquad (4.3a)$$

$$P = A_1\ n(1 + g)^{-1} = A_1\ n(1 + i)^{-1} \quad \text{for } i = g \qquad (4.3b)$$

Note that the preceding formula for the case $i \neq g$ can also be rewritten as

$$P = \frac{A_1\ [1 - (1+ g)^n\ (1+ i)^{-n}\ ]}{[i - g]}$$

EXAMPLE – GEOMETRIC GRADIENT

Scrooge has once again struck it lucky and found a new oil well. Although tiny, the well would bring $100,000 in cash flow in the first year. Scrooge estimates that cash flow should increase 5 % yearly. Assuming an interest rate of 3%,

a. what would be the present value of Scrooge's new oil well over the next eight years?
b. What would be the new present value over eight years if $i = 5\%$ instead?

---

[7] For derivation details, see Newnan et al., *Short Title*.

Step 1: Find the periodic (yearly) interest rate.

In this case, annual $i = 3\% = 0.03$.

Step 2: For (a), find the present value equivalent to a geometric stream of yearly cash flow with geometric gradient $g = 5\% = 0.05$, first period cash flow $A_1 = 100{,}000$ over $n = 8$ years:

$$P = \frac{A_1(1-(1+g)^n(1+i)^{-n})}{i-g}$$

$$= \frac{100{,}000(1-(1+0.05)^8(1+0.03)^{-8})}{(0.03-0.05)}$$

$$= \$831{,}584.85$$

Step 3: For (b), find the present value equivalent to a geometric stream of yearly cash flow with the same interest rate and geometric gradient, $i = g = 5\% = 0.05$, first period cash flow $A_1 = 100{,}000$ over $n = 8$ years

$$P = A_1\,n(1+g)^{-1}$$
$$= 100{,}000*(8)*(1+0.05)^{-1}$$
$$= \$761{,}905$$

EXAMPLE – ARITHMETIC GRADIENT

Betty expects to earn \$4,000 in the first month of selling her fabulously soft and chocolaty cookies. She expects her revenue to increase by \$500 each month for the next eleven months from the top-notch advertising service she has contracted. What is the present value of her expected first year revenue? Use an interest rate of 9%.

$$A = 4{,}000; \; G = 500; \; n = 12$$
$$\text{Monthly } i = 0.09/12 = 0.0075$$
$$\text{Present value} = A(P/A, 0.0075, 12) + G(P/G, 0.0075, 12)$$
$$= 4{,}000(P/A, 0.0075, 12) + 500(P/G, 0.0075, 12)$$
$$= 4{,}000(11.435) + 500(61.889) = 45{,}740 + 30{,}944$$
$$= \$76{,}684$$

Or using the formulae

$$= 4000\,\frac{[(1+0.0075)^{12}-1]}{0.0075(1.0075)^{12}} + 500\,\frac{(1+0.0075)^{12}-1)-12(0.0075)}{0.0075^2\,(1+0.0075)^{12}}$$
$$= 4{,}000(11.435) + 500(61.889)$$
$$= \$76{,}684$$

Breakeven Analysis, Payback Period, and Benefit-Cost-Ratio Analysis

BREAKEVEN ANALYSIS

Entrepreneurs with start-up businesses will often want to know how many products or units you must sell to reach breakeven. **Breakeven** is defined as when income generated from selling some quantity of units, Q, balances the expenses incurred to produce that product. You must realize that there will be some set of fixed expenses that you must pay even if you sell no units at all; we call these fixed expenses, and they consist of things such as rent, salaries, leases, advertising, and so on. Then you have variable costs, which are proportional to the cost of producing a single unit. Of course, equations can be developed that could characterize variable costs as say affected by volume or some other more complicated approach. However, breakeven analyses are generally done to give you a basic idea of how many units you have to sell before you go "cash positive," which is always a very happy day in the life of the company. There is a problem given to reinforce this concept.

PAYBACK PERIOD

When you are writing your business plans, you will want to include the simple calculation of payback period (PBP). PBP is defined as the time for the cumulative net annual profit to equal the investment. So, in simple terms, if you are showing a net profit of $100,000 per year and the total investment was $500,000, you'd report a payback period of five years. Simple enough (of course, this is all prorated based on percentage of ownership and the funds invested by an individual investor). The tricky part (for those studying for the FE exam, and for certain, one of these questions will be on the exam), what is the correct answer when say the investment was $510,000 with the same annual profit. The mathematical answer is 5.1 years, but, if you are asked how many years and the only answers were, say, 4, 5, or 6 years, then you'd have to answer 6, because at the end of 5 years, you haven't paid back the investment, so you'd round *up*, and the answer would be 6 years. The logic to this approach is that revenues may be mainly realized toward the end of the year or some other contrived logic. If you're lucky on the FE exam, the possible given answers will include 5.1 years and *not* 6 years. If both choices given, then round up to 6 years.

Note that when doing a payback period analysis, the effects of inflation and interest (equivalence calculations for time-value effects) are not considered (amazingly).

BENEFIT-COST RATIO (BCR)

The last of the triumvirate here, benefit-cost ratio (BCR). You will not typically use this term in entrepreneurship during your fund-raising events. Once you've launched your business, however, and cash becomes more precious with regard to how to spend it on competing needs, doing a BCR analysis can be useful in evaluating alternatives. The benefit-cost ratio is calculated as

$$BCR = \frac{PW \ of \ Benefits}{PW \ of \ Costs}.$$

Government projects almost always require this calculation be shown.

4.10 Problems

1. Refer back to the last example problem (Betty selling chocolate cookies). From extra word-of-mouth advertising, Betty finds that her actual increase in revenue comes up to be $750 per month instead of the $500 she expected. Still using an interest rate of 9%, what is the present value of her revenue now?

2. Answer the following questions based on the given cash flow:

   a.

Time	Cash Flow
0	1,250
1	1,500
2	1,750
3	2,000

   Does the cash flow have an arithmetic or geometric gradient? What is the value of G or g?

   b.

Time	Cash Flow
0	650
1	695.5
2	744.19
3	796.28

   Does the cash flow have an arithmetic or geometric gradient? What is the value of G or g?

3. James has patented a new tax filling software that solves all income tax return filling woes in New York. Not surprisingly, the software sells extremely well and has brought in $80,000 in the first year. James expects to open branches in other states during the next four years, at a rate of one branch per year. After opening, each new branch is expected to bring James an additional increase in revenue of $30,000. What would be James's effective annual income over the first five years? Use an interest rate of 6%.

4. Rita sees a new opportunity with first mover's advantage to sell ready-made frozen Mexican food in a small university town with a large Hispanic population. She estimates the initial cost of running the business to be $20,000 a year, but estimates that cost will decline for the next two years by $5,000 each year as her staff gained experience and becomes more efficient at their jobs (expenses are $20,000, $15,000, and $10,000). At an interest rate of 10%, what would be the equivalent uniform annual running cost of the business during the first three years?

5. Ms. Li expects to earn $50,000 in her first year of selling imported aromatic teas, a new specialty product. However, she projects that her revenue will decline by $2,000 each year over the next five years after the first year due

to her product's novelty wearing off and the competition. Using an interest rate of 10%, what would be her equivalent uniform annual revenue?

6. Chris earns $60,000 this year from selling apple cider made from his grandmother's old secret family recipe in supermarkets. As long as he can keep the recipe as a well-protected trade secret, Chris estimates his income each year to increase $ 8,000 at least for the next three years ($24,000 total) with his plan to expand his distribution channels to online and small grocery stores. Using an interest rate of 6%, determine the following:
   a. What is Chris's equivalent effective annual income during the four years?
   b. What is the future value of Chris's income during the four years?

7. Peter has hired a new marketing manager who promised to bring him 10% yearly increase in revenue over the next three years by using a clever promotion and product distribution strategy. If the increased yearly revenue was $100,000 in the manager's first year, what is the present worth of the expected additional income over the manager's first three years? Use an interest rate of 8%. (This could be used as a guideline on how to much to salary to pay the marketing manager, as in some fraction of the generated revenue increase to the company.)

8. You are considering different types of distribution channels for your product. You are confident of generating $1 million worth of sales yearly. How much of this sales revenue will you receive if you
   a. sell directly to your customer?
   b. go through a wholesaler?
   c. go through both a wholesaler and a retailer?
      Assume that you have to pay 10% commission to the wholesaler in part (b). For part (c), assume that the retailer takes 5% of generated sales, and that the wholesaler takes 10% of what is left before passing the money along to you.

9. Consider again the situation raised in question 8 with some modifications. Now, if you sell the product to the customers directly, you have to pay for sorting and transportation cost, which accounts for 12% of sales. Going through a wholesaler eliminates these costs because the wholesaler would do these services for you (it's part of its commission and a wholesaler can do so more efficiently). Furthermore, the retailer can increase your total sales by 10% because the friendly staff at the local stores can provide more intimate customer support and promotion. If you aim to maximize profits, which of the three distribution channels should you choose?

10. You want to calculate your breakeven point on the hydroponic business you are considering starting. From the analysis you used in your business plan, here are the basic parameters:
    • unit cost per system of $125 and selling price of $250
    • fixed costs to run the business (rent, salary, overhead, and Selling, General and Administrative costs or SG&A) of $12,000/year.

How many systems to you need to sell to break even? Investors like to know these values. Why?

11. What is the payback period for an investor if he or she receives 4% of the annual profits, his or her investment was $100,000, and the company's net profits (after tax) is $500,000.

12. Your hydroponic business has been successful for two years, so you are looking at doing more of the manufacturing process "in-house." A fiberglass molding machine has an annual cost equivalence of $5,000 (considers equipment life, salvage, up-front cost, etc.) and will result in annual increased profits of $10,000. What is the benefit-cost ratio (BCR) of this choice?

5 Creating Your Team

Associate yourself with men of good quality if you esteem your own reputation. It is better be alone than in bad company.

George Washington

5.0 Entrepreneur's Diary

I was getting pretty close to starting my fish business; I was concluding some final negotiations to secure the capital I needed to implement the business plan strategy (I needed $500,000). As you get involved in this entrepreneurship arena, you'll quickly discover that there is a lot of parallel processing going on. You will need to juggle several activities concurrently in order to pull this business launch off successfully. My situation was that I knew I was going to need a general manager (GM) about the same day I broke ground on the fish production building. So, about three months before I thought I needed my GM, I initiated a concerted effort to find a competent GM, who would be my first paid employee. I spread the word through my contact list and a national online site for aquaculture. I also received a call from Dr. Tom Fields, a colleague at a private fish farm near Saratoga Springs, New York, with whom we (Cornell University) had done business with over the years.

Dr. Fields was downsizing his salmon breeding operation and needed to let go his manager, Brian. Well, I quickly called Brian and told him about the job, and Brian was very excited. I told him that I did not have the capital "in hand," but I expected to shortly. Shortly, soon came! I called back Brian, but he had decided to take an alternative job with a State Hatchery facility (a safe and secure job). But, Brian recommended another person, Mike, with whom he had worked and who was currently the GM at a fish company that was in ownership transition. I called Mike, and along with my Cornell research mentor, Bill, we gathered at the Bob Evans Restaurant in Cortland, New York. Mike was a Cornell graduate, among other things, and had all the necessary work experience, and Bill and I both liked Mike. Now what? I called Brian (our original target for hiring, who said Mike was the best hatchery person he had ever seen) at home and got Brian's wife on the phone. I conducted an extensive pseudo interview with Brian's wife. I was looking for those other personal factors, for example, temper, stability, and honesty. We hired Mike. That was the best business decision I made for my start-up fish business.

Wherever you look in the management literature, almost every piece of advice written about management teams includes the word *balance*. Most experienced entrepreneurs confirm the need for the top leaders in a company to serve as both the visionary and the operational manager – tasks that draw on so many different skills that it is rare to find them in a single person. Although the creation of the business is usually best done by a "big-picture person," the long-term sustainability of the business depends on effective management of the "small details."

Although most experts agree balance is needed on a management team, they also agree it is hard to achieve. During the bootstrapping stages of a new company, you the entrepreneur usually do everything from taking out the garbage to making the high-level strategic decisions. You will develop a strong sense of control over every aspect of your business. When it is time to grow, you need to accept the probability that additional management skills will be healthy for the company, even though it will require you to give up some control. At that point, you need to set aside ego considerations and do a ruthless inventory of your own qualities and abilities in order to identify what is missing and how those gaps may affect the success of the company. Sometimes a good mentor can be a big help with this task because of the ability to bring an objective, outside perspective.

Ironically, if you succeed in creating a balanced team, you will probably also set into motion a level of "creative tension" in the workplace. Leading a company with people who are different from you can be a challenge. The entrepreneur's job is to keep everyone focused on the same set of goals, despite differences in backgrounds, approaches, and working styles.

5.1 The Start-Up Team

You and two or three founding team members will lay the foundation for your entire organization, but how you act as "employees" is, perhaps, of fundamental importance. Your team should share a common vision of the business plan and the methods you plan to use to build a lasting company. Few things impress venture capitalists more than a team that demonstrates they have

1. fully discussed the vital issues facing the start-up company,
2. achieved consensus on the proper courses of action, and
3. internalized this consensus to a considerable degree, so that the level of agreement is very high.

No matter how good your business concept is, you will not raise investment capital unless the investors are convinced that you have a strong management team in place. The best business plan and marketing strategy will fail if management competency is lacking. Some investors simply invest in individuals with a previous successful track record; that is, they will bet on a winning horse.

You may hear investors say they prefer a great team with a mediocre idea to a mediocre team with a great idea. Even if a business is not going to require outside investors, this wisdom still has meaning for every entrepreneur. Leadership of the company is crucial. If potential investors look at the management team and say, "Wow!" – you can be sure they will be anxious to find a way to become involved in the project, even if they think the idea you brought to them is not at the top of their current list of potential investment projects.

Talents and skills that are needed in a start-up team include

1. vision,
2. successful sales, marketing, information technological ability,
3. people management skills/human resources,
4. financial and legal expertise, and
5. operational competence.

A very common dilemma seen in start-ups is that a founder has a strength in one area (e.g., vision) at the expense of other functional areas (e.g., marketing or finance). For example, you can become so involved with product innovation (the idea) that you lose sight of the product's market potential. It is unlikely that you and one or two other founders will have all these skills, so you can balance the company's human resources by hiring the right employees and picking appropriate key consultants for the business. Adding partners and sharing company equity may be necessary to gain special expertise or to access to new markets or other resources. Partners can share risk, reduce the workload, and provide necessary resources (for example, cash, expertise, and network connections). But partnering with the wrong people can lead to disaster and to broken relationships.

When choosing consultants (see Additional Resources at the end of the chapter), be careful to select an individual or firm with experience working with small companies that have goals similar to yours. Ask each consultant for a list of references that includes other start-up businesses and call its references. Remember, if you move toward growth and/or acquisition or an IPO, you will need to think about "graduating" to a different team of consultants, but equally important, you must consider requirements for doing such from the very beginning of your business activities so that legalities are being met as you go forward, for example, needing to file quarterly accounting statements for three or four years.

Evaluation of Start-Up Team

Before you attempt to raise any additional money to expand your start-up, first evaluate the strengths and weaknesses of your team. You will need to evaluate team members by both the strength of their resume (because that is what an investor will do) and by how the team members have performed to date for you, as in actions speak louder than words. How well do the actions of each team member speak to his or her abilities to understand the company's vision, and how well does each team member work with the others? Here are some questions to ask in doing this internal evaluation:

1. Do the people currently on board have the critical experience and skills needed to move your company forward?
2. Is there evidence that the members function well as a team?
3. Will the reputations of the team serve to attract investors and other first-rate employees?
4. Are you missing any critical players?
5. Does your company vision fit with the current management team members?

Friends and Family and the 50/50 Approach

Many start-ups originate as a family-based business or may be financially supported by friends and family. You should be especially careful in crafting and managing the family member relationships that are intermingled with such a business, and it is best articulated in written form in your business's operating agreement. Be certain that all members of management have extremely clear expectations ahead of time about the risk, vision, scope, direction, and leadership structure of the business.

Choosing to partner with someone you know and trust is very appealing. But if leadership is not clearly outlined from the very beginning, the result can be disastrous because expectations of each founding partner may not be well understood by the others. In addition, there is a lot of room for misunderstandings about equity and ownership. If a friend, colleague, or relative helps out in some way during the initial stages of your start-up and expects "sweat equity" in exchange, there will be heartbreak and feelings of betrayal if those expectations are not met. Set clear expectations from the very beginning with everyone involved with your new business. Do not hold out vague promises for profit sharing and equity. Have a clearly defined plan from the very beginning, as defined in your operating agreement about how employees or others might participate in the success of the business.

And the worst scenario in forming your business is a 50/50 ownership structure with your founding partner. This prevents any clear demarcation of who is in charge, making decision making difficult at best. Starting out in a 50/50 ownership structure to prevent squabbling and bickering at the beginning is simply laying the groundwork for greater pain and heartache at a later date. Sure you want a team approach, but someone must be in charge. So, the 50/50 – don't do it.

5.2 The Board of Directors and CEO

The board of directors (BOD) has the ultimate fiduciary responsibility in a company, meaning that they may be sued by the stockholders for negligence or otherwise abdicating this responsibility. They have the responsibility for selecting the chief executive officer (CEO). Once the CEO has been selected, the members of the BOD should function as reviewers or counselors rather than trying to run the company for the CEO.

The CEO may need the BOD's assistance and support in obtaining financing. The CEO may need advice from the BOD on product and marketing mix, selling to key customers, and finding accounting and legal assistance. In these areas, the BOD may take a rather active role, especially because the board, in the seed round, consists of only a few members (say, five) with the key seed-round financiers as members.

You can expect that the early round investors will insist on board representation. For your start-up, try to minimize the number of board members, because each member will require time of the CEO to keep them informed and for the board member to feel a part of the company. The board member wants to be involved because he or she has probably invested money. Recruiting board members with name recognition is a key credibility point for the early start-up.

Choosing board members is important; as some may well be directors for life (four of the eleven directors of Intel have been there at least twenty-five years). When you are seeking investors, remember that the early investors can often help you to find additional capital from their colleagues. Also, they will often have experience in running companies and have other forms of expertise that will be helpful to your company. A good investor contributes much more than just money.

A good investor contributes much more than just money.

Receiving good legal advice during your start-up formation is critical. You may be tempted to include a lawyer (noninvestor) on your board. But, because you should keep your board small (only four or five members total), you may wish to avoid adding the lawyer to your board. Some lawyers are not entirely comfortable with the compromises and consensus building that must occur on a small board.

Consensus of like-minded people may have limited value. Hence, a homogeneous board should be avoided. Rather, people with a range of backgrounds should be selected. The most common trap is to include only people with tech backgrounds 100% aligned with the start-up plus some financial people.

Although the board should not meddle in the day-to-day activities of the company, its members should be able to spend the time necessary to learn about your company's business and understand its products, competitors, and customers. Board members who are merely well known but who lack the time required to learn about the business may be of little value.

Your CEO (probably you) should take a proactive approach to the board, bringing issues to their attention rather than concealing items or taking a defensive posture. The board's expectations should be spelled out clearly ahead of time.

Each board meeting should follow a well-planned agenda, consisting of information about progress and key issues that need to be dealt with. Materials for the board meeting should be given to the board members prior to the meeting so they have time to review and reflect before the actual meeting.

Although other officers of your company are rarely board members, and thus have no vote, you may wish to invite them to the board meeting to encourage an atmosphere of openness. This also gives the other officers an awareness of the board's views on various issues, as well as providing the board insight into the company's management team.

5.3 Building and Managing Your BOD

Board Structure

By definition, every corporation has a board of directors (called managers in a Limited Liability Company, an LLC). The shareholders (called members in an LLC) elect the board and the board elects the operating officers, e.g., the CEO. The minimum legal size of the board varies by state, but in most states, the minimum size is three people (typically a president, a secretary, and a treasurer – also referred to as the officers of the company). Review your state's rules on this matter. Do not confuse a board of directors with a board of advisers or a strategic or technical advisory board. The advisory boards can be very useful, particularly for creating credibility with outside investors.

The BOD must elect a chairman of the board who is often one of the insiders or key founders, such as the president or CEO. However, every case is different, and you may want one of your outsiders to be the chairman, say, in particular when the founders are young compared to some of the others.

Directors will need to be protected from personal liability due to the actions of the company to as high a degree as possible. Although director and officer (D&O) insurance is an option, it is very expensive (but required in an IPO) and generally not used by smaller, privately held companies. Additionally, make sure that the articles of incorporation of your company provide the directors with the highest limitation on liability afforded by the state in which the company is incorporated.

Selecting Your Board

For a start-up company, your board will be what is called a working board. These boards actually participate in the management of the company, particularly on the business management side. Such a board will meet frequently (monthly) and will have some rather spirited discussions as a start-up company is characterized by chaos and the need for rapid assessment of strategy and implementation plans. Other boards are called reporting boards, which meet several times a year to review status of the company. When things go badly, the reporting board will inform the shareholders that a change in company management is needed, generally meaning that the CEO (you) should be replaced. And, finally, there are lame-duck boards, which simply rubber-stamp actions of the CEO or founders.

As you build your BOD, try to expand your management capabilities. Select board members who add a needed dimension, and look for mentoring ability. Don't add members to the BOD who are carbon copies of your own competencies.

For your start-up company, your BOD may be one of your most critical resources. Finding and recruiting members for the BOD can be a dilemma for many start-up entrepreneurs, especially if they are young, because this is primarily a networking task. There is a lot of work in pitching your idea to each potential member. In fact, it may be necessary to have a short written business plan or at least an executive summary to show to prospective BOD members in order to get the right people involved. Sometimes your primary investor may dictate whom they want added to the BOD.

Before you get locked into a BOD (which legally every incorporated business must have), consider putting together an advisory board. In the very earliest stages of the business, the addition of an advisory board to support your core management team is a signal to an investor that you recognize the value of experienced perspectives afforded by a strong advisory board. It also shows that you have the initiative and salesmanship to formalize mentoring relationships.

Building an effective board should be one of your primary early goals. Potential investors will evaluate you partly on how successful you were at this. Assuming that you are the CEO (at least for a while), the CEO's ability to recruit top board talent reflects well on the CEO's ability to manage. Bringing top talent in also demonstrates that the CEO is not afraid of strong leadership and of constructive criticism coming from the board and that the CEO doesn't want to impose a single vision on the company.

Unfortunately, many entrepreneurs are so worried about control of their company being taken away from them by the board that they either opt for no board at all or fill their boards with longtime friends or family members. This is one of the worst-case examples for a management snafu in a start-up company.

A potential investor is always worried about how wisely you'll use their money. In particular, for start-up companies whose founders are young, the credibility of some "gray-heads" can be enormous. Always remember that one of your primary constraints will be lack of capital, and hence, one of your primary objectives is to raise capital in sufficient quantities to move your company aggressively forward. Thus, you must convince investors to invest.

A potential investor must be comfortable with your company before they invest and seeing someone that is thirty to forty years younger than themselves at the reins of the company can be a bit unsettling. The gray-heads provide some comfort to the potential investor. The following list gives some examples of the types of things investors are thinking about and how your board member selection can reduce their fears and/or concerns:

- The founders were able to get Ms. Fish, who has been very successful in the seafood business, so she must be a believer. I can invest with some confidence, because I know that Ms. Fish does not want to be embarrassed by this company's failure.
- Once I invest, Ms. Fish will be there providing valuable input to the young company founders, so my investment will be safeguarded.
- Ms. Fish will provide some sage advice from her years of experience and provide some valuable industry contacts for sales, so the young founders will not squander this money I'm investing (one of their biggest concerns).

And the preceding list could go on. Your job as company founder is to provide comfort to the potential investor that you will be successful because you will be gleaning the experience you lack from Ms. Fish.

How to Recruit Board Members

Your university alumni are a great source for board members. You can probably assume that there is some type of alumni listing of successful types or there may be an entrepreneurial program at the university, for example, the Cornell entrepreneurship program (www.eship.cornell.edu/). Use the business-sector trade journals or local banks to seek out talent. Try to land that first key outside person for your board. Work hard at this. Then the board can become your best salesperson to recruit other board members. Do what you can to arrange interaction between your star board members and a potential board recruit. Let personal chemistry work to your advantage.

Never ask a potential board member to be an investor. Once on the board, the board member will likely become an investor, but by their choice. If a board member does not become an investor after a year or so, you might want to rethink this board member's tenure; that is, why are they not investing? Does this mean they do not think the company is a good investment?

Board Compensation

Board members do not join your board to get rich. They are already rich. However, everyone wants to participate in the upside potential of a start-up. That's what we are all hoping for. Assuming a five-member board maximum, board members should receive 2% to 3% of the company ownership over some vesting period of, say, three years, or use 1% per year for up to five years. There are, of course, variations on this formula. Some immediate vesting should occur upon signing on, for example, 1%. You may want to have a tryout period for the new board member. For example, your pitch to them might be for them to attend two board meetings, and then the potential board member and the company can mutually agree that the person come on the board formally. This gives both parties a look-and-see opportunity. If it is not working, then neither party should feel wronged if the person does not come on. Removing a noncontributing board member once they are on the board can be problematic.

There are ways other than monetarily to take care of your board members. Like us all, they need to be sincerely appreciated for their efforts. Don't forget: they are basically doing you a big favor by sitting on your board (if they are good). Go out of your way to make their visit to the board meeting a pleasant experience. Make their travel arrangements or offer to do so; put them up in your house, and/or feed them home-cooked food. Provide presents from the company supplies (if you are making something that can be easily given) so they can share with their friends. Schedule some type of entertainment, for example, a fishing or golf outing? Have fun together! You will be surprised how much faster the relationships between the board members and the management team evolve and how much more effective they become once everyone is working hard but having a good time together. Most successful people actually like hard work if it is fun and builds toward a common goal.

Managing the Board

A good BOD can be one of your most effective management tools. An effective BOD will consistently bring good ideas to the table and solutions to problems. Bad boards are really bad in more ways than can be mentioned. They can destroy what little momentum your company is building and, worse, can bring failure to your company for a wide variety of reasons, for example, failure to communicate effectively with outside vendors and creditors, bad public relations, working against the interests of the company to save their own personal agendas, or protecting their own reputations (it was their fault). Board shakeups in these cases must occur, and in some cases, they become well known in the local community or public eye.

The following are some suggestions on how to manage and evaluate your board.

MEASURING BOARD COMMITMENT

Although not initially, but during the first two years of membership, a board member should become a significant investor. Some suggest a number of $100,000 as an indication of commitment and belief in the company. There are, of course, other ways to show commitment, for example, bringing investors to the table and/or arranging financing with local banks.

PARTICIPATION

During board meetings, how many questions are being asked by board members? How many motions and strategic concepts are being brought forward by board members? It is hard to give a target number, but one is certainly some sort of benchmark. Be wary of the strong silent types.

COMMUNICATION

The CEO should make a habit of bringing their board members up to speed each week or certainly at least biweekly on the current status of the company. The call can be short, just five minutes, but it keeps the directors involved and feeling that they are valued, which they are! This will not take much time effort by the CEO and is invaluable. Think how you would feel as a director.

AGE AND VITALITY

Be leery of potential directors with the "retired," "former," or "emeritus" in their titles. One is more than enough on your board. It is amazing how quickly these individuals can become out of touch, and they are the last to recognize it. Remember that there is a difference between being "old" mentally and being "old" in years. Some very "young" people are older than seventy years old.

WEEDING OUT WEAK BOARD MEMBERS

Some board members may become weak or ineffective over time. Establish a formal review mechanism for both the CEO and individual board members in which all board members participate. A key here is that the board develops the review criteria for effectiveness. Then, once a board member is not meeting these criteria, replacement time has come.

NEW MEMBERS

For new board members, provide an orientation notebook that includes the following:

1. Company mission statement
2. Corporate and governance essentials
3. Copy of the corporate bylaws
4. Directory of current board members
5. Board committees and their members
6. Directory of company officers and top executives
7. The board's meeting schedule for the next year
8. Copies of the company's annual reports and other important legal filings
9. Corporate code of ethics
10. Director membership criteria
11. Director compensation and benefits information

Automatic Board Seats

Investors making a sizable investment (say, 10% of your company's equity basis) often expect to be placed on the board. Keep your board small, certainly five or fewer members. An investment does not automatically qualify someone to be on your board. The board must be a working board and leverage their contacts and knowledge base for the maximum benefit of the company. Accepting $200,000 from Aunt Mary, who demands to be on the board, may be a poor choice compared to accepting $50,000 from Ms. Fish, who is well connected in her industry.

Company founders may want to place themselves on the board to hold places, but even this is probably not a great idea. Simply do not fill all board seats that were created in your company bylaws, until you have the appropriate people that can help you. You might take a board seat as a means to hold the spot, if you are willing to immediately step down when the right person appears. But sometimes, temporary board seat holders decide they really do not want to step down when the time comes. Don't expect Aunt Mary to step down when it is obvious that she should. Keep people off your board if they are not intended to be long-term placements.

5.4 How to Hire and Retain Good Employees

Most of us will start our first companies while we keep our "real" job (the one that pays us a salary). You'll be working eighty hours or more a week, so be prepared. Soon after or concurrent with your successful fund-raising campaign, you will be hiring one or more employees. If I haven't said it before and I may say it again, remember this adage when hiring decisions are in front of you: Hire slow; fire fast. This means a whole bunch of things, but among others, it emphasizes keeping your fixed-costs overhead as low as possible as long as possible. Why? Because you have to pay fixed costs whether you have income or not, and it is very painful to fire someone through no fault of his or her own when you do not have the cash to pay them.

> Words to the wise:
> Hire slow; fire fast.

You need to have a prudent and careful procedure in your hiring procedures. The objective of the hiring process is to find a person whose skills, abilities, and personal characteristics suit the job. There are two sets of factors to consider:

- Background factors, which include education and experience
- Personal factors, which include intellectual ability, personality, and motivation

Although background factors are important, personal factors may be more important, especially for a start-up. Background factors can be discerned by reading the job candidate's resume, personal factors generally cannot. One needs to collect this information from an interview or by other means (e.g., phoning references,). The last thing you want in a small company is personality conflicts between the members.

Background Factors

EDUCATION

Education is a pretty easy one to check out. How well did the potential employee do as a student and how good of a school did they go to? You should know that some research has found almost no correlation between early-career job performance ratings and prior GPA in school among just-graduated engineers.[1] Still, large corporations go for candidates with high GPAs, probably because this is the simplest approach and is also logical. You can find good employees that have lower GPAs, but you will have to be more careful in making sure these potential employees have what you need, for example, internal motivation and dependability.

EXPERIENCE

Experience is also important, because they will provide a historical record of previous work and will give you a much better idea of how the person performs out of an academic setting. There are three things to consider regarding experience:

1. Functional experience
2. Company experience
3. Level of responsibility

PERSONAL FACTORS

Personal factors include the following three factors:

1. Analytical ability
2. Creative ability
3. Decision-making style

As part of the personal factor evaluation, you will need to make reference checks. It should not surprise you that the people the candidate gives you will provide very positive comments about the candidate; that's why you were given these people as references. You need to be able to check out the candidate from people who interacted with your candidate but were not on his on her list of references. You can do this by telephone, so there is no record. Be very careful not to compromise a person's current position. If other people at the candidate's current employ are unaware that the candidate is seeking other employment, you do not want to compromise this person's current position.

The personal factors will be the most critical when deciding which candidate to choose from a group of nearly equally qualified candidates. There are entire books written on this subject. I'm not sure that after reading them, you are much better off in trying to be a better judge of talent and who will fit best into your company. Simplify the process: analyze beforehand what type of employee you need to fill a job and then choose the one that you think will work best with your

---

[1] Denis M. S. Lee, "Academic Achievement, Task Characteristics, and First Job Performance of Young Engineers," *IEEE Transactions on Engineering Management* EM-33, no. 3 (1986): 127–133.

current employees. Have more than one person interview a potential candidate, and especially have the people who the new person will interact with most participate in the interview process.

Recruitment: Pitching Your Company

Hiring an employee is a two-way street. While you are screening your candidates, they will also be screening you and your company. If you have prepared a full and comprehensive business plan, this part should be quite easy, because your business plan outlines the goals of the business and its strategy to move forward, and you should have an employee handbook detailing rights and responsibilities of your employees. Consider the benefits you will want to provide to both single employees and families with children. If you can provide day care for mothers and fathers with small children, that would give you a large advantage over other companies in hiring quality individuals.

Finding recent college graduates may be easy by pitching the upside opportunities of your company, but you may need to eventually consider people with spouses and families. Then you will have to consider other factors, for example, quality of your schools, quality of life in the community, and access to an airport, among others. Locating in an area where a companion or spouse can find employment easily would be advantageous. You don't have much control over these factors, however, other than when you make a decision on where to site your business, so think about these things before you do!

Common Mistakes

Although we may think we are excellent recruiters and judges of employee talent, we often make many common mistakes. Avoid the following in particular:

1. Political pressure – hiring a person strongly and personally recommended by one of your investors or board members even though they may not have some key qualities.
2. Unrealistic expectations – demanding too many qualities can have the effect of discouraging people to apply.
3. Just-like-me bias – hiring people who are just like you may lead to a lack of people that are qualified in other key areas than your expertise.
4. Delegating – while the CEO should delegate some tasks, hiring key people is not one of them.
5. Unstructured job interviews – this has the tendency to lead nowhere and to turn off the job candidate.
6. Not contacting references – this may lead to hiring people with inflated or invented experience, among other things.
7. Only checking supplied references– the candidate's references will almost for certain be very good.

How to Hire

Define the problem you need to solve through hiring. First, define the job profile:

- Position title – should describe the function and level
- Duties – responsibilities, task, and scope
- Required and desired experience and education
- Intellectual capabilities – analytical, creative decision making
- Personality characteristics – those believed to relate to good performance in the position
- Motivational characteristics – goals, interests, and energy level

Next, generate a list of candidates. Ask for recommendations from contacts, use conventional advertising, hold recruiting events, and so on. Then, methodically evaluate the candidates. Conduct structured interviews in which the candidate's competencies are assessed by asking behavior-based questions. Consider the background of the candidates when drawing up questions so that the questions will be appropriate and ones that the candidate can in fact answer.

Try to use short cases and situation-based questions to ask the candidate how he or she would respond. Contact the candidate's references and ask pointed questions. Have a prepared question list, or at least notes, when you call the reference persons, for example, please describe a particular project Candidate X worked on and describe their contributions and performance. You may have to be somewhat inventive in how you solicit candid opinions from a reference given by a candidate. You might try to go beyond the candidate's reference list to other people, for example, a coworker.

Finally, be sure to be completely familiar with the do's and don'ts of what you are allowed and not allowed to ask during an interview. Illegal interview questions include any interview questions that are related to a candidate's

- age
- race, ethnicity, or color
- gender or sex
- country of national origin or birth place
- union activities or interest
- religion
- disability
- marital or family status or pregnancy

The American Disability Act (ADA) prohibits an employer from asking an applicant whether he or she has a disability or from inquiring into the nature or severity of a disability (however, the employer may ask questions about the applicant's ability to do the job). The National Labor Relations Act (NLRA) prohibits employers from questioning employees about union membership or activities. Neither Title VII nor the Age Discrimination in Employment Act (ADEA) prohibits any specific questions. An employer, however, should not ask questions that may imply discrimination. Moreover, some state laws expressly prohibit certain types of pre-employment questions, such as questions about marital status or number of dependents. Check into these rules further with a local attorney or the community chamber of commerce.

5.5 Contracts

A contract is a legally binding agreement involving two or more people or businesses (called parties) that sets forth what the parties will or will not do. A contract is formed when competent parties mutually agree to provide each other some benefit (called consideration), such as a promise to pay money in exchange for a promise to deliver specified goods or services or the actual delivery of those goods and services.

Contracts can be oral or written; exceptions include contracts involving the ownership of real estate and commercial contracts for goods worth $500 or more, which must be in writing to be enforceable. Unfortunately, with oral contracts, if there is a dispute between the contracting parties, they are difficult to enforce. Remember, ink (written) requires no memory by the participating parties. Some simple advice: try to avoid ever having to use legal means to enforce a dispute. In most cases, you will lose much more than you ever gain. Try to work out some compromise that you can accept, even if you think that you are being taken advantage of grossly.

> A contract in *ink* requires no memory by the parties involved.

Ask yourself, "Why do people have disagreements in the first place?" Probably because they failed to reach a clear and agreed to understanding of what the two parties were to do for each other in the first place. Spend time creating contracts that you clearly understand and will support. And you should probably ***have your attorney review your contracts*** before you sign them to be safe.

5.6 Employee Contracts: Hiring

The most common type of contract that you will need as you begin your start-up is an employee contract. Before hiring the candidate or as part of the employment contract, include the following:

- **Preemployment clearance.** Make sure that the employee has not taken any intellectual property from a previous employer, and make a statement to that effect as a part of the employee's written employment agreement.
- **Confidentiality agreement.** Every employee should sign one as part of the hiring process. If you ask an employee to sign a confidentiality agreement at

a later date, then you should provide some extra compensation, not merely his or her continued employment.

Once you have finished the above two tasks, you can proceed with a formal employee contract. You will want to protect certain interests of your company by including several clauses in your contracts that include the following:

- **Covenant not to compete:** The employee agrees to not compete with the former employer for a period. These are generally not enforceable for periods of more than one year. You can also consider area restrictions, for example, not east of the Mississippi, if you are doing business say in the Northeast. (More detail on this later.)
- **No-moonlighting clause:** Prohibits the employee from participating in any business activities (even after-hours activities) unrelated to the employee's job with current company. Moonlighting should also be covered by an explicit policy in the employee handbook, as in you can't do it.
- **Nondisclosure agreement:** Prohibits the employee from using or disclosing any of the company's trade secrets unless the company authorizes it. The prohibition continues even after the employee quits. (More detail on this later.)
- **Antipiracy or no-raid clause:** This restricts the employee's ability to hire employees from their previous (or current) employer for a period; one year is typical.

Noncompete Clauses

Noncompete clauses are usually only for technical staff and senior management. You should use an attorney to help draw up an enforceable noncompete clause that will vary by state. Generally speaking, noncompete periods longer than two years are not enforceable, but they vary from state to state and will depend on the nature of the business. If you are the employee, then the covenant not to compete should be viewed carefully before you sign your employment contract. You do not want to be burdened by a frivolous lawsuit from your former employer that saddled you with a three-year noncompete clause even though it is probably not enforceable. Don't sign such a contract; seek other employment.

To be binding and enforceable, the covenant not to compete must meet certain requirements:

- Ancillary to some other agreement (e.g., a formal employment agreement)
- Protect only legitimate interests of the employer
- Limited in scope (duration, geography, and affected activities); for example, a noncompete clause restricting competition for longer than two years would usually be thrown out.
- Not be contrary to the interests of the public. This varies from state to state.

BREACH OF A NONCOMPETE OR NONDISCLOSURE CLAUSE

If the court finds that the employee breached a valid noncompete clause, the court will impose liability on the offender, with the most common form of relief being an injunction that would require an employee to stop competing against the former employer. In some cases, actual damages are assessed.

Breaching a nondisclosure clause can result in serious negative ramifications to the employee. For example, Patrick Worthing was employed in the fiberglass research center of PPG Industries, Inc. of Pittsburgh. Worthing attempted to sell Owens Corning, a competitor, numerous types of confidential research information from PPG. The Owens Corning executive immediately informed PPG and the FBI, which led to the arrests of Patrick Worthing and his brother, Daniel Worthing. The grand jury indicted the Worthings who then pled guilty to the court. The court sentenced Patrick Worthing to fifteen months in prison and Daniel Worthing to five years' probation including six months of home detention.

5.7 Employee Contracts: Restrictions During Employment

You may want to start a company while you are still employed to your primary employer. Sometimes this is allowed and other times not. The decision will be related to how your current duties relate to the company you propose to create. In some cases, this can work to the benefit of your primary company (employer), but this decision will depend on the type of employment category you are in. There are three main categories of employment:

1. Key employees (officers, directors, managers)
2. Skilled employees (engineers, marketing specialists, sales representatives)
3. Unskilled employees (draftsmen and -women, technicians, assembly-line workers, custodial staff)

Key employees owe a duty of loyalty to their employer. This duty, which exists regardless of whether or not there is an employment contract, prohibits employees from doing anything that would harm their employer, for example, creating a competitor. Neither key nor skilled employees may compete with their employer or solicit fellow employees to work for their new company during the period of employment. Both key and skilled employees may make plans for a new company, however. Unskilled employees, unless they have explicit no-moonlighting or noncompete clauses, are unrestricted in their off-hour interests unless these activities are detrimental to their employer's interests.

Restrictions by Types of New Venture

Your restrictions on what you can and cannot do will also partly relate to the type of new venture that you are creating.

NONCOMPETING

If the new enterprise is a noncompeting business, the employee (key, skilled, and unskilled) is free to establish and operate the new venture so long as it does not interfere with his or her current job performance or violate any provisions in any employment agreement. The conditions that you must follow if you remain employed with your current employer include the following:

- No employer resources may be used.
- All activities must be conducted after hours.

Subject to these conditions, the employee may

- make telephone calls and emails
- rent an office.
- hire employees (but not coworkers from the original employer)

COMPETING

If your new venture will compete directly with your current employer, your actions are more restricted as follows:

- Key employees and skilled employees may not prepare for and plan the new venture if doing so would interfere with their job responsibilities.
- Under no circumstances may the employee (you the entrepreneur) be involved with the operation of the new business while still employed.

Solicitation of Coworkers

Solicitation of fellow coworkers may be a problem. Say you are a skilled employee. If you hire employees who were under an employment contract for a definite term away from your employer, you may be liable for damages for encouraging the worker to break that contract. Even if you are not under contract and your employment is "at-will" (terminable at any time by either party for any reason), you can still be held liable if your conduct leads the coworker to violate any restrictive covenants, such as nondisclosure, noncompete, or misappropriation of trade secrets.

Key employees are even more restricted. If a key employee induces another employee to move to a competitor and the inducement is wilfully withheld from the employer, the conduct is a breach of fiduciary duty.

5.8 Leaving a Company

You have decided that your start-up company now needs your full-time attention and that it is not ethical to stay in your current employment any longer. You have probably already had some discussions with your current employer about

Table 5.1. Rules to Follow When Leaving an Employer

Number	Rules
1	Take nothing with you.
2	Take no notebooks, meeting notes, schematics, drawings, business plans, or portions thereof.
3	Make no copies of anything, not even magazine articles with your handwritten notes in the margins.
4	Tell your supervisor of the whereabouts of any electronic information storage devices that you had access to or used.
5	Do not write any portion of a business plan until well after you have left your employer, preferably months afterward.
6	Do not meet with cofounders until after all of you have left your employers.
7	Do everything you can to prevent the possibility of a lawyer twisting your actions into a story that makes you look like a sneaky thief.
8	After leaving your employers and upon your first meeting as founders, get your patent counsel and general counsel to write a letter to your former employers stating what business you plan to enter and asking what specific proprietary information the employers may feel you possess.
9	Talk to no one from or at your former employer until your legal counsel gives you the all-clear signal.

Note: This table is based on Nesheim, *The Power of Unfair Advantage*.

your start-up and may have even been giving periodic updates. Sometimes, the employer may actually invest in the start-up, or they may also immediately terminate you.

During these discussions, you may have received verbal assurances that everything is fine and not to worry about such clauses as discussed earlier, for example, noncompete. It is highly likely that if your new company becomes successful and in particular starts to hurt your former company in any manner, your former employer will bring suit against you for damages of some sort. So, be sure to at least ask for a signed bill of good health that you have not violated any of the above restrictions during your employment (see Rule 8 in Table 5.1).

When the time comes to leave a company, you should follow procedures for resigning very carefully. Usually, you must resign in writing to your supervisor with a copy of the letter sent to the director of human resources. After handing the resignation letter to your supervisor, he or she may very likely collect your keys and security passes, summon the security officer to escort you to your office to collect your personal effects, and then escort you from the property.

If you are working for a technology-type company, the company's trade secrets are an important issue, especially if you leave a company to start your own business. You should work hard to "get out clean." Nesheim provides summary guidelines for leaving your company in Table 5.1.[2] Some advice is hard to follow, for example, rule 5, not writing a business plan until after you have left.

[2] John L. Nesheim, *The Power of Unfair Advantage* (New York: Free Press, 2005).

5.9 Additional Resources

General small business resource site

www.nolo.com/lawcenter

CitiBusiness

http://us.citibusiness.com/

Small Business Network

http://www.americanexpress.com/homepage/smallbusiness.shtml?aexp_nav=smbustab

These two sites, by Citibank and American Express, respectively, contain a wealth of practical advice for start-up entrepreneurs. The Citibank site has a section I recommend called Business Builders that includes advice on structuring the business. The Small Business Network has an Ask the Advisor feature where you can ask questions and get feedback.

Choosing and Managing Small-Business Consultants

BIZPLANIT

http://www.bizplanit.com/vplan/management/newsletter.htm

The advice on this site includes an article from a free monthly newsletter related to business planning. Other parts of the site address common mistakes made in writing about the management team in your business plan.

ENTREPRENEUR MAGAZINE: GUIDE TO PROFESSIONAL SERVICES
(1997, JOHN WILEY & SONS)

Authored by Leonard Bisk and published as part of the Entrepreneur Magazine Series (available on Amazon.com as an e-book). This book tells you what every entrepreneur or manager should know about how to find, hire, and manage a winning team of expert advisers and service providers. It includes tips on hiring and managing consultants and is aimed at the practicing entrepreneur.

Assessing Personal Management Abilities

Prentice Hall Website

http://www.prenhall.com/scarbzim/html/check1.html. Entrepreneurship and New Venture Formation, by Thomas W. Zimmerer and Norman M. Scarborough (1996). The checklist on this site can help entrepreneurs understand their own readiness to take risks and can serve as a starting point for a discussion between partners about what they each offer to the management team.

5.10 Video Clips

Directions for viewing clips on the Prendismo Collection (formerly from www.eclips.cornell.edu):

- Visit http://prendismo.com/collection/. (You must subscribe with the site for full access to all features.)
- Click on the **Subscribe** at the top of the site. Students receive reduced rates.
- Once you have subscribed, type in the name or title of the clip in the **Search** at the top right corner of the screen.
- Click on the clip you wish to view.
- Click on the play icon to view the clip. If you have not become a subscriber, you will only view the first 20 seconds of the clip before being prompted to log in.

1. Jessica Bibliowicz (Founder of National Financial Partners)

   - BS from Cornell University
   - CEO of National Financial Partners, an independent financial services distribution system
   - NFP was created with $124 million of Capital from Apollo Management, a leveraged buyout firm
   - Title of Video Clip: "Jessica Bibliowicz Discusses How Management Team Was Built"

2. Richard Hayman (Hayman Systems)

   - Hayman Systems founded in 1938 by Stanley Hayman, as a cash register company
   - Richard Hayman led the company into being a value-added reseller and systems integrator of POS systems
   - Hayman systems was acquired by MICROS systems, Inc. in 1999
   - Title of Video Clip: "Richard Hayman Discusses Method of Hiring New Employees"

3. Kevin McGovern (Chairman and CEO of McGovern & Associates)

   - Discusses professional and educational background and decision to attend law school
   - Discusses international business, entrepreneurship, and specific examples from SoBe Beverage company start-up
   - Undergraduate degree from Cornell University and law degree from St. John's University
   - Title of Video Clip: "Kevin McGovern Discusses Importance of Finding Good People to Support Start-up and Fill In Experience Gaps (Advisory Board)"

4. Richard Saltz (Founder Business Innovation Strategies)

   - Discusses specific challenges facing not-for-profit organizations
   - Discusses obtaining funding, working with venture capitalists, valuation and business planning

- BS and MBA from Cornell University
- Title of Video Clip: "Richard Saltz Shares Thoughts on Creating Board of Directors for Not-For-Profit Organization"

5.11 References and Bibliography

Lee, Denis M. S. "Academic Achievement, Task Characteristics, and First Job Performance of Young Engineers." *IEEE Transactions on Engineering Management* EM-33, no. 3 (1986): 127–133.

Weiss, Alan *Managing for Peak Performance: A Guide to the Power (and Pitfalls) of Personal Style*. Shakopee, MN, Las Brisas Research Press, 1995.

5.12 Classroom Exercises

1. Why do so many people start businesses on a 50/50 basis with their friend?
2. How might you find appropriate board members?
3. What do you think the highest motivating factor can be to join a start-up? Is this motivation different for women and men?

5.13 Engineering Economics

This section of the text is meant to guide you in learning engineering economics, which would be useful for any aspiring entrepreneur beyond the basic objective of passing the FE test. For this chapter, we focus on cash-flow diagrams and incorporate some of the material from previous chapters (future value calculations and arithmetic gradients).

Although drawing a cash-flow diagram may seem trivial to an advanced engineering student who has now mastered upper-level calculus, constructing a cash-flow diagram is a good tool to visually see the sequence of cash events. Once these events are detailed, constructing the equivalence calculations can be done in a more systematic manner, and simple mistakes can be avoided. We will give you some problems to let you master this technique.

Example Problem – Drawing Cash-Flow Diagram

Consider equipment with a purchase price of $3,000. The equipment brings revenue of $600 yearly but costs $100 to maintain each year. Its life span is five years. Assume zero salvage value. Draw the cash-flow diagram.

The initial $3,000 to purchase the equipment is an expenditure represented by a downward arrow, so are the yearly maintenance costs. The yearly income of $600, however, is an income represented by an upward arrow in each period. Following the end-of-year convention, the yearly revenue and income are assumed to be obtained and incurred at the end of each period. Note that the purchase occurs at $t = 0$. There are five periods, each representing a year.

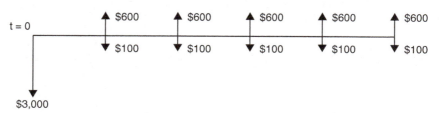

5.14 Problems

1. What does the horizontal axis represent in a cash-flow diagram?
2. How are receipts (cash inflow) represented in a cash-flow diagram? How about disbursements or investment (cash outflow)?
3. Do we need to factor in expenses that occur outside the time period considered for the cash flow diagram?
4. In the preceding problem, how do you address (or account for) sunken costs?

5. You invest $50,000 as a warrant in a start-up company and their business plan shows very high returns over the next five years. So, you structure the warrant that it requires repayment at a rate of 50% interest per year compounded annually. (Note: you will also demand your initial investment also be converted into common stock at the then current share price). Draw the cash-flow diagram.

6. You want to buy a sugar-cone oven for your new organic ice cream store, but you do not have much cash at hand. The seller offers to sell you the oven for a monthly payment of $150 over four months, due at the end of the month. Draw the cash-flow diagram that would represent the present value of the machine purchase price and then calculate the present value using an interest rate of 6% in your calculation.

7. It is summer and your retail shop's business is seeing poor sales (the college population has gone home for the summer). To conserve cash flow, you ask each of your four employees to kindly take a $100 pay cut per person per month for the three months of summer (June to August). You promise to pay them back in December when you would have lots of cash; plus you will give them an additional bonus of $100 per person. Draw the cash-flow diagram; show the future value of the total amount of money you would have to pay back in December to all your employees should they accept your proposal. (Use an interest rate of 12% in your calculations, and assume monthly compounding between the end of August and December).

8. You are recruiting Ms. Smith to be your sales manager. You are pretty confident that because of her previous sales contacts, she will increase your current yearly revenues $100,000 per year for each year starting in year 2 (takes him a while to implement strategy, so sales are static in year 1, then $100,000, $200,000, $300,000, $400,000 increases). Before you negotiate her salary, you want to determine how valuable these added sales are to you. Taking $t = 0$ to be the time Ms. Smith joined, draw the cash-flow diagram representing the *future* worth of Ms. Smith's efforts (i.e., the increase in revenue she brought in) to the company five years after she joined in. Assume a market interest rate of 5%. (a) So, what is the future value of this added revenue? (b) What is the annual equivalent value of this added income over this five-year period? Use an interest rate of 5%.

9. Our organic fish company pays $8,000 in monthly salary to its general manager. The company has a profit-sharing plan that results in the GM receiving an end-of-year bonus totalling $25,000. Draw the cash-flow diagram representing the salary payments if
   a. the salary is paid at the end of month.
   b. the salary is paid at the beginning of the month. In both cases, assume that the bonus is always paid on December 31, and use P for present worth. Determine the equivalent annual salary of the GM if the nominal market interest rate is 10% (nominal rate means on an annual basis;

we don't know if this has some different equivalent rate for compounding on a more frequent basis).

10. Mr. Johnson is considering whether he should employ Ms. Arnold to work as marketing manager in his firm. As a fresh college graduate from a prestigious university, Ms. Arnold would command a starting annual pay of $60,000 paid in equal increments monthly ($5,000/month). Alternatively, he can contract an advertising company to come up with an advertising campaign and conduct it for your company for two years for an upfront fee of $25,000 and $4,000 per month (end-of-month payments) for twenty-four months. The company's internal rate of return is 20%, market interest rates are 5%, and inflation rates are currently 3% per year

Assuming that the marketing manager and advertising contract are perfect substitutes, from a purely economic point of view, would it be better for Mr. Johnson to hire Ms. Arnold or to hire the advertising company? Justify your choice of interest rate used (part of the problem is to determine which interest rate you should use to evaluate the alternatives).

6 Creating Your Company

People are definitely a company's greatest asset. It doesn't make any difference whether the product is cars or cosmetics. A company is only as good as the people it keeps.

Mary Kay Ash (1915–2001; U.S. Business Executive)

6.0 Entrepreneur's Diary

You're making great progress. You've got a board, you are about to hire two key employees, and you are just about ready to start executing your company business strategy. But, you need a company structure to embody these attributes. In fact, you can't even open a bank account without an Employer's Identification Number (EIN). This is no simple decision. In particular, your board (typically made up of "older" types) will probably steer you in the direction of a conventional commercial corporation, aka C-Corp. Why, because they are familiar with it. People just about always think what they are familiar with is best, else why would they have been doing it all these years? I believe that a limited liability corporation (LLC) is probably your best bet at this point of your company history for a variety of reasons we discuss in this chapter. I'm not a lawyer, and this is where you should seek legal advice once you think you know what you want to do. In fact, never go to a lawyer and ask them an open-ended question such as "What should I do?" They will just about always guide you in the most conservative direction legally, and this may not be best for your company.

6.1 Types of Ownership Structures

Before you can decide on an ownership structure for your business, you should learn a little bit about how each structure works. As a good reference on deciding which ownership structure is most suitable for your business, read "Choosing the Best Ownership Structure for Your Business" (go to www.nolo.com/lawcenter).

The most common forms of business structure are the following:

- Sole proprietorship and Partnership
- Limited partnership

- Limited liability company (LLC)
- Corporation (for-profit)
- Nonprofit corporation (not-for-profit)
- Cooperative

Sole Proprietorship and Partnerships

For many new businesses, the best initial ownership structure is either a sole proprietorship or – if more than one owner is involved – a partnership. A sole proprietorship is a one-person business that is not registered with the state as a limited liability company (LLC) or corporation. You don't have to do anything special or file any papers to set up a sole proprietorship – you create one just by going into business for yourself.

Legally, a sole proprietorship is inseparable from its owner – the business and the owner are one and the same. This means the owner of the business reports business income and losses on his or her personal tax return and is personally liable for any business-related obligations, such as debts or court judgments.

Similarly, a partnership is simply a business owned by two or more people that has not filed papers to become a corporation or a limited liability company (LLC). No paperwork needs to be filed to form a partnership – the arrangement begins as soon as you start a business with another person. As in a sole proprietorship, the partnership's owners pay taxes on their shares of the business income on their personal tax returns and they are *each personally liable* for the entire amount of any business debts and claims.

Sole proprietorships and partnerships make sense in a business in which personal liability isn't a big worry – for example, a small service business in which you are unlikely to be sued and for which you won't be borrowing much money for inventory or other costs.

LIMITED PARTNERSHIPS

Limited partnerships are costly and complicated to set up and run. They are not a good choice for the average small-business owner. Limited partnerships are usually created by one person or company, the "general partner," who will solicit investments from others – who will be the limited partners.

The general partner controls the limited partnership's day-to-day operations and is personally liable for business debts (unless the general partner is a corporation or an LLC). Limited partners have minimal control over daily business decisions or operations and, in return, they are not personally liable for business debts or claims. Consult a lawyer on limited partnerships if you're interested in creating this type of business.

Corporations and LLCs

Forming and operating an LLC or a corporation is a bit more complicated and costly, but well worth the trouble for some small businesses. The main benefit of an LLC or a corporation is that these structures limit the owners' personal liability for business debts and court judgments against the business.

> A C-Corp or an LLC limits the owner's personal liability against judgments against the business.

What sets the corporation apart from all other types of businesses is that a corporation is an independent legal and tax entity, separate from the people who own, control, and manage it. Because of this separate status, the owners of a corporation don't use their personal tax returns to pay tax on corporate profits – the corporation itself pays these taxes. Owners pay personal income tax only on money they draw from the corporation in the form of salaries, bonuses, dividends, and the like.

LLCs are similar to corporations because they also provide limited personal liability for business debts and claims. But when it comes to taxes, LLCs are more like partnerships: the owners of an LLC pay taxes on their shares of the business income on their personal tax returns. All income and losses pass from the LLC to the members each tax year.

Corporations and LLCs make sense for business owners who either

1. run a risk of being sued by customers or clients or run the risk of piling up a lot of business debts, or
2. have a good deal of personal assets they want to protect from business creditors. To learn more about forming an LLC or a corporation, see articles on each topic at www.nolo.com/lawcenter.

Nonprofit Corporations

A nonprofit corporation is a corporation formed to carry out a charitable, educational, religious, literary, or scientific purpose. A nonprofit can raise funds by receiving public and private grant money and donations from individuals and companies. The federal and state governments do not generally tax nonprofit corporations on money they make that is related to their nonprofit purpose, because of the benefits they contribute to society.

Cooperatives

Some people dream of forming a business of true equals – an organization owned and operated democratically by its members. These grassroots business organizers often refer to their businesses as a "group," "collective" or "co-op" – but these are usually informal rather than legal labels. The legal structure of cooperatives dates back to 1844, when the Rochdale Equitable Pioneers Society was established in Great Britain. Rochdale Principles are still followed by every cooperative business. Cooperatives typically operate on the principle of one person, one vote or voting by some other proportional means agreed to by its members. Cooperatives will have a board of directors selected from the members and becomes accountable to all the members. Cooperatives can be a group of producers, consumers, workers, or any other group with some common interest. For example, a consumer co-op could be formed to run a food store, a bookstore, or any other retail business, or a workers' co-op could be created to manufacture and sell arts and crafts.

6.2 Sole Proprietorship Basics

Sole proprietorships are so easy to set up and maintain that you may already be one without knowing it. For instance, if you are a freelance writer, a builder who takes jobs on a contract basis, a salesperson who receives only commissions and no salary, or an independent contractor who isn't on an employer's regular payroll, you are automatically a sole proprietor.

Even though a sole proprietorship is the simplest of business structures, there are generally still some requirements to maintain one legally. You may have to comply with local registration and license or permit laws to make your business legitimate. You are totally liable for any debts or judgments against you that happen as a result of your activities. This means all your personal assets are vulnerable to a law suit.

Personal Liability for Business Debts

A sole proprietor can be held personally liable for any business-related obligation. This means that if your business doesn't pay a supplier and loses a lawsuit, the creditor can legally come after your house or other possessions.

EXAMPLE

A. Lester is the owner of a small manufacturing business. When business prospects look good, he orders $50,000 worth of supplies and uses them in creating merchandise. Unfortunately, there's a sudden drop in demand for his products, and Lester can't sell the items he's produced. When the company that sold Lester the supplies demands payment, he can't pay the bill. As sole proprietor, Lester is personally liable for this business obligation. This means that the creditor can sue him and go after not only Lester's business assets, but also his other property as well. This can include his house, his car, and his personal bank account.

B. Shirley is the owner of a vegan donut shop. One day Roger, one of Shirley's employees, is delivering donuts to a client using a truck owned by Shirley. Roger strikes and seriously injures a pedestrian while he was texting on his cell phone (against company policy). The injured pedestrian sues Roger, claiming that he drove carelessly and caused the accident. The lawsuit names Shirley as a codefendant because Roger worked for her. After a trial, the jury returns a large verdict against Roger – and Shirley as owner of the business. Roger had no assets. This means the pedestrian can go after all of Shirley's assets, business and personal.

By contrast, the law provides owners of corporations and limited liability companies (LLCs) with what's called "limited personal liability" for business obligations. This means that, unlike sole proprietors and general partners, owners of corporations and LLCs can normally keep their house, investments, and other personal property even if their business assets do not meet the obligations of a judgment.

Paying Taxes on Business Income

Legally, a sole proprietorship is not separate from the person who owns it. The fact that a sole proprietorship and its owner are one and the same means that a sole proprietor simply reports all business income or losses on his or her individual income tax return – IRS Form 1040 Schedule C.

As a sole proprietor, you'll have to take responsibility for withholding and paying all income taxes, which an employer would normally do for you. This means paying a "self-employment" tax, which consists of contributions to Social Security and Medicare, and making payments of estimated taxes throughout the year.

Registering Your Sole Proprietorship

Unlike an LLC or a corporation (discussed later), you generally do not have to file any special forms or pay any fees to start working as a sole proprietor. All you have to do is declare your business to be a sole proprietorship when you complete the general registration requirements that apply to all new businesses. Most cities and many counties require businesses – no matter how small – to register with them and pay some minimum tax. In return, your business will receive a business license or tax registration certificate. You may also have to obtain an employer identification number (EIN) from the IRS, a seller's permit from your state and a zoning permit from your local planning board. You may think you are too small to worry about these legalities and that you can ignore these requirements. This is a bad idea; if you are caught, you may be subject to back taxes and other penalties.

And if you do business under a name different from your own, you usually must register that name – known as a fictitious business name – with your county. This is also called a DBA, taken from "doing business as." Sometimes it is a good idea to remain a DBA until you are really sure what form of corporate structure you want to register as. Some investors may prefer one form over another, and if you've already filed in a different form, for example, C-Corporation, it can be a bit problematic to re-form in an alternative structure.

6.3 Partnership Basics

By definition, a partnership is a business with more than one owner that has not filed papers with the state to become a corporation or LLC. There are two basic types of partnerships – general partnerships and limited partnerships. This section discusses only general partnerships – those in which every partner has a hand in the management of the business. A general partnership is the simplest and least expensive co-owned business structure to create and maintain.

Personal Liability for All Owners

First, partners are personally liable for all business debts and obligations, including court judgments. This means that if the business itself can't pay a creditor, such as a supplier, a lender, or a landlord, the creditor can legally come after any partner's house, car, or other possessions. The creditors will go after the easiest assets to capture first, so if your partner is out of state and/or has nonliquid assets, then they for sure will go after your assets first.

Second, any individual partner can usually bind the whole business to a contract or other business deal. For instance, if your partner signs a yearlong contract with a supplier to buy inventory at a price your business can't afford, you can be held personally responsible for the sum of money owed under the contract. In effect, each partner is 100% liable for any debt or judgments incurred by any of the partners.

> Each partner is 100% liable for any debt or judgments incurred by any of the partners.

There are few limits on a partner's ability to commit the partnership to a deal – for instance, one partner can't bind the partnership to a sale of all of the partnership's assets – but generally, unless an outsider has reason to know of any limits the partners have placed on each other's authority in their partnership agreement, any partner can bind the others to a deal.

Third, each individual partner can be sued for – and be required to pay – the full amount of any business debt. If this happens, an individual partner's only recourse may be to sue the other partners for their shares of the debt. Because of this combination of personal liability for all partnership debt and the authority of each partner to bind the partnership, choosing a general partnership for your business structure in a start-up is a poor choice.

Partnership Taxes

A partnership is not a separate tax entity from its owners. The IRS calls it a "pass-through entity." This means the partnership itself does not pay any income taxes on profits. Business income simply "passes through" the business to each partner, who reports his or her share of profit/loss on his or her individual income tax return. In addition, each partner must pay quarterly estimated tax payments to the IRS each year or face penalties from the IRS for this as well.

Although the partnership doesn't pay taxes, it must file Form 1065, an informational return, with the IRS each year. This form sets out each partner's share of the partnership profits (or losses), which the IRS reviews to make sure the partners are reporting their income correctly.

Creating a Partnership

You don't have to file any paperwork to establish a partnership, but simply agreeing to go into business with another person will get you started. Of course, partnerships must fulfill the same local registration requirements as any new business, such as applying for a business license (also known as a tax registration certificate). Most cities require businesses to register with them and pay at least a minimum tax. You may also have to obtain an employer identification number from the IRS, a seller's permit from your state, and a zoning permit from your local planning board.

In addition, your partnership may have to register a fictitious or assumed business name. If your business name doesn't contain all of the partners' last names, you usually must register that name – known as a fictitious business name or a DBA, doing business as – with your county.

Ending a Partnership

One disadvantage of partnerships is that when one partner wants to leave the company, the partnership automatically dissolves. In that case, the partners must fulfill any remaining business obligations, pay off all debts, and divide any assets and profits among themselves.

If you want to prevent this kind of ending for your business, you should create a "buy-sell agreement," which can be included as part of your partnership agreement. A buy-sell agreement helps partners decide and plan for what will happen when one partner retires, dies, becomes disabled or leaves the partnership for other interests. One way a buy-sell agreement helps avoid this situation is by allowing the partners to buy out a departing partner's interest so business can continue as usual.

The Uniform Partnership Act

Each state (with the exception of Louisiana) has its own laws governing partnerships, contained in what's usually called "The Uniform Partnership Act" or "The Revised Uniform Partnership Act" – or, sometimes, the "UPA" or the "Revised UPA." These statutes establish the basic legal rules that apply to partnerships and will control many aspects of your partnership's life unless you set out different rules in a written partnership agreement. To find your state's partnership statutes, see Learn about Legal Research at

http://www.nolo.com/lawcenter/statute/index.cfm.

Don't be tempted to leave the terms of your partnership up to these state laws. Because they were designed as one-size-fits-all fallback rules, they may not be helpful in your particular situation. It's much better to put your agreement into a document that specifically sets out the points you and your partners have agreed on.

What to Include in Your Partnership Agreement

While the owners of a partnership are not legally required to have a written partnership agreement, common sense dictates that you construct a written agreement to describe the details of ownership, including the partners' rights and responsibilities and their share of profits. Here's a list of the major areas that most partnership agreements cover. You and your partners-to-be should consider these issues before you put the terms in writing.

1. **Name of the partnership.** Shouldn't be too many arguments here.
2. **Contributions to the partnership.** You and your partners work out and record who's going to contribute cash, property, or services to the business before it opens – and what ownership percentage each partner will have.
3. **Allocation of profits, losses, and draws.** Discuss and then detail in writing.
4. **Partners' authority.** Without an agreement to the contrary, any partner can bind the partnership without the consent of the other partners. So, detail and describe carefully any restriction here.
5. **Partnership decision-making.** You can create types of decisions that require unanimous or simple majorities. You should carefully think through issues like these when setting up the decision-making process for your business.
6. **Management duties.** Basically who does what? Guidelines here are very helpful, so everyone knows basically what his or her responsibilities are.
7. **Admitting new partners.** Agreeing on a procedure for admitting new partners will make your lives a lot easier when this issue comes up.
8. **Withdrawal or death of a partner.** You should set up a reasonable buyout scheme in your partnership agreement; this may be for a friendly or a forced departure of a partner.
9. **Resolving disputes.** You should assume you will need a mechanism to settle disputes. Agree beforehand how to do this. Try to avoid going to court to resolve problems with partners (or anyone else for that matter).

6.4 Limited Liability Company (LLC) Basics

LLCs try to combine the best aspects of partnerships and corporations. An LLC combines attributes from both corporations and partnerships: the corporation's protection from personal liability for business debts and the simpler tax structure of partnerships that pass profits and losses directly to the partners. Although setting up an LLC typically requires more legal paperwork than does creating a partnership, running one is significantly easier than running a corporation.

Number of Members

You can now form an LLC with just one owner (called members) in most states (check local state regulations to be sure). If you want to form an LLC in a state that requires more than one member, you can make your spouse or some other trusted family member your LLC's second member. There is no maximum number of owners that an LLC can have. Single-member LLC's may affect your tax filing status of the LLC; for example, a single owner LLC in New York files as a sole proprietor or partnership.

Limited Personal Liability

Similar to shareholders of a corporation, all LLC owners are protected from personal liability for business debts and claims. This means that if the business itself can't pay a creditor – such as a supplier, a lender, or a landlord – the creditor cannot legally come after any LLC member's personal assets. Because only LLC assets are used to pay off business debts, LLC owners stand to lose only the money that they've invested in the LLC. This feature is the so-called limited liability aspect of this company format.

Exceptions to Limited Liability

Although LLC owners enjoy limited personal liability for many of their business transactions, it is important to realize that this protection is not absolute. This drawback is not unique to LLCs, however – the same exceptions apply to corporations. An LLC owner can be held personally liable if he or she

- personally and directly injures someone;
- personally guarantees a bank loan or a business debt on which the LLC defaults;
- their LLC fails to deposit taxes withheld from employees' wages;
- intentionally does something fraudulent, illegal, or clearly wrong-headed that causes harm to the company or to someone else; and/or
- treats the LLC as an extension of his or her personal affairs, rather than as a separate legal entity.

This last exception is the most important for you to understand. In some circumstances, a court might say that the LLC doesn't really exist and find that its owners are really doing business as individuals, who are personally liable for their acts. To keep this from happening, make sure you and your co-owners do the following:

- Act fairly and legally. Do not conceal or misrepresent material facts or the state of the LLC's finances to vendors, creditors, or other outsiders.
- Fund your LLC's adequately. Invest enough cash into the business so that your LLC can meet foreseeable expenses and liabilities.

- Keep LLC and personal business separate. Get a federal employer identification number, open up a business-only checking account, and keep your personal finances out of your LLC accounting books.
- Create an operating agreement. Having a formal written operating agreement lends credibility to your LLC's separate existence.

Business Insurance

A good liability insurance policy can shield your personal assets when limited liability protection does not. Insurance can also protect your personal assets in the event that your limited liability status is ignored by a court. In addition to protecting your personal assets in such situations, insurance can protect your corporate assets from lawsuits and claims. Be aware, however, that commercial insurance usually does not protect personal or corporate assets from unpaid business debts, whether or not they're personally guaranteed.

LLC Taxes

Unlike a corporation, an LLC is not considered separate from its owners for tax purposes. Instead, it is what the IRS calls a "pass-through entity," such as a partnership or sole proprietorship. This means that business income passes through the business to each LLC member, who reports his or her share of profits/losses on his or her individual income tax return. Each LLC member must make quarterly estimated tax payments to the IRS, when there is positive liability (you made some profit).

Although an LLC itself doesn't pay taxes, LLCs must file Form 1065, an informational return, with the IRS each year. This form, the same one that a partnership files, sets out each LLC member's share of the LLC's profits (or losses), which the IRS reviews to make sure the LLC members are correctly reporting their income.

LLC Management

The owners of most small LLCs participate equally in the management of their business. This arrangement is called "member management." The alternative management structure – somewhat awkwardly called "manager management" – means that you designate someone (or a position) to take responsibility for managing the LLC. The nonmanaging members simply participate in LLC profits/ losses. In a manager-managed LLC, only the named managers get to vote on management decisions and act as agents of the LLC.

Choosing a Name for Your LLC

The name of your LLC must comply with the rules of your state's LLC division. Although requirements differ from state to state, generally

- the name cannot be the same as the name of another LLC on file with the LLC office.
- the name must end with an LLC designator, such as "Limited Liability Company" or "Limited Company," or an abbreviation of one of these phrases ("LLC," "L.L.C." or "Ltd. Liability Co.").
- the name cannot include certain words prohibited or designated for specific businesses by the state, such as *bank*, *insurance*, *corporation*, or *city* (states differ widely on prohibited terms).

Your state's LLC office can tell you how to check if your proposed name is available for your use. Often, for a small fee, you can reserve your LLC name for a short time until you file your articles of organization. Besides following your state's LLC naming rules, you must make sure your name won't violate another company's trademark.

Once you've found a legal and available name, you don't usually need to register it with your state, because when you file your articles of organization your business name will be automatically registered.

Forming an LLC

You create an LLC[1] by filing "articles of organization" with the LLC division of your state government. This office is often in the same department as the corporations division, which is usually part of the secretary of state's office. Filing fees are around $100. Articles of organization are short, simple documents. Many states supply a blank one-page form for the articles of organization, on which you need only specify a few basic details about your LLC, such as its name and address, and sometimes the names of the members. Also, you must name a contact person involved with the LLC (usually called a "registered agent") who will receive legal papers on its behalf. Your registered agent must have a legal address in the state in which your LLC is registered. There are companies that provide this service for a fee of roughly $200 per year or you can use "Uncle Charley or brother Jim" who lives in your state of registration.

In addition to filing articles of organization, you must create a written LLC operating agreement. Although you don't have to file your operating agreement with the state, it's a crucial document because it sets out the LLC members' rights and responsibilities, their percentage interests in the business, and their share of the profits. Some banks may require you to provide them a copy of your LLC operating agreement when you open your business account.

Finally, your LLC must fulfill the same local registration requirements as any other new business, such as applying for a business license and registering your business name.

The following summarizes the steps you take to make your LLC a legal reality:

- Choose an available business name that complies with your state's LLC rules.

---

[1] One of the easiest ways to form your LLC is to use an online service such as the Company Corporation (www.incorporate.com).

- File formal paperwork, usually called articles of organization, and pay the filing fee.
- Create an LLC operating agreement, which sets out the rights and responsibilities of the LLC members.
- Publish a notice of your intent to form an LLC when required by your state and file any required local registration requirements.
- Obtain licenses and permits that may be required for your business. In some states, for example, Arizona or New York, you must publish in a local newspaper a simple notice stating that you intend to form an LLC. You are required to publish the notice several times over a period of weeks and then to submit an "affidavit of publication" to the LLC filing office. Be sure to check if your state requires this as well.

Ending an LLC

Under the laws of many states, *unless* your operating agreement says otherwise, when one member wants to leave the LLC, the company dissolves. In that case, the LLC members must fulfil any remaining business obligations, pay off all debts, divide any assets and profits among themselves, and then decide whether they want to start a new LLC to continue the business with the remaining members.

Your LLC operating agreement can prevent this kind of abrupt ending to your business by including buy-sell provisions, which set up guidelines for what will happen when one member retires, dies, becomes disabled or leaves the LLC to pursue other interests.

Creating an LLC Operating Agreement

Similar to corporate bylaws, an operating agreement governs the workings of your LLC. Even though operating agreements need not be filed with the LLC filing office and are rarely required by state law, it is essential that you create one. An LLC operating agreement allows you to structure your financial and working relationships with your co-owners in a way that suits your business. There are many sources for blank or sample LLC operating agreements, but you must be sure that your operating agreement is drafted to suit the needs of your business and the laws of your state.

You can pay a business lawyer for assistance to form your LLC, which may be a good idea the first time you form an LLC and especially for those that opt to have a special manager or management group run the LLC. If expense is an issue, software that helps you create your own LLC may be your best alternative. This is one area in which you should consult a lawyer if you have any questions.

Protecting Your Limited Liability Status

The main reason to make an operating agreement is that it helps to ensure that the courts will respect your limited personal liability. This is particularly key in a one-person LLC, in which, without the formality of an agreement, the LLC will look very similar to a sole proprietorship. Just the fact that you have a formal written operating agreement will lend credibility to your LLC's separate existence.

WHAT TO INCLUDE IN YOUR OPERATING AGREEMENT

There's a host of issues you must cover in your operating agreement, some of which will depend on your business's particular situation and needs. Most operating agreements include the following:

- The members' percentage interests in the LLC
- The members' rights and responsibilities
- The members' voting power
- How profits and losses and distributions will be allocated.
- How the LLC will be managed
- Rules for holding meetings and taking votes
- Buy-sell provisions, which establish a framework for what happens when a member wants to sell his or her interest, dies, or becomes disabled

Make sure you fill out the particulars in the following key areas.

PERCENTAGES OF OWNERSHIP

The owners of an LLC ordinarily make financial contributions of cash, property, or services to the business to get it started. In return, each LLC member receives a percentage of ownership in the assets of the LLC. Each member is usually given an ownership percentage that's in proportion to his or her contribution of capital, but LLCs are free to divide up ownership in any way they wish. These contributions and percentage interests are an important part of your operating agreement.

DISTRIBUTIVE SHARES

In addition to receiving an ownership interest in exchange for their investment of capital, each LLC owner also receives a share of its profits and losses, called a "distributive share." Most often, an operating agreement will provide that each owner's distributive share corresponds to his percentage of ownership in the LLC. For example, because Mr. Smith owns 35% of his LLC, he receives 35% of its profits and losses. Ms. Jones, on the other hand, is entitled to 65% of the LLC's profits and losses because she owns 65% of the business. If your LLC wants to assign distributive shares that aren't in proportion to the owners' percentage interests in the LLC, you'll have to follow rules for "special allocations." These rules will need to be carefully defined in your operating agreement; here it is best to consult a tax attorney in your state.

DISTRIBUTIONS OF PROFITS AND LOSSES

First remember that profits and losses from a tax perspective are allocated at the end of each tax year. Whether these profits are actually distributed to members is a separate decision (cash might be kept in the company to meet anticipated cash needs). Therefore, in addition to defining each owner's distributive share, your operating agreement should answer these questions:

- How much – if any – of the allocated profits of the LLC (the members' distributive shares which are reported to the IRS) must be distributed to LLC members each year?
- Can members expect their LLC to pay them at least enough to cover the income taxes they'll owe on each year's allocation of LLC profits?
- When will distributions of profits be made? Or are the owners entitled to draw periodically from the profits of the business?

Because you and your co-owners may have different financial needs and marginal tax rates (tax brackets), the allocation of profits and losses is an area to which you should pay particular attention.

VOTING RIGHTS

Although most LLC management decisions are made informally, sometimes a decision is so important or controversial that a formal vote is necessary. There are two ways to split voting power among LLC members: either each member's voting power corresponds to his or her percentage interest in the business or each member gets one vote – called "per capita" voting. Most LLCs mete out votes in proportion to the members' ownership interests. Whichever method you choose, make sure your operating agreement specifies how much voting power each member has as well as whether a majority of the votes, super majority (greater than two-thirds, usually interpreted as 70%), or a unanimous decision will be required to resolve an issue. Generally, super majority votes are required that directly affect membership interests, for example, decision to sell the company. Simple majority votes might define how other issues are resolved. If unclear, this is when you should consult an attorney.

OWNERSHIP TRANSITIONS

Many new business owners neglect to think about what will happen if one owner retires, dies, or decides to sell his or her interest in the company. These concerns may not be on your mind now, but such situations crop up frequently for small-business owners, and it pays to be prepared. Operating agreements should include a buyout scheme – rules for what will happen when one member leaves the LLC for any reason.

Licenses and Permits

After you've completed the steps described earlier, your LLC is official. But before you open your doors for business, you need to obtain the licenses and permits that all new businesses require. These may include a business license (sometimes also referred to as your "tax registration certificate"), a federal employer identification number (EIN), a sellers' permit, or a zoning permit.

6.5 Corporations

A corporation is what most commonly comes to mind when someone refers to a commercial business. In fact, corporations are distinguished by being a separate legal entity, either a C-Corporation or an S-Corporation, named after the rules that created it as described in subchapter S of the tax code. Both of these corporate forms limit the liability of the company's actions against its shareholders by the process of incorporation, which gives the corporation separate legal standing in the eyes of the law. A C-Corporation will issue stock and have shareholders, have a set of bylaws defining company regulation, and officers that run the affairs of the company, for example, a chief executive officer (CEO), a chief financial officer (CFO), a chief operations officer (COO), and so on. Taxation liabilities are different between a C-Corp and an S-Corp. A C-Corp must pay taxes on all profits generated each year, regardless of whether any profits are paid out to shareholders. Losses in a C-Corp can be carried over to subsequent tax years to reduce future tax liabilities. Conversely, all profits and losses in an S-Corp (or an LLC) are passed each tax year down to the shareholders. For more information, see http://www.investorwords.com/1140/corporation.html#ixzz0zujg4cLe.

6.6 Corporations versus LLCs

Now that you've learned something about both LLC's and corporations, how do you know which form is right for your business? If you are doing a start-up, you really only have two choices: an LLC or a regular C-Corporation. For the majority of small businesses, the relative simplicity and flexibility of the LLC makes it the better choice. Taxation differences were just discussed.

An LLC isn't always the best choice. One of the trickier factors is the taxing liabilities associated with fringe benefits. For example, you'd like to provide fringe benefits to one or more of the owners. Often, when you form a corporation, you expect to be both a shareholder (owner) and an employee. The corporation can, for example, hire you to serve as its chief executive officer and pay you a salary as well as pay your fringe benefits (health insurance and direct payment of medical expenses), both reducing the taxable income of the corporation, because, currently, they are not treated as taxable income to the employees, which can be an attractive feature of doing business through a regular corporation. These opportunities for you to receive tax-favored fringe benefits are reduced if you do business as an LLC.

Self-Employment Taxes Can Tip the Balance

One of the subtle tax advantages of an S-Corporation relates to the federal self-employment tax. You know that federal taxes are withheld from employees' paychecks. From 2014, for example, employers must withhold 5.2% of the first $106,800 of an employee's pay for Social Security and 1.45% of total earnings for Medicare taxes alone (so this is 6.65% of the first $106,800 and then 1.45% of any wages earned above that). The employee contributes 6.2% plus 1.45% amount and sends these funds to the IRS. The total sent to the IRS is 13.3% on the first $106,800 of wages and 2.9% on anything above that. A self-employed person pays the **self-employment tax**, which is 13.3% tax on the first $106,800 and a 2.9% tax on earnings above that amount.

For an S-Corporation, you pay the self-employment tax *only* on money you receive as compensation for services, but not on profits that automatically pass through to you as a shareholder. For example, if your tax year share of S-Corporation income is $100,000 and you perform services for the corporation reasonably worth $60,000, you will owe the 15.3% self-employment tax on the $60,000 but not on the remaining $40,000. By contrast, the taxation rules on income for members of an LLC imposes the self-employment tax on an LLC owner's entire share of LLC profits in any of the following situations:

- The owner participates in the business for more than 500 hours during the LLC's tax year.
- The LLC provides professional services in the fields of health, law, engineering, architecture, accounting, actuarial science, or consulting (no matter how many hours the owner works).
- The owner is empowered to sign contracts on behalf of the LLC.

Until the IRS clarifies the rules on self-employment tax for members of an LLC, you *should assume* that 100% of an LLC member's earnings would be subject to the tax, or it would be better for you to consult your local tax accountant for their his or her current interpretation. This then becomes a primary reason why you would choose an S Corporation over an LLC, that is, an S-Corporation shareholder will pay less self-employment tax than an LLC member with similar income. You'll need to decide if this potential tax saving is enough to offset such LLC advantages as less formal record keeping and flexibility in management structure and in the method of distributing profits and losses. This is pretty tricky stuff, so you should *seek consultation* from a tax attorney or certified accountant on this issue for their current interpretation of this particular rule for your state of incorporation.

6.7 Summary of Company Structure Issues

We've now covered the general choices in forming your new company as you begin to launch your venture. Table 6.1 provides a quick summary of attributes for the different company forms for different important issues.[2]

6.8 Additional Resources

General small-business resource site:
www.nolo.com/lawcenter.

---

[2] The Company Corporation (2003). 2711 Centerville Road, Suite 400, Wilmington, DE, 19808 (http://www.incorporate.com)

Table 6.1 Summary of Advantages and Disadvantages by Business Type

The Issue	Business Type				
	Sole Proprietor	**Partnership**	**S-Corp**	**C-Corp**	**LLC**
Liability protection?	No	No; all partners jointly liable for actions of other partners	Yes	Yes	Yes
No. of owners allowed	1	At least 2; no top limit	No more than 75	No limit	No max; 1-person LLC OK in most states
How income is taxed	Owner pays tax on personal returns	Profits flow through to partners, tax paid on personal returns.	Profits flow to partners, tax paid on personal returns	Corporation pays tax on profits; owners pay tax when cash/property distributed	Profits and losses flow to partners; tax paid on personal returns
Deduct losses on personal returns?	Yes	Yes	Yes	No	Yes
Avoid payroll paperwork?	Yes	Yes	No	No	Yes
Special allocations of income or expenses among owners?	Not applicable	Yes	No	No	Yes
Is a written agreement advisable when starting?	Not necessary	Yes	Yes	Yes	Yes

Source: Table taken from The Company Corporation (2003). 2711 Centerville Road, Suite 400, Wilmington, DE, 19808 (http://www.incorporate.com).

Ownership Structure

Foley and Lardner
http://www.foleylardner.com/FILES/tbl_s31Publications/FileUpload137/1024/handbook_entity_2002.pdf

This is a document covering how to choose your business entity. The Foley and Lardner site is packed with handbooks and guides relating to legal issues for start-ups; often information is directed toward those doing business in California.

Clifford, Denis, and Ralph Warner, *The Partnership Book*: *How to Write a Partnership Agreement*. 6th ed. Berkeley, CA: Nolo, 2001.

This is a good self-help book that can help you think through the details and put them in writing.

6.9 Forming Your Company

Well, if you are actually ready to form your start-up company, your first step is to obtain your employer identification number (EIN) from the IRS. You need the EIN number to open a bank account. You can call the IRS Business & Specialty Tax Line at (800) 829–4933. The hours of operation are 7:00 a.m. to 7:00 p.m.

local time, Monday through Friday. An assistor takes the information, assigns the EIN, and provides the number to an authorized individual over the telephone. Here is the link to Form IRS SS-4 (review the questions before you call the 800 number): http://www.irs.gov/pub/irs-pdf/fss4.pdf

6.10 Video Clips

Directions for viewing clips on the Prendismo Collection (formerly from www. eclips.cornell.edu):

- Visit http://prendismo.com/collection/. (You must subscribe with the site for full access to all features.)
- Click on the **Subscribe** at the top of the site. Students receive reduced rates.
- Once you have subscribed, type in the name or title of the clip in the **Search** at the top right corner of the screen.
- Click on the clip you wish to view.
- Click on the play icon to view the clip. If you have not become a subscriber, you will only view the first 20 seconds of the clip before being prompted to log in.

1. Chuck Winship (Founder of Sugarbush Hollow LLC)

- CEO of Sugarbush Hollow LLC – a farm that produces maple syrup and value-added maple products
- Discusses entrepreneurship, business planning and networking
- Received undergraduate degree from State University of New York at Buffalo and master's degree from Cornell University
- Title of Video Clip: "Chuck Winship Shares Thoughts on Legal Organization of Company"

2. Margaret Taft (Founder of Design Works – Calf Cozies)

- Founder of Design Works, a consulting company that provides computer generated patterns to clients
- Create of the Calf Cozy, a fleece blanket for calves that enables them to conserve energy for growth
- Discusses challenges in production, distribution, advertisement, marketing and financing of Calf Cozies
- Title of Video Clip: "Margaret Taft Discusses Moving From Sole Proprietor to Corporation"

6.11 Classroom Exercises

1. Will an LLC or a corporation protect your personal assets from committing negligence?
2. When or for what reasons might you strategically convert from an LLC to a C-Corp?

6.12 Engineering Economics

This section of the text is meant to guide you in learning engineering economics, which should be useful for any aspiring entrepreneur beyond the basic objective of passing the FE test. We'll cover capitalization cost in this section, which is a very common way to describe project cost. This term and concept will be present on the FE exam.

Capitalization Costs

You know what capital is: money. So the simple answer to a project that cost you $1.00 in start-up funds is that the capitalized cost of this project was: $1.00. It gets a little more complicated when you have to determine the capitalized cost of a project where the service or equipment has to last for some indefinite period, like "forever." In this latter case, you need to determine an annualized cost of the service/equipment and then divide that by the interest rate or your cost of capital or your own internal rate of return on what you expect for your investments. We'll still refer to all these different possible interest rates as "i" as we have throughout this text. So, with this context then, capitalization cost, P, is defined as

$$P = \frac{A}{i},$$

where
    P = present worth value (as P is always defined) of the capitalized cost,
    A = annualized cost, and
    i = interest rate appropriate for context of issue.

The other common context of one of these type of problems is where you have an upfront cost for a service or item, P', and then there is some annual maintenance cost or annualized cost, A, to maintain the service or item. Capitalized cost, P, is calculated in the following manner:

$$P = P' + \frac{A}{i}.$$

EXAMPLE. THREE WAYS TO CALCULATE CAPITALIZED COST, P

We'll do a capitalized cost calculation in three different ways, all giving the same result. Use whichever method makes the most intuitive sense to you.

METHOD A:

Device Costs $1.00 = P'$

Replace every five years for infinity and assume interest of 7% (for example only):

$$P = P' + \frac{A}{i}$$

$$A = F\left(\frac{A}{F}, n, i\right), \text{ where } F = \text{future disbursemen}$$

$$A = F(0.1739), \text{ from interest tables}$$

$$P = P' + \frac{F(.1739)}{.07} = P' + F(2.4843)$$

But, $F = P' = \$1.00$, so then $P = \$1.000 + \$1.000 * \$2.4843 = \$3.4843$.

This assumes that the device did *not* increase in cost.

Note that disbursements occur as $1.00 to buy unit at *time* = 0; then, 2.4843 is the annualized cost of the device spread over five years. You then compound this annualized cost for five years (1.00 compounded at 7% for five years = 1.403), so you get the $P = 2.4843 \times 1.403 = 3.4853$, which then allows you to replace the item and start over with the compounding of funds.

METHOD B:

Find A given a present disbursement of cost at time = 0 of $P' = \$1.00$.

$$P = \frac{A}{i}; A = P'(A \mid P, i, n) = P'\left(\frac{A}{P}, 0.07, 5 \text{ years}\right) = P'(.2439)$$

$$P = \frac{P'(.2439)}{.07} = \$3.4843 \ (\text{same answer!!})$$

(Many think this is the simplest and most logical way to think of capitalization costs.)

Note: You start with $3.4843, and then you disburse $1.000, which leaves you with $2.4843 to compound at 7% for five years ($F = P[F/P, 0.07, 5 \text{ yr}] = 1.40255$), so $2.4843 \times 1.40255 = \$3.4843$, and you are back to where you started. So you can do the disbursement and have sufficient dollars to compound to give you the needed money in five years to start all over again.

METHOD C:

Assume the same i for five years, compute equivalent interest for five-year period as if it were a single period (the period is five years as a unit of time), and the cost to replace the device for that period of five years is P' (equivalent to A term in Method B above):

$$i_{5\,yrs.} = (1 + i)^5 - 1 = (1 + .07)^5 - 1 = 1.40255 - 1 = 0.40255$$

$$P = P' + \frac{P'}{.40225} = P'\left(1 + \frac{1}{.40225}\right) = \$1.000 * (3.4842) = \$3.4842$$

Opportunity Cost and Infinite Life Calculation

What is capital cost, P, of a project that costs $1.00 and lasts forever?

$$A = i(P')$$

$$P = \frac{A}{i} = \frac{i(P')}{i} = P' \text{ (your initial cost)}$$

A term you will also hear about is ***opportunity cost***, OC, for a project. This is the cost of obligating your capital to a particular usage/project when you do not have access to this money for something else you want to do. Hence, you have a "lost" opportunity or more simply an opportunity cost. Opportunity cost is defined as

$$OC = i * P'.$$

Example 6.1 – Capitalized Cost

Mr. Wideawake intends to start a new company in the form of an LLC and is considering buying the permanent right to open a popular coffee franchise, Coffee Time, in his hometown. He expects to have a monthly net profit of $ 15,000 from the shops. If the nominal interest rate is 8% (is based upon a yearly period by definition) what is the maximum amount of money (P) that he should be willing to pay for the franchise right?

Solution. In this problem, the amount of money that Mr. Wideawake pays for the franchise right can be seen as the capitalized cost P. In order for Mr. Wideawake not to suffer a loss, the maximum amount of P should be equivalent to the present worth of having a monthly stream of income A = $15,000 for an infinite time. Thus, we can use the capitalized cost equation as follows:

$$A = \$15,000; \ i = 0.08/12 = 0.0066667$$
$$P = A/i = 15,000/0.0066667 = \$2,250,000$$

Therefore, the maximum amount of money he should be willing to pay is $2.25 million. Of course, there would be other considerations, for example, making sure the net profit mentioned was after Mr. Wideawake had paid himself a reasonable salary and also included any tax ramifications.

6.13 Problems

1. Define capitalized cost.
2. Give the formula for the present value (P) of a uniform series of payments (A) with infinite lifetime.
3. Ben is interested in looking into buying common share stock in Flight Technologies Inc., a company that manufactures Boeing aircraft components. Because aircraft construction is a multiyear process, Ben would only be paid his share of the profits once every three years, the first payment occurring at the end of the third year after he bought the shares. The company estimates that Ben would receive $50,000 every three years. Ben

wants to evaluate this choice using a nominal interest rate of 8%. What is the maximum amount of money that Ben should be willing to pay for the shares?

4. PB and J are the founders of a world famous sandwich company (PB&J, LLC). J, Jr., however, is not interested in taking over his father's member units (shares) in the sandwich company because he wants a less risky venture and would like to have a fixed income stream (he's conservative and doesn't like peanut butter and jelly, PB&J, sandwiches all that much). Fortunately, PB and J have an operating agreement that allows and defines how PB buys all of J's member shares. The buyout agreement requires PB&J, LLC to pay J's family $100,000 immediately on sale and an additional $100,000 every ten years as long as the company is still around. The buyout agreement also requires PB to deposit an amount of money in a dedicated trust fund to make the ten-year payments. How much cash does PB need now to fulfil the buyout agreement? The bank offers a commitment of 5% nominal interest (a yearly interest rate) on the deposit.

5. When he takes over his father's C-Corporation, Don decides that he would be better off without double taxation and so wishes to convert the corporation into an LLC run by him and the rest of the former major shareholders. However, one major shareholder, Mr. Tom, is unable to put in the commitment as an active LLC member (other members have chosen to be a member-managed company), Tom has asked the company members to buy his common stock shares, which they do. Because the members feel indebted to Mr. Tom for his unique contributions (it was his patent around which the company was formed), 70% of the members (required in an LLC to make a major decision) agree to pay Mr. Tom an additional $20,000 now, as well as $5,000 every five years as long as the company is operating. What is the present value of all payments made to Mr. Tom and his family? Use an interest rate of 5%.

6. Little Jimmy (a high school entrepreneur) has just opened a gourmet hamburger restaurant and is looking to buy a good grill. The grill he is considering costs $800 to buy and would incur a maintenance cost of $25 every half-year. Use an interest rate of 8%.

   a. Assuming the grill operates indefinitely, what would be the capitalized cost of just maintaining the grill (not including its initial cost)?

   b. If we include the cost of buying the grill, what would be the capitalized cost of the grill if it operates indefinitely?

   c. Instead of the $25 every six months, assume the maintenance cost of the grill is zero in the first year, but increases $10 yearly afterward, what is the capitalized cost of maintaining the grill assuming it operates indefinitely (assume yearly interest of 8% still)?

7. Phil, Dan, and Eve decided to determine the percentage of ownership in their newly formed LLC according to the amount of capital they had

contributed in cash or "in-kind." Phil had contributed $5,000, Dan put in $3,000, and the use of Eve's house, car, and accounting services was equated to $ 2,000. Determine the percentage of ownership of the three owners, rounded to the nearest whole number.

8. The percentage ownership in an LLC is as follows: Amy and Ann each have 20%, Kevin owns 25%, and the remaining 35% belongs to Tim. There is a recent opening for an accountant in the LLC, and after numerous round of interview, the possible candidates have been narrowed down to Mr. A and Mr. B, who are equally qualified. The decision was to be reached by an owners' vote. Amy and Kevin favored Mr. A, but Tim and Ann liked Mr. B better. Who will be hired if

   a. each owner has one vote?

   b. voting rights correspond to each owner's percentage ownership in the business?

9. You want to determine the difference in federal taxes you'd pay for a company you are the primary owner of, depending on whether or not you have incorporated as an S-Corp or as an LLC. Assume your share of "profits" are $150,000 per year. You provide management services to the company valued at $100,000 per year. Calculate the tax you'd pay if you were an LLC and if you were an S-Corp.

7 Financial Accounting

I have no use for bodyguards, but I have very specific use for two highly trained certified public accountants.

Elvis Presley

7.0 Entrepreneur's Diary

It seems as if addition, subtraction, multiplication, and division should be fairly simple. It is, until you want to apply it to accounting and do your tax calculations! There are many ramifications resulting from your choice of corporate structure on how you end up paying taxes. For example, who in their right mind would choose to pay taxes twice on profits you make from your company? Well, if you choose a C-Corp for your company structure, that's just what you'll be doing! But sometimes a C-Corp is the best choice, and you might ask why. I think my accountant was probably the key professional with whom I interacted with during the early days of my start-up. I recommend seeking an accountant's advice early, *before* you create your company structure.

7.1 Getting Started

At this point in the text, we've covered the basic steps of defining your business, developing a marketing strategy, and differentiating your business from the competition. In this chapter, we cover the creation and use of financial accounting statements in the typical business plan, including depreciation and taxation issues. We stress the importance of creating these financial statements from the **top down** versus the **bottom up** by basing the figures on details from the demand side of the equation, that is, sales and the costs of production.

We also discuss the data needed to create the necessary financial statements for your bank or potential investors. Finally, we briefly explain what belongs in the written business plan (see Chapter 8 "Writing a Business Plan," for a detailed discussion) and discuss the expectations of investors.

As soon as an entrepreneur starts to work on the financial statements for a new venture, it may become apparent that much is unknown. How many customers can be expected? How often will they buy? What is the cost of distribution? What are the costs of manufacturing the products and how will these costs change over the first few years? The answers to most of these questions for a start-up business owner are simply educated guesses.

Of course, you don't start a company unless you think you will be successful, so the financial statements for a start-up company are generally pretty optimistic, sometimes unrealistically so. That's why investors often accuse start-ups of "smoke and mirrors" when it comes to the financial section of the business plan. This leads to the question of why even bother with financials? But they are necessary, because valuations are based on them and they serve as an effective means of forcing you to analyze your business strategy. Second, because modern financial statements are spreadsheet based, an advisor or investor will usually be able to adjust a few figures and correct the whole financial accounting section quite easily, as long as the fundamental principles are followed.

Hence, the financial statement section of the business plan has several purposes. First, the creation of the financial statements forces the start-up team to focus on the dollars and cents of the business: What are the expected sales? what are the costs? What are the projected growth patterns?

Second, creating the financial statements may expose some problematic or even deal-killer numbers. For example, the start-up may be targeting a consumer product market and have all of the manufacturing costs pegged to the penny, only to find that to effectively launch the product, a multimillion-dollar advertising campaign is needed. Or, that the negative cash flow is so large that the start-up must raise considerably more investment dollars than they had anticipated, even though positive net cash flow is predicted by the end of the second year.

Third, the financial statements indicate to the outside users, especially investors, whether or not the start-up team has the maturity and knowledge to actually run the business. Reasonable numbers indicate a well-thought-out business plan and marketing strategy. This is especially important to an investor, who is understandably leery of investing in a start-up company with a financially naive management team.

Fourthly, the financial statements constitute an important set of milestones for the start-up's business plan. These numbers will be used in the future to determine whether the start-up is meeting the financial projections or is said to be *on-plan*. If sales for the second year are projected to be $900,000 but actually turn out to be only $100,000, this should cause some immediate recalibration of the company's strategy. If your start-up company operates below your financial projections, or ***below-plan***, your investors and board will usually react by blaming you, which typically results in a new CEO or some other gut-wrenching event to turn around the situation.

If the start-up continues to operate below-plan, raising additional capital will become very difficult and often comes at a dramatically reduced company valuation. A company devaluation will result in original share owners' investments being dramatically diluted unless they choose to participate in the new round of financing. Thus, you should recognize the importance of making sure that the pro forma statements are attainable. Realistic predictions in your pro forma statements will go a long way in establishing your credibility as a leader of your company.

For most mature, established businesses, financial accounting statements are purely historical in nature. That is, the net income for the previous year will be reported as well as the value of the assets at the end of the previous period. For the start-up business, however, the new firm typically has minimal history related to sales and revenue generation. Hence, venture capitalists and other users of financial information will ask to see projected or *pro forma* financial statements. Pro forma just means "projected" or "future estimated." The Latin definition of *pro forma* –"for the sake of form or perfunctory" – is a bit misleading here!

Even though pro forma statements are simply projections, they should be taken very seriously and be carefully constructed. Because you have no past data (or very limited) to analyze, you must estimate or guess at the proper values for all of the numbers. Using these estimates, you construct an income statement, a balance sheet, and a statement of cash flow. Another type of financial statement, the statement of retained earnings, is not usually dealt with when constructing pro forma statements.

Scope of This Issue

A good business plan will require a professional looking set of financial statements. Thus, either someone in your team should be competent enough to construct an accurate pro forma financial statement or you should retain a professional to construct these statements. Don't use financial software that you don't understand. Mistakes will signal incompetence to potential investors.

Angel investors and venture capitalists expect to see business plans with pro forma statements for three or five years out, respectively. The Small Business Administration (SBA) Loan application typically requires pro forma financial statements for three years into the future.

This chapter will help outline key components of the financial section of your business plan, how to write about financials, and mistakes to avoid in working on this part of the business plan. Chapter 9 in this book, "Fund-Raising," discusses funding through outside sources. Even if you need only tens of thousands of dollars for your start-up (not hundreds of thousands or millions), it is still important to think in terms of achieving milestones (sometimes called value inflection points) and to understand the importance of cash flow.

Top-Down versus Bottom-Up Financials

Financials are called "top-down" when you start with the underlying sales assumptions and costs of production and then build the financial model around those clearly stated assumptions. This approach requires you to project sales volume, which in turn requires you to carefully analyze the customer base via market research (see Chapter 3). The market will dictate the average selling price and the volume of sales you can obtain for your business and as such, the expected sales revenue should be an input to the system. This will then drive the rest of the financial planning process. This forces you to do your analysis from the top of the income statement down to the eventual bottom line where your profitability is revealed. We prefer to call this a top-down analysis.

Building your financial model based upon some predetermined level of profit that you need to make your company successful is called the "bottom-up" approach and should be avoided. The bottom-up approach is essentially using reverse engineering by changing sales numbers at the top or other underlying data such as costs until the necessary profitability is obtained. Some would-be entrepreneurs start at the bottom line of the income statement and enter a net income they are trying to achieve to create what they think is a viable company or to support their fixed overhead, for example, salaries. Then they go on to determine the sales volume (the top line) necessary to achieve that desired bottom line figure. Regardless of how you label this approach (top down or bottom up, sometimes labeled both ways), it is obviously flawed and backward and almost never works. When venture capitalists and other investors detect this approach, they will reject the financial projections and the business plan as a whole, instantly.

On the cost side, a top-down approach requires careful research and documentation. This means calling vendors, getting price quotes, and forecasting changes – all these activities help you have a more realistic idea of what it is going to cost to produce and sell a product or service. Estimating costs will also require you to determine your manufacturing and servicing/distribution functions and required support personnel that you will need to run your business, which will determine payroll costs.

The Financial Story

When narrating the "financial story" in your business plan, think about how you can link certain milestones with your funding requests. For example, it is clear that some period of funding is needed to meet expenses until the breakeven point is reached. You can also use graphs to demonstrate the need for funds at specific milestone events, for example, completing prototype testing, breakeven point, reaching target sales levels. These milestones can be discussed in words, but graphs and figures are often used to portray the point in time when the business crosses the breakeven and other key milestones. Regardless of the size of your project, your financial plan for funding must make sense in terms of how the business is expected to perform over time. Before going further into the details of pro forma construction, we introduce some basic financial terms.

7.2 Basic Financial Terms

Capital Expenditure

A *capital expenditure* is an outlay of cash or the issuance of debt or equity to acquire a long-term (greater than one year) asset or sometimes called *noncurrent assets*. These expenses are not included in operating expenses, but rather a portion of these expenses are allocated to future periods as depreciation expense (see the following discussion) in the income statement and the balance sheets.

Fixed, Variable, and Total Cost

A *fixed cost* is the starting or base cost of producing your company's product line, before you produce the first item, or the costs that are incurred without producing any product. The fixed cost does not change with production volume. When calculating fixed costs, you should include overhead expenses that are independent of the level of production such as utilities, telephone and office staff.

A *variable cost* is a cost that depends on production volume. The variable cost includes those costs that vary directly with sales, for example, raw material or part components. For example, if you were manufacturing an electronic device and it costs you $10 to produce each unit (cost of materials and electricity to make one unit), then the total variable cost of producing the device is $10 times the number of devices that you want to produce. A *marginal cost* of a device is the cost of producing one more unit, which in the preceding example would be $10.

The fixed and variable costs can be used together in the cost equation, which computes total product cost depending on the number of units (U) produced.

$$\text{Total Cost} = \text{Fixed Cost} + \text{Variable Cost} \qquad (7.1)$$

If you were producing electronic devices and used a production facility with a fixed cost (annual rent for the factory) of $10,000 and a unit cost per device of $10/unit, then your total cost equation is

$$\text{Total Cost} = \$10,000 + \$10 \ (U).$$

You should notice that the total cost will always be higher than $10,000 unless the fixed cost can be reduced. This is one reason why start-ups should try to avoid fixed costs as much as possible during the early months or quarters of existence, since you'll be stuck with these fixed costs even if your business model changes. Think of trying to outside source or contract as much as possible in your early start-up period.

Breakeven Point

In addition to the income and cash flow statements, investors also like to understand when your start-up expects to breakeven on sales of products, that is, the

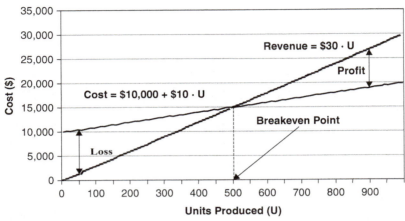

Figure 7.1   Example of breakeven, loss, and profit.

breakeven point. Investors will examine both the level of the breakeven and the timing (how many months or years before breakeven is reached).

The ***total revenue*** from the sales of a product can be calculated from the ***marginal revenue***, or the money gained from the sale of one unit, times the number of units sold. In the electronic device example, if you sold the devices to a distributor for $30 per unit, then the total revenue for selling 1,000 units would be $30,000.

Net income is the total revenue from sales of a product minus the total cost of production. If net income is a positive value it is called ***profit***, and if it is a negative value it is called ***loss***.

The ***breakeven point*** of production is the number of units produced where net income is zero. This means that the total revenue and total cost of production are equal at this point. Producing more units than the breakeven point would create a profit, and producing fewer units would create a loss. Figure 7.1 demonstrates a very simplistic production cost function. In reality, production runs are much more complicated and are rarely produced at a constant rate. Your costing analysis should embed any of these nuances into the cost model when possible or at least qualify any simplifications that are made so not to mislead the reader or investor.

CALCULATING BREAKEVEN

Breakeven can be calculated in terms of units or revenues:

$$Breakeven\,units = \frac{Fixed\,Costs}{Unit\,Price - Unit\,Cost} \tag{7.2}$$

$$Breakeven\,Revenue = Breakeven\,Units \\ \times Average\,Selling\,Price\,per\,Unit. \tag{7.3}$$

Consult the Additional Resources at the end of the chapter to find industry standards on gross margin.

Gross Revenue and Net Sales

Gross revenue is the money generated from your company's sales, that is, the amount of money you collect from customers. If you sell to a distributor, this is the wholesale price. If you sell direct to customers, this is the retail price. *Net sales* are gross revenue after product returns, allowances, and rebates have been subtracted.

Gross Margin and Gross Margin Percentage

Gross margin is defined as the difference between total sales (Sales) and cost of goods or services sold (*CGS*):

$$Gross\ Margin = Sales - CGS. \tag{7.4}$$

We express the ***gross margin percentage*** as a percentage of sales, or

$$Gross\ Margin\ Percentage = \frac{Gross\ Margin}{Net\ Sales} \tag{7.5}$$

Your company's gross margins should be comparable to your industry's norms. Examples of gross margin percentages for some particular companies and types of retailers include the following:

- Dell Computer: 20% to 22%
- Retailer Walmart: ~20%
- Harley Davidson: 34%
- Philip Morris: 42%
- Specialty retail: >50%
- General retail: 50%, sometimes call keystone markup
- Airline industry: 5%
- Software: 90%
- Distributors: 20 to 40%
- Service industry and charities: 100% (no cost of goods)

If your cost of goods sold is zero (such as at a charity bazaar), your gross margin is 100%.

To keep your gross margins up, you will need to either keep your average selling price up (avoid markdowns) and/or keep your CGS down. It is difficult to beat the industry norms on gross margins, but it can be done. One approach is to position the company as a premium seller, such as Nike does with athletic footwear. This approach allows you to charge more than your competition, but this can be very difficult to achieve, particularly for a start-up or a company with a low advertising budget.

Cost of Goods Sold (CGS)

CGS is the amount it costs you to produce or buy the product you have sold. If you are a retailer (clothing store, music store, electronics shop, etc.) this is simply the price you have paid for the merchandise. If you are a manufacturer, this is the total cost you incur to make the product: raw materials, labor, and overhead. If you are in a service industry, then typically CGS is zero, so you'll describe your sales by gross revenue rather than by gross margin.

ESTIMATING CGS

The baseline goal is to keep the cost of goods sold (CGS) at a fixed percentage of net sales that is comparable to the rest of the industry in which you are competing. If the strategy is to wrest market share from competitors by aggressive product pricing, then you'll try to have your gross margin percentage be smaller than the competition. If your strategy is to maintain a cost advantage through proprietary technology or other advances in manufacturing, then the gross margin may be higher than industry norms and may grow over time (but be careful not to be too optimistic). Finally, you might be able to win price concessions from your suppliers, due to anticipated high volume, to achieve a higher gross margin than your competitors do.

There are generally two methods to calculate cost of goods sold:

- The ***unit costing method,*** where

$$CGS = (\text{number units sold}) \cdot (\text{cost per unit}) \qquad (7.6a)$$

- The ***percentage cost method***, used in retail businesses where there are standard markups, calculated as

$$CGS = \text{selling price} \cdot (1 - \text{gross margin desired}), \qquad (7.6b)$$

where

$$\text{gross margin} = 1 - \text{percentage markup}.$$

SGA or *SG&A*

SGA are the expenses incurred in the course of ordinary activities of a business, which include selling, general, and administrative expenses (SGA or SG&A). It does *not* include cost of goods sold (CGS), interest, income tax expenses, depreciation, debt service, principal payments, or loan fees. It does include such things as sales commissions, advertising, wages, office supplies, rent, maintenance, utilities, insurance, travel and entertainment, real estate taxes, and miscellaneous expenses. Sometimes advertising costs will be shown as a separate line item on the income statement, particularly if they are significant part of your marketing strategy.

Profit Margin

The ***profit margin*** calculates what percentage of your sales generates net profit. This indicates the extent to which your company is making money from its product sales. The net profit is the incoming money to a company after all costs have been subtracted to generate the sales. The profit margin tells you what percentage of the selling price of your product results directly into money being available to the company for other activities, for example, shareholder profits.

$$\text{Profit Margin} = \frac{\text{Net Profit}}{\text{Net Sales}} \qquad (7.7)$$

A high profit margin has positive consequences for both employees and investors. The profit margin is affected by both the selling price and the efficiency of costs in producing a product. For example, say company A and company B make the same product, and company A sells the product for $5/unit with a profit margin of 10% and company B sells the unit for $6/unit with a profit margin of 8%. Even though company B is making more revenue from each unit sold, company A is making $0.50/unit net profit whereas company B is only making $0.48/unit. Therefore, if both companies sold approximately the same number of units, then company A would be the more profitable company because it was able to reduce the cost of its operations and subsequently make more money while beating out company B's product with a lower selling price.

Sunk and Opportunity Costs

Several other costs that are not used in the formal financial statements can have a bearing on other decisions and discussion. Two examples of such costs are sunk costs and opportunity costs.

Sunk costs pertain to money that has already been spent. Because the money has been spent and cannot be recovered, it should not affect the decisions made about what to do in the future.

Opportunity costs represent the money foregone or given up by making an alternative decision or said another way, the cost of passing up the next-best choice when making a decision. For example, let us say your company owns a parcel of vacant land. You can either build a warehouse on the land or sell the land for cash. If you build the warehouse, then you have spent the opportunity cost of that land. The opportunity cost here would be what you could have sold the land for. It would be wrong to say that the land was free since you already owned it and weren't using it. Opportunity costs do not appear on pro forma analyses. Opportunity costs (*OC*) are also defined as the cost of obligating capital for some service, project, or other investment. The *OC* is calculated as

$$OC = i \times C, \qquad (7.8)$$

where i = interest rate appropriate for the situation, for example, market interest or marginal rate of return:

C = initial cost of item or project or service.

7.3 Pro Forma Income Statement

The income statement shows the revenues and expenses of the business. It is also called the profit and loss statement, or simply the P&L statement. It is usually the first financial statement that is written.

$$Net\ Income = Revenue - Expenses - Adjustments \qquad (7.9)$$

Revenue is the amount of money collected from your customers paying money directly to you. It is not the income generated by some third-party retailer selling your product after they have purchased the product from you at wholesale or below market price. Expenses are the cost of goods sold and costs incurred in the revenue-generating process, including rent, wages, raw materials, utilities, and advertising. On the income statement, depreciation and interest expenses are not considered operating expenses but are adjustments subtracted from net income before calculating taxes, which are then subtracted to produce after tax net income.

Forecasting Revenue

For a start-up, an income statement requires a forecast of revenue or sales. You will want to keep track of both the number of units sold and the total sales revenue in dollars. Total sales revenue is

$$Sales(\$) = (Number\ of\ Units) \times (Average\ Selling\ Price). \qquad (7.10)$$

The average selling price is what the customers are willing to pay for your product. It has nothing to do with how much it costs your company to make it. Again, if you are selling products through a third-party retail channel, remember to use the price you obtain from the sale, for example, what is commonly called a wholesale price. Alternatively, if you are fortunate enough to be able to sell directly to customers, then you can collect the retail price directly and use the retail price in the sales equation. Your market strategy might be a combination of both. A comparison of your average selling price to the selling prices of your competition is also valuable information to provide your potential investors.

Remember that you must adjust your sales revenue for allowances of normal returns and discounts. For start-ups, returns and allowances can be large. Estimating a return rate up front is tough, so the best you can do is to base your rates on those occurring for similar products in similar industries. Your initial return rates might be higher at the beginning as you refine your product and manufacturing processes.

If your start-up sells more than a single model or version of your product, or has a diverse product line or lines, you will aggregate the purchases by each customer type over a given period. Based on your market research, you should define the customer profile and then estimate how many customers fall into each broad category. Then, decide on the geographic scope of your marketplace and estimate the total number of customers within each category in the relevant area.

Finally, calculate what share of the market you think you can capture. Your market research provides the basis for this analysis.

Although only the aggregate number goes in your business plan, building the sales estimate from a realistic starting point will increase your credibility with the investor. The background spreadsheets go in an appendix, and your narrative "story of the business" should explain how you went about estimating sales. You should estimate monthly sales for the first year and then make annual estimates (generally) or quarterly (for a seasonal business) depending on what your investor requires.

SALES REVENUE PER EMPLOYEE

Being aware of how your revenue per employee compares to others in your industry is important for you to determine whether this statistic is one of your company's strengths or weaknesses. This statistic is commonly expressed as sales revenue per employee; that is,

$$\frac{\text{revenue}}{\text{employee}} = \frac{\text{Total Sales}}{\text{Total Headcount}}.$$

The revenue per employee should be in line with industry norms. Of course, this depends on the industry. It is unusual to see a vital business with annual revenues of less than $60,000 per employee. Conversely, a business with annual revenues of more than $600,000 per employee is rare (Table 7.1).

Table 7.1. Revenue Statistics for Various Types of Companies

Company	Revenue ($ M)	Employees	Revenue per Employee ($)
The DIRECTV Group, Inc.	17,246	11,300	$ 1,526,195
Steel Dynamics, Inc.	8,081	5,940	$ 1,360,357
Google	21,796	20,222	$ 1,077,814
Apple	32,479	32,000	$ 1,014,969
Costco	72,483	75,000	$ 966,440
Amazon.com	19,166	20,700	$ 925,894
NVIDIA Corporation	4,098	4,985	$ 822,038
Dell	61,133	80,800	$ 756,597
Biogen	3,172	4,300	$ 737,586
QUALCOMM	11,142	15,400	$ 723,506
Research In Motion Limited (USA)	6,009	8,387	$ 716,514
Broadcom Corporation	4,658	6,853	$ 679,720
Microsoft Corporation	60,420	91,000	$ 663,956
Lam Research Corporation	2,475	3,800	$ 651,292
Juniper Networks	3,572	5,879	$ 607,651
Cisco	39,540	66,129	$ 597,922
eBay	8,541	15,500	$ 551,049
Yahoo!	7,209	14,300	$ 504,091
Apollo Group, Inc.	3,141	17,736	$ 177,093
Paychex, Inc.	2,066	12,700	$ 162,702
Starbucks	10,383	172,000	$ 60,366
Infosys	4,176	103,078	$ 40,513
Cognizant	2,136	59,500	$ 35,892

Predicting Expenses

Your *average cost* for your product is the total fixed and variable costs of producing a specified number of units divided by that number of units. If you were producing electronic ticket scanners and your company's fixed cost were $10,000 and had a marginal cost of $10/unit, the variable costs for 1,000 units is $10,000 and total cost would be $20,000, making the average cost of $20 per unit.

Similar cost-scaling advantages occur when using an outside contractor to manufacture your product. Typically in a start-up, you will first have to test market your product to a limited number of customers. But whether you want 10 units or 10,000 units produced, there is an initial setup charge by the manufacturer. Try to avoid the temptation of lower average cost for a much larger production run, because almost always, you will make changes to the prototype after your first customers test your product. It is pretty frustrating at that point if you have a large inventory of now obsolete nonsaleable items. You'll also look stupid to your investors

Typical Categories in the Income Statement

COST OF GOODS SOLD (CGS)

The direct costs incurred to produce your products or services and does not include any SG&A.

SG&A OR SGA

These are the expenses incurred in the normal course of ordinary activities of a business and include selling, general, and administrative expenses (office staff). Although administrative expenses are expected to grow as the business grows, they should track with headcount, not with sales. However, projected headcount figures are sometimes hard to come by. Perhaps a good rule of thumb is that administrative expenses should grow at half the rate of sales growth. Hence, if sales are expected to increase by 40% per year, assume that administrative expenses will increase by 20% per year. Showing rapidly growing administrative expenses (that may outstrip sales growth) is a red flag to investors and potential investors.

SELLING EXPENSES

These expenses depend on your compensation scheme for your sales force. If you pay a straight commission, this will be a percentage of sales. If your sales people require a fixed salary, then you would define a fixed cost for these expenses. For your pro forma projections, assume a percentage of sales unless you have good justification for some other scheme. Note that this item may be combined with the SG&A expense (discussed earlier). If your product success is highly dependent on a sales staff, it is probably a good idea to show it as a separate line item expense, partly to show your investors that you are taking sales and the need to generate them seriously.

ADVERTISING/PROMOTIONS

Advertising and promotions are usually considered to be a percentage of sales. Again, if advertising is a significant part of your marketing strategy, you should probably show this as a separate line item. Advertising will probably be at some minimal level and then track with sales as sales volume grows to significant levels. Determining the minimum level of funds allocated for advertising will take some serious strategy discussions and is often a critical ingredient of your company's success.

LEGAL AND ACCOUNTING FEES

Legal and accounting fees should be higher at the outset and then decline. At the worst, they should be flat. Legal and accounting fees should not track with sales.

RENT

Rent should be flat on a month-to-month basis initially. This will increase if the business expands and needs additional space.

UTILITIES

Utilities will likely track with sales (when you have a manufacturing process that uses electricity). Note that you should not include any utility expenses that were included in the CGS category.

R&D (RESEARCH AND DEVELOPMENT)

R&D is only appropriate for high-tech firms, not retailers or those in similar areas, and is usually a certain percentage of sales. R&D costs should be minimized as much as possible in a start-up until significant profits levels are being reached or if the business depends on the R&D effort to move the company to the next round of funding or some milestone event.

INTEREST EXPENSE

Interest expense does not track with sales, but it is a certain percentage of outstanding long-term debt. This is often zero for start-ups.

DEPRECIATION EXPENSE

The cost of depreciation is used in an income statement and balance sheets and is the decrease in the worth of equipment or capital asset as it ages and is utilized. Depreciation is used to decrease taxable income. Depreciation costs are not part of your cash flow statement as they do not affect your cash balance directly. A depreciable asset cannot be expensed in a single year, but must be expensed over its depreciable life. Depreciation is explained in detail later in this chapter.

INCOME TAX EXPENSE

Income tax expense occurs if the company is profitable; Federal rates are typically about 35% of net income before taxes; remember that LLCs pass profits to members without being taxed in the LLC.

Start-Up Costs

The number one rule in a start-up business is to not run out of money. Although your income statement might show a profit in the first year, your cash flow statement will probably have some initial months in which negative cash flow occurs. These funds must be available from cash reserves, which are typically cash your investors have provided. Start-up costs define how much money must be raised to open the doors of your business and remain open till positive cash flow commences. Your start-up costs and needed upfront cash will likely include the following items (see Additional Resources at end of chapter):

- Payroll expenses prior to start-up
- Payroll expenses before breakeven is reached
- Legal/professional fees
- Insurance
- Licenses/permits
- Advertising and promotions
- Building construction or remodeling
- Newly purchased equipment
- Rent and utilities
- Inventory
- Supplies

The *number-one rule* in a start-up business: Do not run out of money!!

Notice that these are all costs incurred before your doors are opened or before a product is sold. You'll need to tailor the list to your particular company. Building inventory on a product that takes months to produce can be a severe cash-flow demand for which you must be prepared. And what happens if production is slightly delayed for reasons such as the weather, growth rates, and so on? Also remember that sales do not result in instant cash, because most customers will require thirty days or longer terms to settle their account. Running out of money can occur even if you have a positive income statement if you cannot pay expenses from your account receivables fund.

Some good advice from the Small Business Administration is that you should have at least three months' worth of operating costs on hand, which constitutes your contingency fund. Your particular business might require many additional months. If your sales cycle and collection of funds turns out to be longer than anticipated or other unexpected surprises drain your available cash, the contingency fund can be used to maintain your company's ability to continue to produce product and services that are being sold. It would be a total disaster to be unable to meet customer demand for your products because of a lack of cash to produce these products.

EBITDA

If you look at an income statement, you will see the infamous term EBITDA. *EBITDA* is an acronym for earnings before interest, taxes, depreciation, and amortization. EBITDA is highlighted because the taxable earnings of a company, that is, after taxes and interest, can be greatly affected by how much debt financing the company has incurred and the amount of depreciation a company is charging against income. If your company was producing income using rented space and rented equipment, the income being generated is against a much smaller amount of invested capital than if you had purchased what you are renting and had built on borrowed money. In fact, one investment strategy is to maximize growth and income as much as possible on borrowed capital under the assumption that when the company is sold, the buyer will inherit all the debt liabilities and your selling price will have been minimally compromised (reduced) by these liabilities being present. Of course, the downside is that operating your company with a large amount of debt will also increase the risk of financial failure (and the banks will take all the assets you pledged when you borrowed the money). Note that income statements represent profits and losses; the cash flow statement (covered next) is only cash flow (what's in your checking account "now") and does not represent impact of depreciation for example.

7.4 Pro Forma Statement of Cash Flow

A *cash flow statement* accounts for changes in cash position. Cash is defined as (cash equivalents are easily accessed, short-term investments):

$$Cash = Currency + Checking\ Accounts + Cash\ Equivalents. \quad (7.11)$$

You can see now that cash is something you can use now to pay bills and meet immediate financial demands. If a business does not have the cash to pay salaries, purchase raw materials, and cover its regular bills, the owner will be faced with closing the doors, even if the prospect of sales is very positive. Hence, the expression that *cash flow is king*. Although this saying has become a cliché, it will still be your top challenge of running a new business. Understanding the realities of how cash flows in and out of the business may be the most important part of the work you do on your financial model. The reality for many small businesses is that expenses will have to be paid before revenues are flowing in. How will this be handled in the business? What sources of cash are available and when will they be used?

Accounting methods use either an accrual or a cash method to keep track of cash flow. Accrual accounting reports revenue when it is *earned* (whether or not it is actually received as cash) and expenses when they are *incurred* (whether or not they are paid). Income statements and balance sheets are based on accrual accounting procedures. However, these two financial statements will not necessarily give a clear picture of the cash "on-hand" status of your company, because you may not have received or expensed significant cash events that have occurred.

The formal statement of cash flows reconciles the procedures of accrual accounting that include noncash events (like depreciation) to actual changes in available cash for your company in order to report the actual flow of cash. The cash flow statement provides an instantaneous accounting of your firm's cash position at the end of some particular period, for example, a month or a year.

Generally, you will construct a monthly cash flow statement for year 1 of your start-up in addition to the required yearly income statements covering three to five years. The reason you do this is because you could run out of cash the first year even though you could have a positive net income shown on your income statement. Cash flow statements beyond year 1 are not generally provided unless your negative cash flow extends beyond the first year or your investor or banker requires it.

The cash flow statement can be constructed using the *direct method* (similar to a checkbook) or *indirect method* from the Income Statement, which adjusts income based on the rules of accrual accounting, and the balance sheet data, which shows changes in accounts receivable, accounts payable, and inventory. Thus, all the pro forma statements are interrelated. The direct method simply tabulates the cash coming into the company and subtracts the cash leaving the company. Incoming cash comes from three sources:

1. Operating activities: any cash transaction related to the company's ongoing business, usually is the producing and delivering goods and providing services
2. Investing activities: buying or selling of depreciable assets; purchase or sale of stock, bonds, and securities; or lending money and receiving loan payments
3. Financing activities: borrowing money and repaying money, issuing stock in exchange for cash, and paying dividends

An example of why the cash flow statement in addition to the income statement is needed is described in the following example. Almost all customers (except retail) will insist on paying for your products on a "net 30" basis. That is, they will send you a check thirty days (more or less) after they have received the shipment from you. Selling goods through a retail channel can be even worse, with payment coming sixty or even ninety days after the goods have been sold. On the balance sheet, you show these sales as accounts receivable, and on the income statement, you will show them as part of sales revenue, even though you have not received the money. On the cash budget, however, you do not show this as a cash flow until actually receiving the check.

As a start-up company, many of your expenses will have to be paid in cash. Typically, companies will want references and/or financial statements before they extend trade credit to you. It is likely that the companies that supply raw materials and supplies to your company will insist on cash payments for the first few months at least.

So you can have a cash-flow problem. Because your customers pay on a delayed basis, yet you must pay your suppliers in cash right away, you will need to find another source of cash, for example, your own personal funds, cash from investors, or borrowing. For some companies, this is not too big of a problem. For you it might be. Think ahead. Again, even if you have a product designed and ready for production, with ready and willing customers to buy it, if you must sell this product through retailers and wait sixty or ninety days for the cash, you may not be able to survive. In fact, some start-ups have found it necessary to raise another round of venture capital for just this purpose, or you will need to have a very friendly banker to supply this cash shortage.

Sources and Uses of Funds

Without a doubt, once you construct your cash flow statement, you will find that you are short of needed capital, and therefore need to raise cash from outside sources (one of the most important reasons you will write a business plan). Somewhere in your business plan, investors will want to see where this need for cash is coming from and how you intend to deal with it, that is, debt or equity financing. Using a brief format, your business plan should show where money is coming from (personal funds, equity investment, loans) and how it will be used. This is a checkpoint to make sure that funding needs are covered by some identified source. Using an investor's capital for depreciable assets is generally a lot more attractive to them than using their cash to pay salaries, because there is some salvage value in a depreciable asset.

7.5 Pro Forma Balance Sheet

The balance sheet lists **assets** of the business (such as cash, buildings, equipment, etc.) and **liabilities** (money that is owed, such as bank loans). The total assets (A) minus the total liabilities (L) equals the **stockholders' or shareholders' equity** (SE) of the business, or the residual value owned by the stockholders of the company. Stockholders' equity will consist of contributed capital, which is the cash actually contributed by investors, and retained earnings. We usually write this as

$$A = L + SE \qquad (7.12)$$

or

$$SE = A - L.$$

Assets will include the following:

- **Cash** – checking account balances plus currency
- **Accounts Receivable** – money owed to us from customers using trade credit. Often tracks with sales. A typical accounts receivable balance is 10% of annual sales.

- **Inventory** – the unsold goods that we must carry. Usually tracks with sales. The level of inventories is usually 10% to 25% of annual sales.
- **Property, Plant, and Equipment** – usually does not track with sales directly. Depreciation expense (on the income statement) is a percentage of this.
- **Patents and so on** – carried at cost less amortization. Amortization expense is recorded on the income statement. This can be a large expense if one has purchased the patents for a large sum.

Liabilities will include the following:

- **Accounts Payable** – this is the money we owe suppliers, typically 10% to 15% of the amount we make in purchases. If our company is a retailer, this is a percentage of cost of goods sold.
- **Leases** – financial obligations for use of property paid over some period. Can be thought of as rental payments that are obligated for some period, sometimes multiple years.

  All leases are not liabilities. Almost all vehicle leases are operating leases, and no liability is recorded on the balance sheet. For a lease to be recorded as a liability, one of the following conditions must be true:
  - Ownership transfer is an option.
  - A bargain purchase option.
  - The lease term is greater than 75% of the asset's useful life.
  - Present value of future lease payments is greater than 90% of asset's fair market value.
- **Bonds and Notes Payable** – all forms of borrowed money; does not track with sales.

Retained Earnings are the net income in past periods that have been retained in the business. It will be a positive number if the preceding years have been profitable and a negative number if the company endured losses over the preceding years.

When a company generates a profit, management has one of two choices: it can either pay it out to shareholders as a cash dividend, or retain the earnings and reinvest them in the business (or a combination). When the executives decide that earnings should be retained, they have to account for them on the balance sheet under stockholders' equity. This allows investors to see how much money has been put into the business over the years. Once you learn to read the income statement, you can use the retained earnings figure to make a decision on how wisely management is deploying and investing the shareholder's money. If you notice a company is plowing all of its earnings back into itself and isn't experiencing exceptionally high growth, you can be sure that the stock holders would be better served if the board of directors declared a dividend. Ultimately, the goal for any successful management team is to create $1 in market value or more for every $1 of retained earnings. Retained earnings (RE) are related to contributed capital and shareholder's equity (SE) as follows:

$$SE = Contributed\ Capital + RE. \qquad (7.13)$$

Let's look at an example. Microsoft had retained $18.9 billion in earnings over the years as of end of year [EOY] 2001. It had more than 3 times that amount in stockholder equity ($59 billion), no debt. Currently, Microsoft (as of 2009) has a market cap (market cap meaning the total value of issued shares of a publicly traded company) of $223 billion and a return on equity of 38%. The company appears to be using the shareholder's money very effectively.

Lear Corporation is a company that created automotive interiors and electrical components for everyone from General Motors to BMW. As of 2001, the company had retained over $1 billion in earnings and had a *negative* tangible asset value of $1.67 billion dollars! It had a return on equity of 2.16%. The company was astronomically priced at 79 times earnings and had a market cap of $2.67 billion. In other words, shareholders had reinvested $1 billion of their money back into the company, and what had they gotten? They owed $1.67 billion. That is a bad investment. Lear filed for Chapter 11 bankruptcy protection in July 2009.

7.6 Example: Pro Forma Analysis

For a real-life example from my own start-up fish company that we will call NEWCO, we develop a pro forma analysis for three years using the following information:

1. NEWCO raises $655,000 from investors all as common stock (share price of $10 per share, so 65,500 shares issued to members).

2. NEWCO borrows $400,000 from local bank at 10% APR to build a single purpose agricultural building; loan is on an interest-only basis for first three years.

3. NEWCO equips the building using $210,000 from cash (the money raised from selling stock). Assume equipment has zero salvage value and a life of seven years.

4. NEWCO has initial legal and accounting expenses of $45,000. (This occurs in year "0" and must be amortized over fifteen years as a start-up cost, not expensed.)

5. For year 1, NEWCO builds fish inventory to 200,000 pounds (fillet equivalent) for the first six months and inventory will stay at this level forever. Inventory is valued at CGS per pound times the weight (fillet equivalent) of the inventory.

6. NEWCO commences sales in the seventh month at 40,000 pounds fillet per month. Selling price is $2.60/lb. All sales are collected two months after sales are made.

7. NEWCO has variable costs (CGS) of $2.00 per pound of product sold (selling fresh fillets) that includes the feed, oxygen, processing, packaging,

Table 7.2. Income Statement for Example Problem

	Year 1	Year 2	Year 3
REVENUE[c]	$ 624,000	$ 1,248,000	$ 1,248,000
COST OF GOODS SOLD[b]-	$ 480,000	$ 960,000	$ 960,000
GROSS MARGIN	$ 144,000	$ 288,000	$ 288,000
Percentage (G.M. / Rev.)	23.1%	23.1%	23.1%
Fixed Operating Expenses	120,000	$ 120,000	$ 120,000
Rent	$ –	$ –	$ –
Salary/Wage[e]	$ 168,000	$ 168,000	$ 168,000
Marketing & Advertising[a]	$ 60,000	$ 60,000	$ 60,000
G & A[d]	$ 62,400	$ 124,800	$ 124,800
OPERATING EXPENSES	$ 410,400	$ 472,800	$ 472,800
OPERATING PROFIT	$ (266,400)	$ (184,800)	$ (184,800)
As a Percentage of Revenue	-42.69%	-14.81%	-14.81%
INTEREST INCOME/EXPENSE	$ (40,000)	$ (37,490)	$ (34,729)
AMORTIZATION EXPENSE[f]	$ (3,000)	$ (3,000)	$ (3,000)
DEPRECIATION (Buildings)	$ (26,667)	$ (26,667)	$ (26,667)
DEPRECIATION (Equipment)	$ (30,000)	$ (30,000)	$ (30,000)
DEPRECIATION EXPENSE TOTAL	$ (56,667)	$ (56,667)	$ (56,667)
NET INCOME BEFORE TAXES	$ (366,067)	$ (281,957)	$ (279,196)
TAX PROVISION (35%)	$ –	$ –	$ –
NET INCOME	**$ (366,067)**	**$ (281,957)**	**$ (279,196)**
ASSUMPTIONS			
Gross Margin	77%	Selling price, $/lb	$ 2.60
Notes		CGS, $ /lb	$ 2.00

Notes: [a] Usually advertising will be approximately 5% of gross sales.
[b] Calculate CGS as a percentage of revenue based on GM.
[c] Second-year revenue is much higher because there are a full twelve months of sales.
[d] General & Administrative is usually a percentage of gross sales; assume 10%.
[e] Includes benefits and payroll taxes.
[f] Amortization of Legal Expenses and so on are SL over fifteen years.

and shipping costs. (Gross margin should be used to calculate costs based on sales revenue.)

8. NEWCO has fixed costs (labor, utilities, rents, travel, insurance, mailing, professional fees per month of $10,000). NEWCO has salaries of $14,000 per month, which includes 35% of salaries in employee benefits and payroll deductions (FICA and worker's compensation).

9. NEWCO will have a marketing and sales budget of $5,000 per month (usually this is approximately 5% of gross sales, but we have simplified here.)

We have constructed the income statement for the above information and present the results in Table 7.2. Looking at this data in Table 7.2, you should see a "glaring" problem with cash flow. Inspection shows that there is a negative Cash balance in year 1, which means you ran out of money, which is the number one rule not to break in a start-up. You'd need to either raise more money by selling a larger percentage of your equity, or you borrow more money from the bank to retain more company equity, but you now have additional debt to service! The remainder of this problem is assigned as one of the problems at the end of the chapter.

7.7 Typical Pro Forma Mistakes

There are several common mistakes in creating a business plan *pro forma* for an investor audience that you need to avoid:

- Financials that are *not* based on justifiable levels of expected sales
- Financials that do not reflect statements made elsewhere in the plan, for example, "We will succeed through exceptional customer service," but we will use low-paid employees, no incentives and no training costs.
- No justification for sales growth, for example, you simply following a linear growth pattern without explanation.
- Too much detail is included in the body of the plan (excessive pages of spreadsheets disgust most readers – if supporting spreadsheets are important, include them in an appendix)
- Unrealistic return on investment for the industry (without any explanation or justification)
- Lack of necessary detail to support assumptions
- No benchmarking of industry standards (see http://www.census.gov/csd/bes/)
- No allowance for contingency funds to account for cash-flow needs during start-up period and/or building of inventory`
- Statements are presented in a non-professional format with unconventional jargon and categories

7.8 Tax and Economic Issues

Depreciation and Amortization

Whereas money will generate interest over time, tangible items will decrease in value as they age. This is true for all tangible assets other than land. Even if an item shows no signs of wear or deterioration, its value should be decreased over time. This decrease has to be accounted for as an expense on the income statement, be recorded on the company's tax returns in some way, and be reflected on the company's balance sheet. This reduction in value for tangible assets such as equipment and vehicles is called depreciation.

A similar mechanism is used for intangible assets, such as patents or goodwill. Certainly patents only have value until they expire. Hence, the value of the patent is gradually reduced from its initial value to zero over time. This is called amortization.

Goodwill is defined as the extra money that your company might pay to buy another company over its actual book value, and this extra money has to be accounted for in some way on tax forms and balance sheets. In general, goodwill simply remains at its original value, without amortization, for the purposes of financial accounting (taxes are different). However, the company will need to examine the goodwill asset each year to check if it has become impaired.

To check for impairment, companies are required to determine the fair value of the goodwill item, using the present value of future cash flows, and compare it to its carrying value. If the fair value is less than the carrying value, which indicates impairment, the goodwill value will need to be reduced to its fair value. The impairment loss is reported as a separate line item on the income statement, and the new adjusted value of goodwill is reported in the balance sheet. You may have to consult an experienced accountant to make sure you execute this calculation correctly.

Goodwill is amortized over a fifteen-year period for tax purposes. Because tax returns and financial accounting records use different procedures, this can get complicated, and reconciling the two may require some deferred liability or deferred asset accounts. Again, you may need to consult an experienced accountant.

DEPRECIABLE PROPERTY

Depreciable Property is property with a useful life of more than one year that is used for your trade or business for income-producing activity. Property with the following characteristics cannot be depreciated:

1. Has an infinite expected life. For example, land is not depreciable property because it has an indeterminate lifetime.
2. Property acquired through an operating lease cannot be depreciated. Note that property acquired through a capital lease is depreciated.
3. Part of the sellable product. Factory inventory and shipping containers are not depreciable.
4. Intangible property such as patents or trademarks,. These properties are *amortized* rather than depreciated (using straight line as opposed to using MACRS, discussed later).
5. Is used for less than one year in operation or has a depreciable life of less than one year. Such items are expenses or direct costs and are not depreciable.

Property is grouped into three basic groups, two groups of tangible property and one group of intangible property. The first group of tangible properties is called *real property* (similar to real estate) and includes land and anything built on, grown on, or attached to the land. The second group of tangible property is called *personal property*, and this includes equipment, vehicles, office machinery, and so on. And finally, *intangible property* includes all intangible assets such as intellectual property.

HOW DEPRECIATION WORKS

Depreciation is used as a deduction against taxable income on a tax return, but it is not "real" money that you spent that year. The amount of depreciation claimed each year depends on the method of calculation, but regardless of method, it depends on the ***initial cost*** (also called ***cost basis***) of the property and the ***salvage value*** at some future time. The salvage value is what you expect the property to be worth when you take it out of service or replace it with some other piece of property. The salvage value for many tangible properties and all intangible property is usually set to $0 at some future time. For example, the value of a patent will go to $0 twenty years after filing when the patent expires. However, just because something has zero worth does not mean that it can no longer be used in the business or has no value to its owners.

The current ***book value*** of depreciable property is the initial cost of the property minus the total depreciation that has been accumulated up to that point. The salvage value is therefore equal to the book value in the last year of depreciation for the property. Therefore, the book value at any given year is

$$BV_t = C - \sum_{n=1}^{t} D_n,$$
(7.14)

where

BV_t = book value of property at year "t,"
C = initial cost of the property (cost basis), and
D = depreciation for given year.

There were historically three main methods used to depreciate property: straight line (SL), sum of years digits (SOYD), and declining balance (DB) methods. In 1986, the modified accelerated cost recovery system (MACRS) was introduced and is now used to calculate all depreciation schedules. MACRS combines aspects of both the SL and DB methods. MACRS is discussed in more detail later in this section. Straight line is usually used for financial accounting.

STRAIGHT LINE DEPRECIATION

Simple, linear depreciation is given the name ***straight line depreciation*** (SL). This form of depreciation is done just the way that it sounds. The depreciation each year (D_t) is the initial cost (C) minus the salvage value (S) divided by the depreciable life (N).

$$D_t = \frac{\text{Initial Cost} - \text{Salvage Value}}{\text{Lifetime}} = \frac{(C-S)}{N}$$
(7.15)

And the book value in any year is

$$BV_t = C - \frac{(C-S)}{N} t.$$
(7.16)

EXAMPLE 7.2

A machine costs $10,000, with a lifetime of four years and a salvage value of $2,000. Compute the SL depreciation and the book value at the end of each year.

Solution

$$C = \$10,000 \ S = \$2,000 \ N = 4 \ years$$
$$D = (C - S)/N = \$8,000/4 = \$2,000/yr$$

Year	SL Depreciation	Book Value
1	$2,000	$8,000
2	2,000	6,000
3	2,000	4,000
4	2,000	2,000

DECLINING BALANCE DEPRECIATION

Another depreciation method is *declining balance method* (DB), which depreciates property at a fixed rate that is higher than is straight line depreciation. Declining balance forms an important role in MACRS and was the official method of depreciation before 1986, when it was modified to the MACRS system. There are two main types of declining balance: 150% declining balance (150% DB) and double declining balance (DDB or 200% DB). The difference between the two methods is the rate of depreciation being used. For this purpose, we define a variable $\alpha = \textbf{1.5 or 2}$ as the rate of depreciation, where $\alpha = 1.5$ is used for 150% declining balance, and $\alpha = 2$ is used for DDB. The formula for declining balance depreciation is

$$D_t = \frac{\alpha}{N}(BV_{t-1}) = \frac{\alpha}{N}\left(C - \sum_{n=1}^{t-1} D_n\right) = \frac{\alpha C}{N}\left(1 - \frac{\alpha}{N}\right)^{t-1} \tag{7.17}$$

and

$$BV_t = C\left(1 - \frac{\alpha}{N}\right)^t. \tag{7.18}$$

Because declining balance depreciation *ignores* the salvage value of property, you must choose from one of four strategies to deal with this problem for the last year's depreciation:

1. The salvage value may happen to equal the book value at the end of the last year. You could also define the salvage value as the book value from declining balance depreciation. In order for this strategy to work, you must set $BV_N = S$, or

$$S = C\left(1 - \frac{\alpha}{N}\right)^N. \tag{7.19}$$

2. You can stop depreciating a property at the point when the book value equals the salvage value. Unfortunately, however, this may leave you with a different depreciable life, and you will probably be paying more per year than you should or pay longer than you have to for the property.

3. In the last year, you can add (or subtract) the difference between the final book value and the salvage value to the depreciation amount.
4. You can use DB with the MACRS system, which converts to SL depreciation at a given time and uses a $0 salvage value. The salvage price is instead treated as a capital gain (selling an asset on the books when its book value is currently zero). Declining balance depreciation is almost always used within MACRS, so the problem of a salvage value is not encountered.

EXAMPLE 7.3

Compute the DDB and the 150% DB depreciation for a machine that costs $10,000, with a five-year lifetime and a $1,000 salvage value.

Solution

For DDB:

$$D_t = 2(10,000)/5 * (1 - 2/5)^{t-1} = 4,000(0.6)^{t-1}$$
$$\text{In year 5: } D_5 = (10,000 - 1,000) - (D_1 + D_2 + D_3 + D_4)$$

For 150% DB,

$$D_t = 1.5(10,000)/5*(1-1.5/5)^{t-1} = 3,000(0.7)^{t-1}$$
$$D_5 = 9,000 - (D_1 + D_2 + D_3 + D_4).$$

Year	DDB Depreciation	150% DB Depreciation
1	$4,000	$3,000
2	2,400	2,100
3	1,440	1,470
4	864	1,029
5	296	1,401

SUM OF YEARS DIGITS DEPRECIATION

The *sum of years digits depreciation* (SOYD) changes along an arithmetic gradient with "G" being the fraction of the sum of years. For example, if the depreciable life is four years, then the sum of years is $1 + 2 + 3 + 4 = 10$ and $G = 1/10$. The general sum of years digits equation is

$$SOYD = \frac{N}{2}(N+1) \tag{7.20}$$

$$G = \frac{1}{SOYD}. \tag{7.21}$$

This kind of a gradient causes larger depreciation values than does SL for the first few years and smaller depreciation values than does SL for the last few years. The general equation for the annual depreciation payment is

$$D_t = \frac{2}{N}\left(1 - \frac{t}{N+1}\right)(C - S). \tag{7.22}$$

And the equation for the annual book value is

$$BV_t = C - D_1 t\left(1 - \frac{t-1}{2N}\right),$$ (7.23)

where

$$D_1 = \frac{2}{N+1}(C - S).$$ (7.24)

EXAMPLE 7.4

Calculate the annual sum of years depreciation and end-of-year book value for a machine that costs $10,000, with a five-year life and a $1,000 salvage value.

Solution

$$C = \$10,000 \; S = \$1,000 \; N = 5 \text{ years}$$
$$D_t = 2/5(1 - t/6)(10,000 - 1,000) = 600(6 - t)$$

Year	SOYD Depreciation	Book Value
1	$3,000	$7,000
2	2,400	4,600
3	1,800	2,800
4	1,200	1,600
5	600	1,000

MIDTERM CONVENTIONS

In general, an item being depreciated is not acquired exactly at the beginning of a year. Therefore, you must adjust the *recovery period*, or the actual period of analysis, so that it starts in the middle of a period (month, quarter, or year). So we use a mid-period convention for these cases; for example, half-year, mid-quarter, or mid-month conventions must be used.

MID-MONTH CONVENTION

Use this convention for all nonresidential real property and residential real property, which employ the SL method under MACRS. With *mid-month convention*, you treat all property placed in service or disposed of during a month as placed in service or disposed of at the midpoint of the month. This means that a one-half month of depreciation is allowed for the month the property is placed in service or is disposed of.

MID-QUARTER CONVENTION

Use *mid-quarter convention* if the mid-month convention does not apply and the total depreciation bases of MACRS property you placed in service during the last three months of the tax year (excluding nonresidential real property, residential real property, and property placed in service and disposed of in the same year) are more than 40% of the total depreciable bases of all MACRS property you placed in service during the entire year. Under this convention, you treat all property placed in service or disposed of during any quarter of the tax year as placed in service or disposed of at the midpoint of that quarter. This means that 1.5 months of depreciation is allowed for the quarter the property is placed in service or is disposed of.

HALF-YEAR CONVENTION

Use *half-year convention* if neither the mid-quarter convention nor the mid-month convention applies. Under this convention, you treat all property placed in service or disposed of during a tax year as placed in service or disposed of at the midpoint of the year. This means that a one-half year of depreciation is allowed for the year that the property is placed in service or disposed of.

One of the three conventions (mid-month, mid-quarter, half-year) **must** be used when calculating MACRS depreciation. You cannot start depreciation at the beginning of a *calendar year.*

The easiest example is the half-year convention. This works by simply cutting the first year of depreciation in half, and then doing the calculations for all of the other years normally. For example, let's say that you have an initial cost of 100%, and you are depreciating it to 0% using DDB and the half-year convention (for simplicity, we are not switching to SL in this example). Then your calculations for a five-year piece of property would be the following:

You can see that using this convention, the depreciation doesn't start out as

Year	D_t Calculation (%)	D_t (%)	Book Value (%)
0		0	100
1	(2/5)100% / 2 = 20	20	80
2	(2/5) 80% = 32	32	48
3	(2/5) 48% = 19.2	19.2	28.8
4	(2/5) 28.8% = 11.52	11.52	17.28
5	(2/5) 17.28% = 6.91	6.91	10.368
6		10.368	0

large as without the convention, but the full value of the property is still accounted for in the last half-year (year 6). The remaining value of the property will always have to be summed into the last period in order to account for exactly 100% of the value of the property.

Similar calculations can be done for mid-month and mid-quarter conventions, but there are also tables that are provided for these calculations (see the Appendix for these details).

Modified Accelerated Cost Recovery System

The *modified accelerated cost recovery system* (MACRS) was put into effect in the Tax Reform Act of 1986 as the official cost recovery (depreciation) system to be used for computing taxes. The system effectively combines the DB and the SL depreciation into a recovery framework based on the depreciable life and the character of a property. The system is called *modified* because it resulted from a change in the accelerated cost recovery system (ACRS), which was used from 1981 to 1986 and differed only slightly.

Table 7.3. Allowable Depreciation Methods To Use for Specific Property Types

Method	Type of Property
GDS using 200% DB (to SL)	• Nonfarm 3-, 5-, 7-, and 10-year property
GDS using 150% DB (to SL)	• All farm property (except real property)
	• All 15- and 20-year property
	• Nonfarm 3-, 5-, 7-, and 10-year property[a]
GDS using SL	• Nonresidential real property
	• Residential rental property
	• Trees or vines bearing fruit or nuts
	• Water utility property
	• All 3-, 5-, 7-, 10-, 15-, and 20-year property[a]
ADS using SL	• Listed property used 50% or less for business
	• Property used predominantly outside the United States
	• Tax-exempt property
	• Tax-exempt bond-financed property
	• Imported property
	• Any property for which you elect to use this method[a]

[a] You may elect to use this method instead for these properties.

Depreciation with MACRS assigns different ***property classes*** for properties, and these classes determine the depreciable life and depreciation method for each property. There are two class-life systems, the first is called the General Depreciation System (GDS), which is the accepted MACRS system for determining class lives, which are three, five, seven, ten, fifteen, and twenty years. The alternative class life system is the Alternative Depreciation System (ADS), which uses the actual lifetime of the property as the class life. In general, ADS is only used for property that is tax exempt, is used less than 50% for business, is used outside of the United States, or is used in farming enterprises. You can use ADS even if the property qualifies for GDS, but once you use either method, you cannot switch back to the alternate. So, be careful.

There are MACRS tables available in this chapter and in the Appendices for calculating depreciation. In the following, we also outline the basic method that was used to create the tables. MACRS uses either 200% DB to SL, 150% DB to SL, or just SL, depending on the convention called for. Table 7.3 specifies what method should be used for different properties.

MACRS always sets the salvage value to $0 (with the GDS class lives), and this is one of the main differences between MACRS and ACRS. Because the asset is always depreciated to $0, then any fully depreciated property when sold is counted as a capital gain.

EXAMPLE 7.5 – HOW MACRS WORKS

Suppose you have a property that has an initial cost of $10,000 and zero salvage value, and it has a GDS class life of five years. The 200% DB, 150% DB, and SL depreciation values for these methods (without a midterm convention, which means full years of depreciation at the beginning and at end of life) are shown in the following table:

Year	200% DB	150% DB	SL
1	**$4,000**	**$3,000**	$2,000
2	**2,400**	**2,100**	2,000
3	1,440	1,470	**2,000**
4	864	1,029	**2,000**
5	776	1,681	**2,000**

The bold values show the path in which the declining balance should be changed to the SL method at the point when the SL method yields higher values. The path only tells you *when* to switch to SL, however, and the bold values for the last three years are *not* correct because SL depreciation is used for the *remaining* book value of the property and the remaining years of property life. Therefore, the SL value for 200% DB would become $1,200, and for 150% DB, it would be $1,633. Then the completed depreciation schedule would look like the following table:

Year	200% DB + SL	150% DB + SL
1	$4,000	$3,000
2	2,400	2,100
3	1,200	1,633
4	1,200	1,633
5	1,200	1,633

For MACRS (see next example), you are required to use a midterm convention for the first year, for example, property placed into service starting the seventh month of the year. Now, you switch from declining balance to straight line as we did in the preceding example. The only difference is that with the mid-year convention, you would use a different method for finding the SL depreciation of the remaining book value, since you are dividing by the remaining life of the property for the SL depreciation component. See the following example.

EXAMPLE 7.6 – HOW MACRS WORKS, MID-YEAR CONVENTION

You have a property with an initial cost of $100 (so you can think in terms of percentage) with a GDS class life of five years and want to find the DDB depreciation schedule for this property with a half-year convention. The DDB and SL depreciation with half-year convention would be as shown in the following table:

Year	200% DB	SL	MACR Choice	Cumulative Depreciation	Book Value
0				0.00	$100.00
1	½(2/5)(100 − 0) = $20.00	½(100 − 0)/5 = 10.0	20.00	20.00	$80.00
2	2/5(100 − 20) = 32.00	(100 − 20)/4.5 = 17.78	32.00	52.00	$48.00
3	2/5(100 −52) =19.20	(100 − 52)/3.5 = 13.71	19.20	71.20	$28.80
4	2/5(100 −71.20) = 11.52	(100 − 71.20)/2.5 = 11.52	11.52	82.72	17.28
5	2/5(100 −82.72) = 6.91	(100 −82.72)/1.5 = 11.52	11.52	94.24	5.76
6	Remaining Balance = 10.37	½ (100 −94.24)/0.5 = 5.76	5.76	100.00	0.00

Table 7.4. Deprecation Rates by Recovery Period Using DDB

Year	Depreciation Rate for Recovery Period					
	3 Years	5 Years	7 Years	10 Years	15 Years	20 Years
1	33.33%	20.00%	15.29%	10.00%	5.00%	3.750%
2	45.45	32.00	25.49	18.00	9.50	7.219
3	15.81	19.20	17.49	15.40	8.55	6.677
4	7.41	11.52	12.49	11.52	7.70	6.177
5		11.52	8.93	9.22	6.93	5.713
6		5.76	8.92	7.37	6.23	5.285
7			8.93	6.55	5.90	5.888
8			5.46	6.55	5.90	5.522
9				6.56	5.91	5.462
10				6.55	5.90	5.461
11				3.28	5.91	5.462
12					5.90	5.461
13					5.91	5.462
14					5.90	5.461
15					5.91	5.462
16					2.95	5.461
17						5.462
18						5.461
19						5.462
20						5.461
21						2.231

Notice that the SL calculation is applied to the remaining book value, having chosen the larger of the two depreciation values from the alternate methods. Value tables for 200% DB, 150% DB, and SL depreciation schedules for different class lives using the half-year, mid-month, and mid-quarter conventions are shown in the Appendix. Depreciation schedules using DDB with the half-year convention for the six GDS class lives are shown in Table 7.4.

To use this table, simply multiply the initial cost of the property by the percentage given for a year to get the depreciation for that year. For example, a property with an initial cost of $12,000 and a GDS class life of ten years would depreciate in the third year by

$$D_3 = \$12,000 \times 15.40\% = \$1,848.$$

Amortization

For intangible property, regular depreciation (MACRS) cannot apply because the property has no determinable depreciable life or rate of depreciation. Instead, this property must be amortized. Rules for amortization are similar to those for depreciation. The property must be owned by the person claiming the amortization, it must be a property that is in use for more than one year, and it must be used in connection with the production of income for a business. Acquiring an intangible property also sets the initial worth of that property. For example, when you amortize your own patent, you can only amortize the direct costs incurred in creating the patent (mostly attorney fees). But if a company bought that same patent from you for $100,000, then it would be able to amortize the patent by using an initial value of what it paid you for it.

There are two major categories of amortized property, and each category is assigned its uniform amortizable life. The categories and their amortizable lives are shown in the following table (prior to 2004, these two categories had different amortizable periods):

Amortization Category	Amortizable Life
Start-Up and Organizational Costs of a Corporation or Partnership	15 years
Intangible Property	15 years

Intangible Property includes the following:

- Going concern value
- Workforce in place
- Business books and records, operating systems, or any other information base (such as a customer list)
- A patent, copyright, formula, process, design, pattern, know-how, format, or similar item
- A customer- or supplier-based intangible
- A license, permit, or other right granted by a governmental unit or agency
- A covenant not to compete entered in connection with the acquisition of an interest in a trade or business
- A franchise, trademark, or trade name
- A contract for the use of any item on this list

All intangible property except for the last four on the list must be acquired by the owner rather than created, which is similar to inventory not being deducted for depreciation, because the owner creates the inventory.

To calculate amortization, use SL depreciation using the initial cost and salvage value of the property over a period of fifteen years (check current tax codes to verify the period). The amortization must begin in the first month that you purchased or acquired the property.

Of special significance to an entrepreneur is the amortization of start-up and organizational costs. A *start-up cost* is a cost that you made to start your company; otherwise, your company would not be functional. This includes costs made before the existence of the business and during its start-up. The start-up costs do not include research and experimental costs (to develop your product), interest, and taxes, but they do include things such as survey costs, legal advice and accounting, advertising, employee hiring and training, and travel expenses. Keep in mind that capital expenses made in starting a business (i.e., building and equipment) must be depreciated rather than amortized. **Organizational costs** are similar to start-up costs because they include payment for things that are needed to organize the start-up of a company. Organizational costs can only be incurred during the first tax year of a company's existence. The tax laws continue to change on how organizational and start-up costs can be applied, so consult a tax attorney or an accountant in this regard before starting your company.

7.9 Income Taxes

Introduction to Taxes

Income tax has a long and tumultuous history in the United States. Tax codes seem to be mostly driven by politics and government agendas compared to logical economic principles. The original income tax was created by Abraham Lincoln and Congress to help pay for the Civil War. This tax was eventually repealed, but was revived under Grover Cleveland in 1896. The Supreme Court ruled this tax unconstitutional, which paved the way for the ratification of the 16th Amendment in 1913, which gave the U.S. government power to tax the incomes of its citizens. The income tax has developed considerably since 1913 (including the creation of the IRS in the 1950s), and in 1998, Congress passed the "IRS Restructuring Reform Act." This act has led to the current division of the IRS into four main parts: Wage/Investment, Small Business/Self-Employed, Large/Mid-Size Business, and Tax Exempt/Government Entities. The IRS[1] employs about 106,000 employees who processed 234 million tax returns and collected more than $2.4 trillion in revenue in 2011.

Taxable Income

All personal and corporate income is subject to taxation. In this section, income means the profit to an individual or company. Taxes are only taken if money is being made. For example, nonprofit companies do not have to pay taxes because, by definition, they do not make any money. The employees of a nonprofit company, however, *do* have to pay taxes, because they are being paid an income.

Tax Period

Two types of taxation periods are used. The taxation period is always one year, but this year can either be a *calendar year* that starts on January 1 and ends on December 31, a *fiscal year* that starts on the first day of any other month than January, or a *52–53 tax year* that does not start at the beginning of a month. Whatever method you use, individual taxes are due to the government on April 15 (unless a weekend or holiday) of each year for your given tax period, and this tax period must always remain the same without accounting for extra taxes due.

---

[1] http://en.wikipedia.org/wiki/Internal_Revenue_Service; retrieved June 8, 2013

Taxing Methods

Besides federal and state taxes that are based upon an individual's taxable income, tax is often collected in other ways, for example, sales, property, excise, luxury, gift and estate taxes. These taxes can be considered progressive because they only tax the people who buy these products. These taxes differ widely between different areas and states and therefore may affect your purchase or sales decisions.

Withholding

The most common way to deduct income tax is through **withholding**, which is when the money is automatically withheld from an employee's paycheck by the government. Withholding counts as an income tax payment and can be deducted from each employee's annual tax payment. This is really the second tax imposed on an employee's pay, because the company itself has to pay taxes on the income that has gone toward salaries. The IRS W-4 Form determines how much money is withheld from your salary, and this form is filled out by all new employees to a company or by employees that have changed the number of allowances or deductions that they are owed. Allowances can be made for employees with dependents, who are single, who have more than one job, who live in an area with high state and local taxes, or who own a home.

A portion of withholding is put toward the Federal Insurance Contributions Act (FICA), also known as Social Security. This money is put into a government trust fund, which provides retirement benefits for people registered with Social Security.

Calculating Income Tax

There are two methods for a business to calculate its taxes, depending on the type of company, cash, or accrual methods, discussed earlier in this chapter.

Calculating Gross Income

Taxable income is calculated using your **gross income**. Your gross income is first subjected to your allowed deductions to become the **taxable income**. The taxable income is used to compute the amount of taxes to be removed from your gross income, which then becomes your **net income**, or net profit.

FOR AN INDIVIDUAL

The gross income of an individual is the gross (total) salary that they earn plus interest income, dividends, and other sources of income. The taxable income is therefore the gross income minus allowances (exemptions) or deductions that are allowed by the government for certain factors such as disability, Social Security, pension plans, retirement plans, charitable contributions, medical expenses, home mortgage interest, reported theft or losses, other non-income tax, and car or business expenses. People cannot claim depreciation for themselves unless they are including tax for a business (i.e., sole proprietorship) in their personal income tax. The standard deduction and personal exemptions that a person may subtract are computed from lines 38 and 40 of the IRS Form 1040, which is used to compute individual income tax.

For an individual, gross income is calculated as follows:

$$Gross\ Income = Total\ Salary + Interest + Dividends + Other\ Income \quad (7.25)$$
$$Taxable\ Income = Gross\ Income - Allowances - Deductions \quad (7.26)$$

FOR A COMPANY

For a company, the gross earnings or income of a company is simply the total earnings, or sales. The taxable income, or net earnings, is the gross income minus the operating expenses, interest payments, nonincome taxes, depreciation, and amortization. The cost of capital is only included once in depreciation, and there should *not* be an initial cost of capital included in the cost of sales or in operating expenses. For a company, the gross income is calculated as follows:

$$Gross\ Income = Total\ Sales \quad (7.27)$$
$$\begin{aligned} Taxable\ Income = {}& Gross\ Income - Operating\ Expenses - \\ & Interest\ Payments - Non\text{-}Income\ taxes - \\ & Depreciation - Amortization \quad (7.28) \\ = {}& EBITDA - Interest - Non\text{-}Income\ Taxes - \\ & Depreciation - Amortization = Net\ Earnings, \end{aligned}$$

where the acronym **EBITDA** stands for Earnings Before Interest, Taxes, Depreciation, and Amortization.

Taxing Companies

Tax laws differ for the five general different types of company structures:

1. Sole Proprietorship
2. Partnership
3. Limited Liability Company (LLC)
4. S Corporation (S-Corp)
5. Corporation (includes C-Corp)

The taxes that apply to each of these company structures may form a substantial role in what type of company you should be. The structures of these business types were described previously in this text. A quick summary of the main taxation differences are shown in Table 7.5.

Business Taxes

There are four main types of business taxes for any business:

1. Income Tax
2. Self-Employment Tax
3. Employment Tax
4. Excise Tax

Table 7.5. Main Taxation Differences by Company Type (repeated from Chapter 6, Table 6.1; from the Company Corporation)

	Corporation	LLC	S-Corp	Partnership	Sole Proprietorship
Personal Liability?	No	No	No	Yes	Yes
Double Tax?	Yes	No	No	No	No
Tax Individual Shareholders?	Yes	Yes	Yes	Yes	Yes
Self-employment tax?	No	Yes	No	Yes	Yes
Able to use lower corporation tax rate?	Yes	Yes	No	No	No
Can use cash method even if sales are greater than $5,000,000?	No	Yes	Yes	Yes	Yes

Table 7.6 Federal Income Tax Rates for Single and Married Filers (2012 Tax Year)

Tax Bracket	Married Filing Jointly	Single
10% Bracket	$0–$17,400	$0–$8,700
15% Bracket	$17,400–$70,700	$8,700–$35,350
25% Bracket	$70,700–$142,700	$35,350–$85,650
28% Bracket	$142,700–$217,450	$85,650–$178,650
33% Bracket	$217,450–$388,350	$178,650–$388,350
35% Bracket	Over $388,350	Over $388,350

INCOME TAX

Federal income tax is a "pay-as-you-go" tax, which means that a company must pay the tax as it earns or receives income during the year. This tax is usually paid through *withholding*, as described earlier, but can also be paid by *estimated tax* in payments throughout the year. All businesses except partnerships must file an annual tax return, and tax payments are only required on a corporation if the total taxes are expected to be greater than $500 or on any other business type if the total taxes are expected to be greater than $1,000.

In general, the IRS Form 1040 is used for personal income tax returns, and the IRS Form 1120 is used for corporate tax returns. The tax rates for individuals continually change as Congress adjusts these rates as political parties change power. Go to www.irs.gov to see current rates.

The taxable income values listed in Table 7.6 reflect reduction in taxable income for either itemized deductions (requires IRS Forms Schedule A and B) using a standard deduction of $5,950 single filer or $11,900 for family filer and a per-person exemption of $3,800 for each qualifying individual (Table 7.6).

Corporate taxes are paid only by corporations and electing LLCs. This tax and the individual taxes of stockholders create the two sides of the *double tax*. Corporate tax rates also continue to change, so consult www.irs.gov for current percentages. As a general guideline, companies having taxable incomes over $100,000 per year will pay about 35% of their net earnings in taxes.

SELF-EMPLOYMENT TAX

This tax gives an individual working for himself or herself (i.e., sole proprietorship, partner in a partnership, or member in a LLC) the Social Security and Medicare benefits that they are entitled to have. Social Security, or more properly FICA-Federal Insurance Contributions Act, provides retirement, disability, survivor, and hospital insurance benefits. People working for themselves are considered to have a business if they make $400 or more a year in net revenue for themselves from their business. In 2012, the rate of self-employment tax is 13.3% of net earnings for the first $106,800 of taxable income and an additional 2.9% Medicare tax on earnings above this level. A person's net earnings is normally 92.35% of his or her total earnings (this is related to the fact that companies paying half of the 15.3% FICA tax on all employees, so half of 15.3 is 7.65% and 100% − 7.65% = 92.35%). These rates and limits change often (depending on the political needs of the time and which party is in power), so check www.irs.gov for up to date information.

EMPLOYMENT TAX

These taxes are what your corporation must pay for having employees. There are three main things covered in this tax: employee income tax withholding, Social Security and Medicare taxes, and federal unemployment (FUTA) taxes.

EXCISE TAX

Excise taxes apply only to specific types of businesses or purchases. The main types of excise taxes are environmental tax; communications and air transportation tax; fuel tax; tax on the retail sale of heavy trucks, trailers, and tractors; luxury tax on passenger cars; and different types of manufacturers taxes. Instructions on these types of taxes are described in IRS Form 720.

Cash Flow with Taxes

Cash-flow analysis including tax effects can be broken down into five basic components:

- Before Tax Cash Flow (BTCF) – this is your raw income minus all costs and payments. It is also known as your gross income.
- Deductions – this is all of the exceptions and deductions that the government may allow you, but it is labeled as the depreciation and amortization deductions that you have claimed for the period.
- Taxable Income – this is the BTCF minus Deductions, or the amount of income for which you will be taxed. This value is what you insert into the tax schedule tables given in the previous section. (Table 7.6)

- Tax – this is the dollar amount of your taxes, and is computed using the tax schedule tables (Table 7.6).
- After Tax Cash Flow (ATCF) – this is the BTCF minus tax, known as your net income for the given tax year.

Your net income is the *bottom-line* value that you are trying to find. The easiest way to compute this value for multiple periods is by using a table. For example, suppose your corporation has

- a gross income of $10,000 in year 1 increasing at a rate of $30,000 per year,
- depreciation of a property worth $12,000 (no salvage value) over a three-year period using straight line depreciation,
- an initial nondepreciable payment of $30,000 in year 0, and
- a tax rate of 25% (for simplicity)

Assuming no deductions other than depreciation, the net income of the corporation would be computed using the following table (tax rate changes with BTCF value from 15% to 17.4% for year 3):

Year	BTCF	Deductions (Depreciation)	Taxable Income	Tax	ATCF
0	$ –30,000	$ 0	$ 0	$ 0	$ –30,000
1	10,000	4,000	6,000	1,500	9,100
2	40,000	4,000	36,000	9,000	31,000
3	70,000	4,000	66,000	16,500	53,500

This is also a good example of where there could be a big difference between being an LLC (profits and losses pass through to the members each tax year) versus a C-Corporation, in which losses can be carried forward to succeeding tax year. In the preceding example for a C-Corp, the $30,000 loss in year 0 could be applied in year 1 to offset the $6,000 of taxable income (making that year have $0 taxable income) and in year 2, the remaining $24,000 of losses could be applied to reduce the C-Corp's taxable income from $36,000 to $12,000. In an LLC, the $30,000 loss would have been passed that tax year to the LLC members. Whether they could take advantage of this loss would be on a case-by-case basis; that is, there are additional IRS rules on applying losses on an individual tax return. (See Section 7.10, "Rules for Deducting Losses" from an LLC, taken from www. smartmoney.com, at end of the chapter for more details.)

EXAMPLE 7.7

An unmarried individual has an adjusted gross income of $62,400. He has a standard deduction of $5,950 (2012 for single person who decides not to itemize deductions) and a personal exemption of $3,800 (2012 for single person). Compute the federal income tax using the 2012 tax table for single individual (use tax rates given in Table 7.6).

Solution

$$\text{Taxable income} = \text{Gross Income} - \text{Deductions} - \text{Exemption}$$
$$= \$62{,}400 - 5{,}900 - 3{,}800$$
$$= \$52{,}700$$
$$\text{Income Tax} = 10\% \times (8{,}700) + 15\% \times (35{,}350 - 8{,}700)$$
$$+ 25\% \times (52{,}700 - 35{,}350)$$
$$= 870 + 3{,}998 + 4{,}338 = \$9{,}206$$
$$(14.8\% \text{ of taxable income})$$

Capital Gains and Losses

When you sell property, your taxes must consider the gain or loss that you realized on that sale based on the original price of the property for which you bought it. A gain or loss from the sale or exchange of property is recognized as a gain or loss for tax purposes. Recognized gains must be included in gross income. Recognized losses are deductible from gross income. However, only certain property can be recognized as having a capital gain or loss, and any losses sustained from property held for personal use is not deductible. Gains on personal assets, for example, stock, of course, are taxed, but if the property has been held for more than one year, it is taxed as a long-term capital gain

A net gain from all of the year's transactions can be classified as either *ordinary income* or *capital gain*. The difference between these two classifications is that capital gain income is often taxed at a lower rate than is ordinary income for personal income tax purposes. Corporations do not enjoy this privilege, but pay the full tax rate on capital gains. A loss from the year's transactions is deductible and called *capital loss*. However, a limit of $3,000 is allowed for an annual capital loss. A loss on property can include loss due to theft, repossession, or condemnation in addition to a loss due to the un-recovered value of the property. All gains and losses are reported for an individual person or shareholder on Schedule D of the IRS Form 1040.

The gain or loss that you compute takes into account all maintenance, renovation, and other money put into the property between the time you bought it and when you sold it. Thus, it is possible to sell a property for more than you paid for it but to still claim a capital loss for your taxes. The value of the property after changes have been made is called the *adjusted basis* of the property.

Capital gains or losses can be claimed on most personal capital items, called *capital assets*, which include things such as stocks and bonds, owned homes, household furnishings, coin or stamp collections, and gems or jewelry. Capital gains and losses are divided into two categories. A *short-term capital gain* or *loss* is generally a property that is held for one year or less. A *long-term capital gain* or *loss* is a property held for more than one year. Some properties (i.e., Section 1231 property) are considered to have long-term capital gains or losses regardless of how long the property is owned before being sold. The distinction between short-term and long-term gains and losses is sometimes important in computing the net capital gain or loss for a given tax year.

Business property is broken down into three sections for reporting capital gains and losses. For each section, the ***net gain*** for that section is the total gain from all property in that section, and the ***net loss*** is the total loss from all section properties. IRS Form 4797 can be used to compute the capital gains or losses for sales of business property.

The three types of business property for capital gains and losses are described next.

Section 1231 Property. This includes real property, leaseholds, cattle, horses and other livestock, unharvested crops, timber cutting, condemned property, and any casualties or thefts of property. A net loss of Section 1231 property is an ordinary income loss, and a net gain is considered as a long-term capital gain.

Section 1245 Property. Includes tangible and intangible personal property, other tangible property besides buildings, certain real property, agricultural or horticultural structures, and storage facilities (not buildings) used for petroleum. Income from the sale of Section 1245 property is treated as ordinary income to the extent of depreciation allowed or allowable on the property, which is the lesser of

1. the depreciation and amortization allowed on the property and
2. the gain realized on the sale of the property minus the adjusted basis of the property.

If the income from the sale of Section 1245 property is greater than the lesser alternative stated earlier, then the extra income is treated as a Section 1231 capital gain.

Section 1250 Property. Includes all real property that is subject to an allowance for depreciation, and is not, nor has it ever been, a Section 1245 property. IRS Form 4797 can be used to compute the ordinary income and capital gain for a Section 1250 property.

Tax rates on capital gains are changed from year to year. Consult www.irs.gov for current rules and rates.

EXAMPLE 7.8

Your company bought a plot of land for $100,000, built a $1,000,000 factory on it, and purchased $500,000 worth of food handling equipment. Annual EBITDA is $400,000, and annual interest payments amount to $50,000. Assume that the company pays no non-income taxes and has a corporate tax rate of $22,250 for the first $100,000 of income and then pays 39% on income above that level; capital gain tax rate is 20%. Assume that the factory has a GDS class year life of fifteen years and equipment has a GDS five-year class life; you put both the building and the equipment into service mid-year (so use half-year convention); at the end of year 6, you sell the equipment for a capital gain of $200,000 (ignore the fact you'd purchase other equipment to replace the old equipment). Determine the following:

a. Develop the depreciation schedule for the building and equipment showing book value at the end of each year and total depreciation that can be claimed each year using a MACR depreciation schedule.
b. What is the after-tax cash flow for the first six years?
c. How much income tax must the company pay in year 6?
d. What is the rate of return from the first six years of after-tax cash flow?

Solution

PART A

Factory (15-year GDS life)

Year	Book Value	150% DB	SL	MACR Choice	Cumulative Depreciation
0	1,000,000	–	–		–
1	950,000	50,000	33,333	50,000	50,000
2	855,000	95,000	65,517	95,000	145,000
3	769,500	85,500	63,333	85,500	230,500
4	692,550	76,950	61,560	76,950	307,450
5	623,295	69,255	60,222	69,255	376,705
6	560,966	62,330	59,361	62,330	439,035

Equipment (5 year   GDS life)

Year	Book Value	200% DB	SL	MACR Choice	Cumulative Depreciation
0	500,000				
1	400,000	100,000	50,000	100,000	100,000
2	240,000	160,000	88,889	160,000	260,000
3	144,000	96,000	68,571	96,000	356,000
4	86,400	57,600	57,600	57,600	413,600
5	28,800	34,560	57,600	57,600	471,200
6	–	28,800	28,800	28,800	500,000

PART B+ C	Tax Rate 22.5% for first $100,000 and	39% for additional income						

Year	Cash Flow BT (BTCF, not including gains)	Interest Expense	Deprec Expense Bldg + Equip	Taxable Income	Capital Gains	Income Taxes (IT)	Capital Gains Tax 20%	After Tax CF
0	(1,050,000)	–	–	–	–	–	–	(1,050,000)
1	400,000	50,000	1 50,000	200,000	–	6 1,250	–	288,750
2	400,000	50,000	2 55,000	95,000	–	3 7,050	–	312,950
3	400,000	50,000	1 81,500	168,500	–	4 8,965	–	301,035
4	400,000	50,000	1 34,550	215,450	–	6 7,276	–	282,725
5 6	400,000	50,000	1 26,855	223,145	–	7 0,277	–	279,723
	400,000	50,000	91,130	258,871	2 00,000	8 4,209	40,000	425,791

Taxable Income = BTCF − Depreciation

Tax = 22,250 + 0.39(Taxable Income − 100,000)s

ATCF = BTCF − Tax

Year 1: Income Tax$_1$ = 22,250 + 0.39 * (400,000 − 100,000 − 50,000 − 150,000) = \$61,250;

do similar calculations for years 2 through 5);

For year 6,

Tax = 22,250 + 0.39(400,000 − 100,000 − 50,000 − 91,130) + 0.2(200,000)
    = 22,250 + 61,959 + 40,000 = \$124,209

ATCF = 400,000 + 200,000 − 50,000 − 124,209 = 425,791

c. How much income tax is paid in year 6

   Capital gains of \$40,000 +

\$84,209 in regular income tax = \$124,209 total

d. Rate of return

   NPW = −1,050,000 + 288,750(1 + i)$^{-1}$ + 312,950(1 + i)$^{-2}$ + 301,035(1 + i)$^{-3}$ + 282,725(1 + i)$^{-4}$ + 279,723(1 + i)$^{-5}$ + 425,791(1 + i)$^{-6}$ = 0

   Iteratively solve for "i" until NPW = ~ 0:

   ROR = 19% (You can solve this in Excel using the IRR function.)

7.10 Deducting Losses from an LLC

If you are part of a limited liability company (LLC), there are special rules as to how you deduct income losses.[2] Remember that in an LLC, profits and losses pass directly through to its members. You may have created your own LLC where you are its only member, and in this case your LLC is considered a disregarded entity for tax purposes and your losses (or income) will be directly reported under your social security number as regular income/loss (Schedule C). A multi-member LLC will issue Schedule K-1s to each member. You must treat the K-1 loss or income based upon your activity being classified as either passive or active.

Passive activity losses can generally be used only to offset passive income. Passive losses in excess of passive income for the year are carried forward to future years. You can deduct the losses in future years when and if you have passive income or when and if you sell or otherwise dispose of the activity that generated the losses. You can avoid the passive activity loss rules if you have ***materially participated*** in the loss-generating activity and become an active participant by meeting IRS regulation standards for such classification by passing any one of several tests. One of the following two tests is probably the easiest to meet.

[2] This is a reproduction of the following article: Bill Bischoff, "Deducting Losses from an LLC: New Wrinkles," *Smartmoney*, October 29, 2009, http://www.smartmoney.com/taxes/income/smallbiz-update-deducting-llc-losses/.

- Substantially All Test: You pass if your participation (time spent) in the loss-generating activity during the year in question constitutes substantially all participation that is collectively contributed by all individual members. In other words, you basically do all the work, even if this requires very few hours per week.
- More-Than-100-Hours Test: You pass if you participate in the loss-generating activity for more than 100 hours during the year, and no other individual participates more than you.

And as always, please consult your tax advisor on this matter to be sure you are following the proper guidelines established by the IRS.

7.11 Additional Resources

IBM's Guide to Financials

http://www.prars.com/ibm/ibmframe.html

If you want to learn how to interpret financial reports, this site has a good tutorial.

Business Services, BC-Canada

http://www.smallbusinessbc.ca/workshop/cashflow.html

This site includes an excellent discussion of how to create your cash flow statement, including step-by-step instruction on how to estimate sales. The description includes many numerical examples if you need help understanding any of the terms used in this chapter. You can also download a publication called *Business Planning and Financial Forecasting* from this related URL:

http://www.smallbusinessbc.ca/pdf/bpff2002.pdf

BizPlanit

http://bizplanit.com/vplan/financial/basics.html

In the section called Virtual Business Plan, this site features a concise description of what goes into the financial section of the business plan, and a list of common mistakes made by entrepreneurs.

American Express

http://home3.americanexpress.com/smallbusiness/Tool/biz_plan/fin/index.asp

This site discusses the following aspects of the financials: risks, cash flow statement, balance sheet, income statement, and funding request. Each section includes an explanation of the topic and tips for preparation.

Business Owner's Toolkit

http://www.toolkit.cch.com/tools/buspln_m.asp

This site features examples of financials from different kinds of businesses.

Small Business Administration

http://sba.gov/starting/suchecklist.html

A worksheet for estimating start-up costs is included in this site. You can also find a complete downloadable start-up kit that has valuable worksheets and an explanation about business planning at http://www.sba.gov/gopher/Business-Development/Business-Initiatives-Education-Training/Business-Plan/.

U.S. Census Bureau

http://www.census.gov/csd/ace/

http://www.census.gov/csd/bes/

These two sites have government-tracked data regarding business expenses (assets and expenditures).

U.S. Internal Revenue Service

http://www.irs.gov

This is a source for tax tables and rules for calculating taxable income for individuals and corporations.

7.12 Video Clips

Directions for viewing clips on the Prendismo Collection (formerly from www.eclips.cornell.edu):

- Visit http://prendismo.com/collection/. (You must subscribe with the site for full access to all features.)
- Click on the **Subscribe** at the top of the site. Students receive reduced rates.
- Once you have subscribed, type in the name or title of the clip in the **Search** at the top right corner of the screen.
- Click on the clip you wish to view.
- Click on the play icon to view the clip. If you have not become a subscriber, you will only view the first 20 seconds of the clip before being prompted to log in.

1. Jeff Parker (Founder of Technical Data Corp and of First Call Corp; both sold to Thompson Corporation)

- Founder of CCBN.com, provider of Internet-based investor and corporate communication services
- Bachelor of Science from Cornell University in Engineering, master's in Business & Engineering
- Title of Video Clip: "Jeff Parker Discusses Challenge of Documenting Realistic Cash Flow in a Business Plan"

2. Elizabeth Ryan (CEO of Breezy Hill Orchards)

- CEO of Breezy Hill Orchards
- Breezy Hill is a fruit and value added producer for New Yorkers

- Aggressively raised private equity to launch a national product
- Created a hard cider division of the company
- Title of Video Clip: "Elizabeth Ryan Shares Thoughts on Financial Analysis"

3. Scott Stewart (Founder Stewart-Peterson Advising Group)

- Cofounder of Stewart-Peterson Advising Group
- Stewart-Peterson Advising Group is one of the largest agricultural advising firms in the country
- Discusses partnership, entrepreneurship, business planning, and management
- Graduate of Purdue University
- Title of Video Clip: "Scott Stewart Shares Thoughts on Importance of Cash Flow"

7.13 Bibliography

Adams, Rob. *A Good Hard Kick in the Ass*. New York, NY: Crown Business Publisher, 2002.

Adelman, P. J. and A. M. Marks. *Entrepreneurial Finance: Finance for Small Business*. 2nd ed. New York: Pearson Prentice Hall, 2001.

Leach, J. C. and R. W. Melicher. *Entrepreneurial Finance*. New York: Thomson South-Western, 2003.

7.14 Problems

1. What is GAAP?
2. Why do we do pro forma analyses if they are not accepted as a GAAP?
3. Why are projections in pro forma almost always on the high side?
4. Why is so much "goodwill" on the books of many companies?
5. What items appear in an income statement that would not appear on a cash flow statement?
6. You have a very valuable patent. Someone offers you (in writing) $10,000,000 to purchase it. Can you now start to depreciate this item with an initial basis value of the $10,000,000? Why or why not?
7. A Caribbean cruise ship has the following data for a three-week cruise:

Charge per room	$400 per week
Fixed costs	$90,000 per cruise
Average Variable cost per room	$100 per week
Ship Capacity	150 rooms

    a. Develop a formula for both total cost and total revenue for the ship.
    b. What is the total number of rooms filled that will allow the ship to break even?
    c. What is the profit or loss for the three-week cruise if the ship operates at 80% capacity?

8. A rural car wash still uses a paid worker to dry off cars that have been washed, for which he or she gets paid $1.00 per car. The owner of the car wash is considering installing an automatic drying system, which costs $8,000 with energy/maintenance costs of $0.20 for each car. If the dryer has a useful life of four years, then what would be the minimum number of cars that would have to be washed each year to justify the purchase of the drying machine?

DEPRECIATION

9. A machine can be obtained for $100,000. The machine has a class life of five-year property; GDS for the property is five years. Compute the depreciation and the book value schedule for the machine using the following (assume mid-year convention for machine being put into place and the machine has no salvage value):
   a. straight line depreciation (assume an ADS five-year life), still assume put into service mid-year.
   b. the MACRS depreciation schedule.

TAXES

10. Brian Duke buys a lot and building in a busy downtown area for $1,000,000. He plans to tear down the building immediately and construct an apartment building in its place, which would cost him an additional $1,375,000. The apartment building would be finished in a year after purchase of the property, and he could then rent the apartments (which can be considered to be filled immediately) for a net income before tax of $500,000 per year. The apartment complex is depreciated using a straight line method with a 27.5 depreciable life and zero salvage value. Mr. Duke plans to sell the apartments in fifteen years and retire. Assume that Mr. Duke has a 34% income tax rate and that the capital gain from selling the apartments would have a 20% tax (instead of being taxed as regular income). When Mr. Duke sells the land and apartments at the end of fifteen years, what should be the selling price if he wants to get a 12% (after tax) rate of return on his investment?

11. In the example pro forma problem given in Section 7.6, the $400,000 loan was considered an interest only loan for the first three years. Also, the business had a negative cash flow during year 1. Make an adjustment to the business in terms of equity capital raised and then show the necessary changes on the pro forma sheets to reflect the bank loan being paid back over ten years at 10% nominal interest (assume payments are made at end of each year, not monthly). Use the Easy Sheets or NEWCO Excel workbook posted on the course website, and construct
    • the income statement,
    • the balance sheet, and
    • the cash flow statement.

8 Business Plans, Presentations, and Letters

The road to success is always under construction.

Arnold Palmer, famous golfer

Genius is 1% inspiration and 99% perspiration. Accordingly a genius is often merely a talented person who has done all of his or her homework.

Thomas Edison

8.0 Entrepreneur's Diary

Launching your business beyond the friends and family stage will require you to rewrite your business plan, probably for the umpteenth time! I will guarantee you one thing: the circle of professional acquaintances that you had thought might invest in your company … well, most won't … and the ones who do invest, will have a lot of advice for you, and probably more than you want! Advice can be a good thing, but when one investor says, "Go right," and the next investor says, "Go left," you'll likely find yourself wondering which direction is correct. But, you must do *something*!

I have been pretty successful in raising capital for my start-ups. My business plans were a critical component of my successful fundraising. Were they perfect? No, but these plans raised money! In fact, one of the VC investors said several times that she thought my business plan for my aquaculture tilapia start-up was one of the best business plans she had ever read. Her VC firm required a supporting partner (a proponent was required before this venture firm would make an investment) to make a personal investment in any company that is submitted to the VC group for consideration. After the initial acceptance, the tilapia business several years later ended up under new management and despite the earlier VC's praise for the original plan, one of the first objectives of the new management team was to write a "really good" business plan. My "best she'd read" plan got tossed out the window. What does this tell you?

8.1 The Business Plan – A Necessity

As an example, let's say you see a new business opportunity at your current employer. This large publicly traded company (Big-Co, Inc, we'll call it Big-Co) had developed a piece of technology. You tried to convince Big-Co to pursue this opportunity, but Big-Co was not interested, because you told it that maximum potential sales were in the range of $3 million to $5 million over the next three to five years. Big-Co said that type of cash flow was insignificant to it (review the lessons in Chapter 2, where disruptive technologies are discussed and why new technologies are often ignored). Well, to you, $3 million or $5 million seemed like a pretty significant cash flow in your personal bookkeeping system. So, you decide to pursue this opportunity. You make the smart move of negotiating rights to this neglected technology from Big-Co, for a modest royalty of 2% of your gross sales.

Because you are currently working in the field in which the new technology will be applied, you have a solid knowledge base about the market and the competition and basic technical information on supply chains, distribution channels, and manufacturers. So, what else do you need? Two things: the capital to launch your business and the business plan that presents your idea to potential investors so you can secure their cash investments.

A business plan is more than a necessary means to an end. Your business plan is a way for you to organize your thoughts, plan your strategies, and analyze your financial returns *before* spending any significant capital. You should develop your first business plan before spending any significant capital, even when it is out of your own pocket. Based on this first business plan and probably using your own money, you need to get to at least some minimal level of "real," meaning that you have the semblance of a prototype, a computer model, simulation – something that actually demonstrates your concept so you can show it to someone. Having done that, you can then revise your business plan. In fact, you will constantly be revising your business plan, strategies, and funding needs as you go forward.

You will use your business plan to raise capital, either equity or debt or a combination of the two. Some beginning entrepreneurs have the misguided view that a first-class business plan isn't really needed until they are seeking funds beyond the family arena. After all, they are your family, and they probably love you deeply. So, why shouldn't several of them basically just give you $10,000 or $25,000 each on good faith? Well, it just doesn't work that way. And, furthermore, the worst thing you can do is to proceed without a well-thought-out and -written business plan at any stage of your business venture.

Writing your business plan requires a plan (ironic, huh?). As I mentioned earlier, there are what seem to be countless "best ways" to write an effective business plan. Some of the plethora of recommendations arises because different business types (high tech, service, food, manufacturing, etc.) and company history (start-up, expansion, profitability) will necessitate different styles and strategies. This text focuses on technology driven start-ups that are in their early stages of development. This really relegates you to writing a business plan for the angel-type investor (see Chapter 9, "Fund-Raising," for further discussion of angel investors), and this is where our focus will be. For example, the exit strategy for an angel business plan will be different from that for a VC.

Before you try to raise capital from angel investors or essentially beyond the friends and family stage, you probably will have formed a legal company (most likely an LLC, see Chapter 6, "Creating Your Company," for discussion of legal forms of a business). Because you are a start-up, you will have little history and certainly will not have any financial history that could impress an investor. However, having just one customer is a critical threshold of success that will impress an angel investor. Conversely, writing a business plan for an existing business that needs capital for expansion, needs to show your previous ability to run a profitable company as demonstrated in your previous cash flow statements. But for your start-up, your ability to raise money will be critically linked to impressing the investor with the potential for success and high financial returns as described by your business plan.

The major thrust of your business plan will be a thorough analysis of the market that identifies the opportunity you see. Here, you should fall back on the six-word definition of marketing: *marketing means solving customers' problems profitably.* You must show potential investors that you are addressing a problem identified in your market analysis and that you can solve it in a profitable manner. This means that you share these profits with your customers (so they will buy your product), and your investors (so they will invest in your company in the first place).

To summarize the above and to be totally clear, there are two key points in a business plan to describe to the reader:

1. the opportunity and
2. your company's unique capability to address this opportunity

Supporting these fundamental determinants of viability are other necessary components that serve to round out your business plan into a complete package, for example, marketing and business strategy, financial analysis, management team, funds needed, and exit strategy. More details are provide on these topics later in this chapter.

Remember, your business plan may be the only window for potential investors to see the opportunity you have to offer. Don't ruin your opportunity by giving them a poorly prepared document. They will judge you by your attention to detail and their first glimpse in this regard will be by reading your business plan. Angels and venture capitalists will often have hundreds if not thousands of business plans presented to them over the course of a year. For yours to be selected for follow-up contact will require that you make a positive impression with this first and only chance. Make sure your business plan is packaged attractively. Your business plan when opened should "lay flat" when on their desks, so if they put it down to take a break/phone call – it will still be open to the spot where they left off (trivial detail, but important).

Your goal and hope is that the business plan – once read by the investors– will lead to further discussion and negotiation (see Chapter 11, "Negotiation," for help here), and ultimately for them to make an investment in your company. Simply put, your business plan should be a short document of ten to fifteen pages (not counting the table of contents, legal restrictions, figures and tables, and appendices) that describes the market you are about to enter and identifies an unmet need (solves a problem) in what you anticipate to be a growing and profitable market (upside opportunity). Let's get going!

8.2 Strategies for Investor Type

VC investors will be extremely careful in how they select companies in which to invest; they even have rules and guidelines on how they make their decisions. As a result of VCs' full vetting, they fully expect you to succeed. Conversely, angels have a different perspective or motivation on investing (see Chapter 9, "Fund-Raising"), and are willing to take more risk and invest smaller sums of money in start-up ventures than will a VC. Never make the mistake of assuming that angel investor Ms. Bigbucks really won't care if you lose her money, because she has so much. Your investors, no matter how wealthy, will expect you to treat their investment as if you were using your own money, and you should.

Angel investors want to make a return on their investment just as much as any VC investors do. An angel, however, is much less motivated by the size of the return than is the VC. In fact, an angel may be reasonably happy that your company simply survives. Your business plan should tell an angel investor very specifically when you expect to reach breakeven, or when they say, "When does the bleeding stop?" In fact, simple survival can be a very good result because it will probably mean that the angel investor will be able to stay involved and continue to help and mentor you towards success. Don't get me wrong here. An angel investor will be happy (thrilled) with a quick and large success, also, but this is not their primary motivation or expectation.

The angel investor is really looking to provide input and advice as your company matures towards financial success. Since the angel expects to be providing some management input, your need to have management experience will be less critical. To a VC, management skills and history are everything. To an angel investor, personal chemistry between your team and the angel investor is more important. The angel investor expects to enjoy being in your company and interacting with you, probably even on a social basis.

For an angel-, your business plan should demonstrate that you address management voids via your board members, which often are filled with investors; that's why you have to look for more than just money from investors. For example, what else does this person bring to the table? The angel investor is much more interested in being involved and adding their skills and creativity to the company than is a VC. The angel investor realizes how difficult it is to launch a successful company and that he or she is intended to fill management voids that typically exist. Finally, because angel investors really want to be actively involved, they look for something that is pretty much in their own backyard. Angel investors like an opportunity where they can get in their car and drive to see you in an hour or so (certainly within six hours of where they live).

The VC investor is motivated almost solely by financial return, which is qualified by the Value Equation (see Chapter 10, "Investor Principles"):

$$V = M \times P \times S,$$

where
  M is the quality of the entrepreneurial team (management),
  P is size of the problem, and
  S is the elegance of the solution.

You must have the first two components of the Value Equation (Team and Problem) for a VC to even be interested in looking at your company as a potential investment opportunity. This means that your assembled team must have a strong previous record of successes and that the P term must be very large ($0.5 billion to several billion dollars). If you have these two requirements, then the elegance of your solution (S) to the problem is the final requisite component to convince a VC to invest in your company. Yes, an angel investor is also interested in these things, but places different values on each.

Besides the Value Equation, I find that investors in start-up ventures are even more interested in what I call the Opportunity Equation (see Chapter 10, "Investor Principles"). This equation is closely related to the Value Equation, but it includes what is probably the key factor for whether or not an investor will invest (besides all those other determining factors); the current level of service to a market opportunity:

$$O = E \times T \times S,$$

where
  O is the perceived value of the opportunity,
  E is whether the market is emerging,
  T is the timing of your entry, and
  S is the elegance of the solution (your competitive advantage).

You can see the similarity between the Value and Opportunity equations. There are many emerging markets in the world as income levels rise in less-developed countries for many products and services. But the key question for the entrepreneur is whether any of these emerging markets are currently underserved, which must be clearly addressed in the market analysis section of your business plan. The more a market is already being addressed, the less likely it is a quality opportunity for a start-up. Start-ups have to identify those opportunities that no one else has currently "seen" or at least have not attracted many servers.

As you develop your business plan, keep the Value and Opportunity equations firmly in your mind but emphasize different components in the business plan based on to whom you are pitching your proposal. As a guideline, these distinctions are summarized in Table 8.1.

Table 8.1 Relative Importance of Business Plan Sections to Different Audiences

Section	Self	Family & Friends	Angels	VCs
1. Business Profile	5	20	20	10
2. Market Analysis	50	50	40	25
3. Market Strategy – 4P's	25	20	20	15
4. Business Strategy	10	15	15	10
5. Operations Plan & Management	5	5	10	30
6. Financials	5	5	5	10
Total Score	100	100	100	100

Note that the executive summary is not shown in the table. This section should be written keeping in mind the preceding distributions of relative importance. The executive summary should be the last section that you write and it is also the most important section of the entire business plan because it is the first section the potential investor will read to determine if they want to read any more of your business plan. Spend a lot of effort in writing this section. It is critical!

8.3 Recommendations on Writing Strategies

So, back to your great idea. A first step in writing a good business plan is to describe your basic ideas about your product and how you think a business can be built around it. If you are basing your business around a piece of intellectual property/invention, seek out the inventor and have an in-depth discussion with them. Try to learn as much about the invention as possible and for whom the inventor thought the invention would be most useful; that is, what problem was being solved? The inventor will have a perspective that no one else has. This knowledge will help you describe your strengths as reflected in your product; this was part of your SWOT analysis that was described in Chapter 3, Section 3.2: SWOT Analysis.

Now, seek out a favorite mentor or board member and discuss a strategy of how to write the plan. Make sure you know who your target investor audience is and then write to that person (as mentioned before), for example, a friend, an angel investor, a banker, or a VC. Next, try to obtain an existing business plan that has already been written in the same general area in which you will be competing. Also, one of your board members or mentors may have a business plan that they particularly liked. Imitate it! Once you become familiar with what the "written plan" is supposed to look like, you'll have a much better chance of producing something similar. This textbook also supports a website that has several sample business plans (see "http://blackboard.cornell.edu/; course name BEE4890"[1]). Note that the posted business plans were written by students and

---

[1] For the textbook and non-Cornell users, see separate website with supporting materials for the text such as sample business plans, www.Cambridge.org/Timmons

are not "perfect" in all respects but are considered good examples of the concepts presented in this chapter.

Become familiar with all the necessary components of the BP you intend to write. Pay particular notice to how various sections should link to each other; for example,

Market Analysis → Market Strategy →
Business Strategy → Operations Plan.

Before writing any section of the business plan, construct an outline of the entire plan. For some sections, diagram the subsections in which details will be important. Having constructed an outline, you should write the business plan section by section and have someone who has not been involved in writing the draft do a review and critique. I recommend writing the market analysis first recognizing that without a significant market, you really don't have a business. Why waste your time if you don't have a viable market? After you've written a section and you have received feedback on its content and style, then rewrite this particular section and have your mentor review it again to see if you both are "on the same page" of what you're trying to achieve. Now, you can write the next section. Repeat this process of write, review, and rewrite until you've completed the first draft of the business plan. Then, have someone else review the entire plan for their impressions and to assess how well you are conveying your message of opportunity. Finally, make sure you do a last check on spelling and syntax. Nothing ruins your credibility any more than having an obvious misspelling – particularly early in the document. Now, after you package it nicely (cover, *binding that allows the plan to lay flat when opened*, paper quality, etc.) you are ready to show it to a potential investor.

Final Tip

MULTIPLE VERSIONS NEEDED

You may want to create several versions of your document. For example, you might want to have a half-page e-mail version (serves as your "pitch" document to get them interested), the executive-summary version, a five-page version, and the full-document version. Once you begin sharing the plan with others, expect the need to make changes and that the document will continue to evolve. So keep the plan in a form that is flexible for revision, because the plan will continue to change as you move forward.

Help in Writing Your Business Plan

You can find additional sources of information on how to write a business plan at the back of the chapter (see the Additional Resources later in this chapter). Additional support is also provided at the Cornell website that supports this text, see "http://blackboard.cornell.edu/; course name BEE4890" and at the Cambridge University Press website for this book, see www.Cambridge.org/Timmons.

8.4 Writing Your Business Plan

When completed, a well-written plan is about ten to fifteen pages (4,000 to 6,000 words) not counting your appendices, cover sheets, and legal restrictions. Even if you are not ready to seek outside capital, you still need to write a business plan that will analyze your opportunities, do a financial analysis, and assemble your management team. Failure to do these things is almost always expensive, and you will incur avoidable risk. The process of writing a business plan will help you understand the key success and risk factors you will face starting a new or expanding an existing business.

Following is a suggested content outline for the elements that need to be addressed in your plan and what I think is a logical progression between these sections. You may want to make some adjustments to this order and perhaps add some additional sections or subsections under a major section heading, for example, under Financial Analysis, you could have a subsection called Revenue Model.

1. TABLE OF CONTENTS

Definition: The table of contents should serve as a navigation tool for the plan, including numbering of all sections and appendices. Here is our suggested content outline:

1. Cover page
2. Legal Qualifiers and Disclaimers
3. Table of Contents
4. Executive Summary
5. Mission Statement
6. Business Profile
7. Market Analysis
   a. Business Opportunity
   b. Problem to be Solved
   c. Competition
8. Market Strategy
   a. Product, Place, Promotion, and Price
9. Business Strategy
   a. Key Milestones
10. Operations Plan
11. Management and Key Personnel
12. Financial Analysis and Exit Strategy
    a. Funding Sources and Uses
    b. Revenue or Business Model

      c. Pro Forma

      d. Exit Strategy

   13. Appendices

Purpose: Most readers of business plans will not read the plan in order. Each reader has their own style and approach to reviewing business plans. Some may start with the management team section, others may quickly flip to the financials. The table of contents will help the investor quickly find the sections of most interest to them. To avoid common mistakes, be sure of the following:

- Keep the table of contents simple and clear.
- Coordinate the headings in the table of contents with any tabs marking various sections.
- Number the sections (it is easier to refer to them)
- Number the pages for all sections, including the appendices.
- Make the amount of text reasonably proportional to the importance of a section.

2. EXECUTIVE SUMMARY

Definition: The executive summary is a one- to two-page synthesis of your entire plan; try really hard to make it one page only.

Purpose: The executive summary is the very first part of the narrative and is the front door to your plan. For many investors, it may be the ***only thing they read***. Make sure that the executive summary sells your business plan to the investor. In fact, make sure its first paragraph sells the business plan to your investor. If the hook is not there, they may not even read the whole executive summary. The executive summary serves as your "elevator pitch" to the reader and should include:

- a general idea of where you are as a company and any significant achievements accomplished to date.
- succinct proof that the market needs your solution to an identified problem.
- a clear explanation of your competitive advantage and how you will sustain it over time.
- a brief financial summary of the business (how you make profits, how much capital you need, how it will be used, any important and identifiable milestones).
- information on why your management team is the right one to make this business succeed.

If you cannot articulate in a single page the focus of your business, your sustainable competitive advantage, the strength of your management team, the existence of a proven need in the market place, the likelihood of profitability, and your funding request, then you can forget about the remaining ten to fifteen pages even being read by the investor. A typical investor will not go beyond the executive summary if that single page is not convincing. To avoid common mistakes, be sure of the following:

- Identify themes with headings, for example, Competitive Advantage and Market Analysis. Organize your summary in short paragraphs, with concise language, and use bullets to emphasize key points.
- Include a clear statement of what you need from the investors, how the money will be used, and what the investor can expect in terms of an exit strategy.
- Keep the summary short; two pages are the maximum, and one page is the ideal.
- Proofread your work for spelling and/or grammar mistakes; careless errors communicate to the investor that you cannot pay attention to detail.

3. MISSION STATEMENT

Definition: A mission statement is a short declaration of the business's "reason for being." An effective mission statement will answer three questions (see Chapter 3, Section 3.1 Mission Statements for a complete discussion and examples):

- What is your product?
- Who is your target market?
- What is your special competence, strength or competitive advantage (read differentiation)?

Purpose: Putting the mission statement at the beginning of the business plan is a way of setting the tone for the rest of the document. It is unlikely that an investor will spend a lot of time examining the mission statement, but it is still important because it shows whether the entrepreneur has a clear and focused vision for the business. To avoid common mistakes, be sure to do the following:

- Avoid using the jargon and buzzwords.
- Avoid hyperbole, for example, fantastic, huge, best, extraordinary, massive, or world-class.
- Keep the mission statement relatively short (twenty to forty words, preferably one or two sentences).
- Check the tone to make sure it is business-like.

4. BUSINESS PROFILE

Definition: A business profile (sometimes called the overview) provides a short overview of your company describing its current status and future direction.

Purpose: The business profile will detail the following:

- who you are and the legal form of your company
- where you are doing business and how long you have been doing it
- description of your product and target market
- your "value proposition"

- major "strengths" or unique characteristics you currently possess and your sustainable competitive advantage

A major goal of this section is to establish credibility in the mind of the investor. This section helps the reader understand how the company will make money. Make sure you tell readers any major positive attributes associated with your company. Particularly, tell them if you already are generating sales, which is one of the best ways to establish your company's credibility. This is a good spot to describe any special *value networks* you have with suppliers or partners and the added value these relationships bring to your company. Also, tell the investor about any strategic assets you own, for example, a patent or a license. The combined value of these special networks you have accessed outside your company boundaries and your strategic assets inside your company represent your *value network*. Closely related to your value network is your *value proposition*. A value proposition makes or saves your customers money by solving a problem they have. You want the investor to grasp your compelling advantages early when reading your business plan, so present them here in the business profile section.

> You want the investor to grasp your compelling advantages early, so present them in the business profile.

To avoid common mistakes, be sure to do the following:

- Include the specific location of the start-up operation (or explain where operations are being handled if it is a "virtual company" and the legal form of the company.
- Paint a visual picture of how and why customers buy the products or services (your value proposition).
- Avoid excessive operational details; those belong in the operations plan.

5. MARKET ANALYSIS

Definition: The market analysis is an analysis of the industry, competitor, and customer.

Purpose: As a whole, the market analysis section is designed to document the fact that there is a demonstrable need for your product and that industry and competitive forces are well understood. After an investor reads this section, you will have clearly identified the business opportunity and the problem or customer need you will solve. There should almost be an "aha!"-type reaction from the reader because these two important facets are clearly shown. To help the reader, use subheadings to identify "Opportunity" and "Customer Problem." In the market analysis section, you also will want to identify your competition and any of their weaknesses.

Earlier in this chapter, we talked about the Value Equation and the Opportunity Equation (also, see Chapter 10, "Investor Principles"). Here, you can use some components of the opportunity or Value Equation as subheadings, for example,

Size of the Problem, Current Status of Market Opportunity. It can often help the investor to provide a description of a typical ideal customer profile and how many of these customers exist and where. This is an ideal spot for a table or a chart.

Inexperienced entrepreneurs often confuse or fail to recognize an important distinction when defining markets. Ideally a market will be **underserved** (or else why "go after it") as well as **emerging** (meaning growing, and is a key component of the Opportunity Equation). Recognizing this difference is part art and part science. Let's look at my favorite market – seafood. The market demand for seafood has existed since the beginning of humankind; however, demand for tilapia (an emerging market over the last several years) emerged because of an inability of the wild seafood catch to supply the demand (on a continuous basis) for mild whitefish fillets. Now another, unrelated example: Ethanol fuel has been around forever, but the demand for ethanol as a fuel oxygenate is now booming and has been an emerging market, due to banning of MTBE, high crude prices, government support in form of tax relief, emphasis on sustainability, and carbon sequestration. See the pattern?

> This brief period of a market being underserved is your entrepreneurial opportunity.

Inexperienced entrepreneurs may be tempted to target a market only because it can be defined and/or only because it is growing. However, my advice is that, ideally, a market should be both emerging and underserved (the latter most likely being a temporary state). This brief period when a market is being underserved is your entrepreneurial opportunity. This is where the first mover's advantage term was coined.

In writing this section, you should demonstrate to the investor that you understand the distinctions and importance between emerging and underserved markets and where you are seeing an opportunity. To avoid common mistakes, be sure to do the following:

- Demonstrate that you are starting a business that can be successful given the specific trends and general health of your industry (sustainable advantage).
- Provide both numerical and qualitative proof that consumers not only desire but also are willing and able to pay for your product or service (customer need).
- Relate the findings of each subsection to your specific business idea and approach. You may want to have a text box at the end of each section labeled "Implications for the Business" in which you summarize what your findings imply about your start-up strategy.
- Use graphs, tables, and bullet points to break up the narrative and to clarify your key points.
- Show proof that you have spoken directly to potential customers.
- Use a mixture of quantitative data and some specific detail. For example, you might indicate the quantitative size of the industry and the number of

consumers in your region and then go on to cite some details of specific conversations with consumers.

- When you are reporting on questionnaires or focus groups, keep only the relevant information in the narrative, providing the specific details in an appendix.

6. MARKETING STRATEGY

Definition: Your marketing strategy is how you address the market opportunity that was described in the market analysis section.

Purpose: Marketing strategy is your specific ideas, plans and actions that outline and guide your company's decisions on the best way to create, distribute, promote, and price a product or service. Marketing strategy is the process that creates customers. Your marketing strategy should have a sustainable competitive advantage.

The marketing strategy is linked to the market analysis by explaining how your business addresses the 4P's of marketing: product, place, promotion, and price strategies. You must articulate clearly and convincingly that you understand how to compete in the market you have described and that you have a specific strategy to do so.

You have identified (or should have) the customer problem in your market analysis. Remember the six-word definition of marketing: marketing means solving customers' problems profitably. So there is no confusion to the investor, clearly re-identify the problem you are solving in the marketing strategy section (briefly). You might even start out the marketing strategy section with a definition of the problem you are solving (probably with a reference for more detail back to the marketing analysis section). You will describe how you make money solving the problem in the financial analysis section presented later in the business plan.

At some point in this section of your business plan, you will address the 4 P's of any marketing plan or strategy. Often, your product has already been described in an earlier section, so it may not need to appear here in detail. You might use a very brief description of the product as a lead in to the other three of the 4 P's. As a quick review of the 4 P's, a brief description of each of the 4 P' is given here and they are given in much more detail in Chapter 4.

PRODUCT

The product description will be brief and not be overly repetitive of details already provided in the business profile section (one of the first parts of the business plan). You need to identify and stress the market advantage that you possess and even label a section as Competitive Advantage. You might elaborate here on what your product does better than your competition, which in part describes your differentiation. The product description should also describe why competitors cannot easily replicate your proposed advantage and differentiation that you have in solving your customer's problem. This might have been previously described in your business profile, so be careful about over duplication here.

PRICE

Price strategy will depend upon what value customers recognize in your solution to their problem (this might be covered under the product subsection). Justifying this value should be done here in this section. To highlight the value description and argument, you could label a subsection as value-added proposition- You need to price your product/service that is based on the value provided, competition, alternatives, and so on. Value is the key concept here. Your pricing strategy as reflected in a revenue model will be presented in your financial analysis section.

PLACE

The Place (distribution) description will identify the strategies used to reach different target customer groups and through what means, for example, direct sales, wholesaler, and so on. Different marketing strategies may be required to reach each segment. The different customer groups will need to be clearly identified and whether different channels or combinations of channels are required to reach them. These channels need to be clearly described as to how they deliver product from first manufacture to intermediate and/or end users.

PROMOTION

Promotion programs are typically directed at the product's end user so that demand is created, but your distribution method may have several levels between you and the end customer. Promotion programs and advertising can be very costly and dominate your overall budget, so be careful in being too aggressive here, particularly for start-ups that have minimal cash. More elaborate and expensive promotional campaigns should be relegated to future phases, which you could mention in your overall strategy description.

To avoid common mistakes in writing your marketing strategy section, be sure to do the following:

- Avoid duplication of materials given in the market analysis.
- Make a clear connection between your market analysis and the direction of your market strategy.
- Clearly identify your value added proposition and then reflect this value in your pricing scheme.
- Identify any threats to your success and your strategy to minimize their impact on your company's success.

7. BUSINESS STRATEGY

Definition: The business strategy section is how you implement your marketing strategy.

Purpose: The business strategy will identify steps and associated costs your business will take to achieve your marketing strategy and their timing and sequence. These steps will probably be described as a series of phases or rounds, for example, phase 1, phase 2, and so on. Your funding requests by round will be determined by the activities conducted in each phase. It is customary that investors want to know what funds are required to go beyond the current phase or round in implementing your overall business strategy as this can affect their decision to invest in the current round.

For your current round of financing, you will need to identify specific *milestones* and the funds needed to achieve these milestones. Typical milestone achievements with their target date of completion could include the following:

- When your product will be entered into the marketplace
- Specific revenue milestones and dates
- Staffing expansions and capabilities added
- Key alliances achieved (when necessary)
- Future funding rounds and amount of funds to be secured

You need to be convincing that you have a realistic view of the funds required to achieve the milestones comprising the current phase. Credibility is enhanced when supplier quotations for significant cost items and any outside contracted services/inputs are included in your appendix. These details are generally needed only for the current round of financing, but they may be appropriate for some specific items required in future rounds that are not part of the current round.

Providing a chart showing milestones and the target dates to achieve them creates a responsibility on your part. A milestone chart makes it easier for your investor to understand your strategy and its timing. Your ability to meet time lines and reach milestones will determine whether you can raise the necessary funds for your next round of financing. An inability to reach target milestones satisfactorily will diminish any interest your current investors have in making further investments. So, be sure your milestones are as reasonable as possible but still aggressively move your company forward.

The milestone projections presented in the business strategy must be consistent with the projections made in the financial analysis section. You will need to be aggressive in making these projections, but you have to be credible, and credibility comes from making a sound case as to why you will be able to do what you say you will do. If you are too conservative, then you will be undervaluing your company when you enter into negotiations. Showing milestone events that demonstrate your company growing at unrealistic rates simply demonstrates to the potential investor that you have no basic business realism and, therefore, should be avoided.

> Showing milestone events that demonstrate your company growing at unrealistic rates simply demonstrates to the potential investor that you have no basic business realism and, therefore, should be avoided.

To avoid common mistakes, be sure to do the following:

- Eliminate duplication and coordinate discussion among the marketing strategy, the business strategy, and the operations plan (following section) sections.

- Distinguish between the funds required for the current round and future rounds of financing.
- Make the business strategy only a page or two so that your plan for conquering the market is very clear to the investor.

8. OPERATIONS PLAN

Definition: The operations plan defines the processes, facilities, and personnel that are required to deliver your product/service to a customer.

Purpose: The operations plan is what transforms your business concept from a plan to a reality. Typically, your operations plan will describe

- facilities needed
- type and numbers of employees needed
- responsibilities of the management team
- tasks assigned to each division within the company
- logistical support functions or tasks required for your company to deliver your product to your targeted customer or customer groups

The operations plan will concentrate on the current phase for which you are trying to raise money. For the typical start-up when very limited funds are available, for example, less than $100,000, many of the preceding tasks will have to be done by one or two people. Strive to minimize your fixed overhead costs in your first phase of your business strategy. You might have just one paid employee (you) who does everything including taking the trash out! However, provide some detail on how you expect the various operations that constitute your company to change as you move to subsequent phases of your business strategy, for example, when you switch from contracted manufacturing services to in-house manufacture, add direct sales to end customers when you start out using a wholesaler, expand your geographic sales territory, or add other product lines and services.

If your company is a service or product provider, then show how you execute delivery to the customer. If you are creating a product, describe how you will design, build, and prepare your first demo product or, if you have a prototype product, how you will create your first supply of products for customers and how many units will you require for the current phase. Some of these first customers may be receiving the product for free or at a reduced price so you can obtain feedback about any needed changes before producing a larger batch of products for sale.

Charts and figures are always helpful to demonstrate product flow from obtaining product inputs for manufacturing, producing the final product, and delivering it to the end customer. You need to document the supply side of the production equation and how well you expect this to scale as you increase product sales. This information will be used in your revenue model presented in the financial analysis.

Anyone can have a great idea or concept, but transforming that idea into reality is what counts. Investors invest in reality. Reality is convincing your potential investors that your operations plan can be executed in the timeframe and for the cost you have predicted. To avoid common mistakes, be sure to do the following:

- Include a diagram to show the flow of operations and important milestones.
- Avoid elaborate hierarchical charts about management and employees when you are a start-up with only two or three employees or just yourself.
- Avoid overlap with the financial analysis section in which you will provide details on your cost of goods, company overhead, and management team.

9. MANAGEMENT TEAM

Definition: This section of the business plan informs the reader about the core leadership team for your business, which includes the board, advisors, and any special employees.

Purpose: After convincing your investor of a valid opportunity, you must also convince the investor that you and your team can successfully address it. Investors consider three major factors to predict success: people, people, and people (this is similar to the three most important things in real estate: location, location, location).

Fundamentally, anyone who evaluates your business idea will be evaluating the team that is going to try to execute your business strategy and operations plan. Put simply, can you make it, can you sell it, can you manage it – and what evidence can you provide to support your claims? Remember that some of your management strengths and expertise may come from your board of directors or from a technical advisory board that you establish. For young entrepreneurs, the quickest way to establish your credibility to potential investors is through your board members.

Although the management team is probably the most important aspect of the business plan to secure investors, you can describe your team fairly briefly. Anything more than two or three lines on any member is probably not needed. In your appendix, you can include short resumes for key team members and refer the reader to them. Investors will be impressed by recent successes and experiences of team members who directly relate to the future success of your company. These strengths are so important that they should probably be mentioned in the business profile section and the executive summary.

To avoid common mistakes, be sure to do the following:

- Identify experiences that directly relate to demonstrating your capability to "make" the product and "sell" the product rather than simply listing credentials (degrees).
- For start-ups, minimize fixed overhead as much as possible.
- For start-ups, minimize payroll for as long as possible (part of fixed overhead). Have only the necessary paid staff; use contract services as much as possible instead of staffing in-house capabilities.

- Avoid the use of multiple part-time people in management roles. This is a typical error for start-ups; someone needs to be responsible and 100% dedicated to the success of the start-up.

10. FINANCIAL ANALYSIS AND EXIT STRATEGY

Definition: The financial analysis and exit strategy analyzes the costs that will be incurred to achieve the milestones previously described in your business strategy and operations plan. You should establish a current valuation of your company (pre-value) and your predicted value in three to five years, which is supported by your revenue model, breakeven analysis, and pro forma statements. Your exit strategy describes how your investors will recover their initial investment.

Purpose: The financial analysis and exit strategy lay the groundwork for discussions between you and your potential investors or bankers that will supply capital to the business. This section will describe the total funding being requested and will tell an investor what percent equity is received in exchange, which will be a negotiation between you and the investor about your pre-value claim. Once pre-value is established, the percentage of equity obtained for an investment is simple arithmetic. So be careful here not to compromise your ability to make the best deal for yourself by divulging too much information regarding valuation or suggesting there is no flexibility at all here. Minimally, this section will tell an investor what their expected internal rate of return (IRR) on their investment will be and an exit strategy defining how they will cash out.

Most business plans will flesh out the financial analysis and exit strategy by having the following distinct parts: (1) sources and uses of funds, (2) business or revenue model, (3) proforma statements, and (4) exit strategy. Each is described below in further detail.

SOURCES AND USES OF FUNDS

The financial analysis and exit strategy should begin by clearly stating

- how much money is being sought in the current round
- if any time constraints are imposed on the round
- if there is a minimum level of investment that must be obtained to execute the succeeding phase, or else the funds are returned

These constraints should be consistent with the information previously presented in the operations plan that gave milestones, phases of the business strategy, and funds required by phase. Having identified the level of funding required, you then detail where you expect to source these funds, for example, investors (equity financing), bank loans (debt financing), founders (self), and grants. The intended uses for these funds are then described. If you have not secured funds from a granting agency, then it is not a current source (list under future activities).

Be very careful showing any funds going to research and development (R&D) purposes. There are special cases in which an investor would look favorably on this; for example, you need limited research funds to refine an already-existing working prototype so it can then be released to the market place. Mostly, however, using investor funds for R&D is similar to asking them to give you money to go to Las Vegas to gamble with the hope that you will multiply their investment. R&D being funded from generated profits may be looked at on favorably (in lieu of passing these profits to shareholders), if these funds are being provided from operating profits and the R&D is needed to maintain company momentum or current market share.

To avoid common mistakes in the sources and uses section, be sure to do the following:

- Clearly show from where funds will be sourced and how and when they will be spent.
- Never show that funds will be used to pay back earlier investors or to cash them out.
- Avoid using funds to support high salaries for founders; typically, management salaries are below market
- Be careful showing any funds going toward R&D.

BUSINESS OR REVENUE MODEL

A business model defined only in financial terms is a revenue model, which is how your business will generate revenue and profits. Here is your chance to actually show an equation, which engineering entrepreneurs love. Verbally describe your revenue model including each area in which the company derives revenue, for example, leases, royalties, or sales to different customer types. Having a breakeven analysis is a good idea to include in this section and then compare the number of sales required for breakeven to your projected yearly sales volumes. You will need to explain and justify important assumptions you are making in the revenue model. Sometimes, it is a good idea to create a table of key assumptions and of how these assumptions change over the first three to five years of your pro forma statement.

PRO FORMA STATEMENTS

Once you have developed your revenue model, then build your pro forma model for statements of income, balance sheet, and cash flow. These numbers must flow easily from the arguments and data that you have presented in earlier sections of the business plan. For example, the percentage of market share you are obtaining as shown by sales revenues (or units sold) should be supported by your market analysis section.

Key assumptions used in the business model should be shown in your income statement so that the investor can easily see them and that the sensitivity of these variables can be easily tested. The pro forma model should make it possible to discuss "what-if" scenarios so that if you are wrong about any major assumptions, the implications for the business will be clear. A sensitivity analysis, favored by some, is generally not necessary. More people now simply prefer most likely revenues and costs and a calculation of a breakeven point. Check with whoever will be reading your plan to see what expectations they have about this type of analysis.

Your pro forma model should be demonstrated for three to five years; angel investors usually are happy with three years and bankers want to see five years. You should show a monthly breakdown of cash flows for year 1 in addition to the three- to five-year summaries. A summary of key financial projections should be presented in the text portion of the plan, whereas full projections should be placed in the appendix. Showing the expected rate of return on equity investments (internal rates of return, IRR) is a value that many investors will want to see. The IRR will have to make economic sense to investors when compared to alternative and generally safer investments that are available to them; for example, IRRs much larger than 10% and as high as 100% per year are often expected.

To avoid common mistakes in the pro forma section, be sure that

- your financials are in a format considered credible by financial analysts for the industry in question.
- you include adequate documentation of your cost and revenue estimates.
- the twelve-month cash flow statement is realistic in terms of when revenues are being generated first, and it agrees with the yearly summary of cash flows.
- the twelve-month cash flow does not reveal a negative cash or dangerously low position, even though the end-of-year cash position is satisfactorily positive.
- revenues are proportional to the number of employees based upon industry standards; for example, revenues are approximately $100,000 to $200,000 per employee depending on the type of industry.
- the balance sheet actually balances.

EXIT STRATEGY

Your plan must address how an investor can cash out of their investment. For most companies, the most logical and probable exit strategy is that your company will be acquired by a competitor or a larger company. This is an excellent strategy. You might have a strategic partner that could become a likely buyer of your company. You have to aggressively manage your company toward continued growth and profitability, because these are the type of companies that are often targeted for acquisition by larger companies or other industry players.

An initial public offering (IPO) is not a realistic exit strategy. First, there are typically less than fifty public offerings in a year (Table 8.2), whereas there are generally well over 500,000 small businesses created each year in the United States! Typical cash flows for IPO candidate companies are $20 million or more in revenues. Now it is a good idea to strive for such a financial goal, and then you will probably become a target for acquisition without an IPO process.

To avoid making common mistakes when writing your Exit Strategy, be sure to

- not naively suggest an IPO and
- paint a clear and realistic picture of how investors can expect to "cash out" (the so-called exit strategy).

Table 8.2 Initial Public Offerings by Year (National Venture Capital Association, 2013)

Year	Number of IPO's	Offer Amount ($Mil)	Median Amt ($Mil)	Mean Offer Amt ($Mil)
1985	48	763	13	16
1986	104	2,414	14	23
1987	86	2,125	17	25
1988	43	769	15	18
1989	42	873	16	21
1990	47	1,108	20	24
1991	120	3,726	27	31
1992	150	5,431	24	36
1993	175	6,141	24	35
1994	140	4,004	24	29
1995	184	7,859	36	43
1996	256	12,666	35	49
1997	141	5,831	33	41
1998	79	4,221	43	53
1999	280	24,005	70	86
2000	238	27,443	83	115
2001	37	4,130	80	112
2002	24	2,333	89	97
2003	26	2,024	71	78
2004	82	10,032	70	122
2005	59	5,113	68	87
2006	68	7,127	85	105
2007	92	12,365	97	134
2008	7	765	83	109
2009	13	1,980	123	152
2010	68	7,609	93	112
2011	51	10,690	106	210
2012	49	21,451	89	438

Summarizing, you should avoid the following general mistakes when writing your financial analysis and exit strategy:

- Keep the narrative focused on key issues (sources and uses of funds, key assumptions, breakeven analysis, profitability milestones, rates of return).
- If you state your pre-value position, then give your rationale for its basis.
- Address the primary risks facing the business and how you are financially prepared to weather these challenges.
- Request enough money to start the business with an adequate safety net, but without an unreasonable amount of extra cash.

11. APPENDICES

Definition: The appendices to a business plan often include any details on financial assumptions, any additional information needed about market research, and perhaps information on patents or intellectual property rights or other details specific to certain businesses.

Purpose: Appendices can provide credibility by showing the degree of research done by the entrepreneur. They also help provide an outlet for material that would be too long to include in the narrative but that the entrepreneur wants to include for the reader. A potential investor will generally give your business plan to an outside expert, or experts, to review for technical and scientific merit. Make sure there is enough science in your business plan to satisfy these types or to at least generate pertinent questions in a face-to-face meeting with the investor.

To avoid common mistakes, be sure to do the following:

- Label and number your appendices.
- Refer to each part of the appendix somewhere in the narrative of the business plan.
- Be extremely careful about confidentiality issues when reporting on customer, industry, or competitor intelligence.
- Ask your prospective reader if they would prefer to have the appendices bound separately from the plan itself.

8.5 Reference Styles

As you write your business plan and make use of literature information to support an argument, for example, size of market, your credibility will be greatly enhanced by proper use of references. This also means you should use a recognized format for citation. A variety of reference styles can be used. One set is provided in the following. The key is to be consistent throughout the document that you are composing; do not mix styles.

1. All publications cited in the text should be presented in a list of references following the text of the manuscript. The manuscript should be carefully checked to ensure that the spelling of author's names and dates are exactly the same in the text as in the reference list.

2. In the text, refer to the author's name (without initial) and year of publication, followed if necessary by a short reference to appropriate pages. Examples: Peterson (1988) has shown that fish swim upstream; or, Eating salmon has been shown to reduce heart disease by 25% (Kramer 1989, 12–16).

3. In the text, if a publication references was written by four or more authors, the name of the first author should be used followed by "et al." This indication, however, never should be used in the list of references. In this list, names of first author and coauthors must be listed.

4. References cited together in the text should be arranged chronologically. The list of references should be arranged alphabetically on author's names, and chronologically per author. If an author's name in the list is also mentioned with coauthors the following order should be used: publications of the single author, arranged according to publication dates – publications of the same author with one coauthor publications of the author with more than one coauthor. Publications by the same author(s) in the same year should be listed as 2009a, 2009b, and so on.

5. Use the following system for detailing your references:

    a. *For periodicals*

       Hopkins, J.S., Sandifer, P.A., Browdy, C.L., 1994. Sludge management in intensive pond culture of shrimp: effect of management regime on water quality, sludge characteristics, nitrogen extinction, and shrimp production. Aquacult. Eng. 13. 11–30.

    b. *For edited symposia, special issues, etc. published in a periodical*

       Benzie, J. A. H., Bailment, E., Frusher, G., 1993. Genetic structure of *Penaeus monodon* in Australia: concordant results from mIDNA and allozymes. In: G.A.E. Gall, H. Chen (Eds.), Genetics in Aquaculture IV. Proceedings of the Fourth International Symposium, 29 April – 3 May 1991, Wuhan, China. Aquaculture, 111, 89–93.

    c. *For books*

       Gaugh, Jr., 1992. Statistical Analysis of Regional Yield Trials. Elsevier, Amsterdam, 278 pp.

    d. *For multi-author books*

       Liao, L. C., Chen, S., 1992. Marine prawn culture industry of Taiwan. In: A.W. Fast, L.J. Lester (Eds.), Marine Shrimp Culture: Principles and Practices. Elsevier, Amsterdam, pp. *653–75.*

    e. *For web articles*

       Varian, H. R. (1997, June 11). The future of electronic journals. Paper presented at the 1997 Scholarly Communication Conference. Retrieved June 27, 2013 from http://arl.cni.org/scomm/scat/varian.html

6. Abbreviate the titles of periodicals mentioned in the list of references according to the *International List of Periodical Title Word Abbreviations* or spell out the journal in total.

7. In the case of publications in any language other than English, the original title is to be retained. However, the titles of publications in non-Latin alphabets should be translated, and a notation such as "(in Russian)" or "(in Greek, with English abstract)" should be added.

8. Work accepted for publication but not yet published should be referred to as "in press."

9. References concerning unpublished data and "personal communications" should **not** be cited in the reference list but may be mentioned in the text as a footnote; provide some of the qualifications of the quoted person to create credibility.

8.6 The Pitch: Creating and Giving Oral Presentations

Content of Your Presentation Pitch

A presentation to a group of potential investors is called a "pitch." I love this term, *pitch*, because it is analogous to a baseball player throwing a ball toward home plate and the batter swings. The difference is that you *hope* the batter hits the ball! You want to throw a pitch that the batter's eyeballs light up as they anticipate smashing the ball over the fence for a homerun. Pow!! So, that is your goal in presenting your business opportunity to your audience. It only requires that one person listening to the opportunity in front of them is attracted to it. But you must convince at least one person, or else you have no new investor.

The common error committed by almost all first-time presenters is that they construct a pitch presentation as a mini-version of their entire written business plan. *No*, no, no!! What you are trying to do is to convince the listener that you have identified a real problem, that you have a solution to this problem, and that a significant number of people with the same problem and are willing to pay someone to fix their problem. Of your presentation, *80% or more* should focus on these key points. Who will care about your management team or financial analysis if they are not convinced that there is a real problem?

Having just said that, there are three major components to your pitch, there are no hard rules on the sequential order of your presentation content or even what the content should be. If your business start-up has a serial entrepreneur as one of your key team members, then you would be foolish not to make a point of this. Make sure that your strengths are highlighted. Often, a good start consists of giving a single sentence that defines your company, service, or product. Say it slowly to be sure everyone gets it. Then you might follow that with a description of the problem your company is solving and how big that problem is. Last, make sure the audience knows how much money is required to implement phase 1 of your business plan. Angel- and family-and-friends-type investors are most concerned about phase 1 financing because you have to move successfully through phase 1 to get to phase 2. The angel investor will want to know how much money is needed before the bleeding stops, that is, when will positive cash flow begin? Leave no doubt in the listener's mind about the answers to these questions.

How you order these presentation components is what creates a successful experience for the audience. As a general guideline, the following questions are usually addressed in the presentation (but remember this is partly art and partly theater!), but where you locate them will depend on the theatrical objectives of your team:

- What problem are you solving in the market? What do you do? What product or service do you, or will you, offer to the marketplace? Are there any critical R&D milestones? Avoid the natural tendency to provide very detailed or overly technical product descriptions. Although it's important to provide a sufficient overview of a particular product, it will be more important to provide an understanding of the product's impact in the marketplace and how customers view the offerings.
- Who are you? If you are the entrepreneur, or the leader who will guide the enterprise through inevitable adversity, state very briefly why an investor should have confidence in you, but do not appear to be bragging or embellishing your accomplishments.

- What is your company's background? Presumably you are from a company that is either operating or is newly formed. If you are from an operating company, describe your company and provide a brief history (described most succinctly by historical financial results in abbreviated form). If your company is a start-up, what are your management team's relevant qualifications?

- What is your distinctive competence? What are the distinct competitive advantages of the product or service for which you would like to obtain financing, that is, your differentiation?

- How do you sell your product or service? What is your marketing strategy? Specifically identify your target market and channels of distribution.

- Who are your competitors? Who else is selling something that fills the same need or performs the same function? What are their relative advantages and disadvantages? Because you will be challenging the best of your potential competitors, be very knowledgeable about them.

- How much can you sell? How big is the market, and how fast is it growing? How much money can you make from your sales? Condensed sales and expense figures are in order, both past and future. If you project these numbers on the screen, keep in mind that they should be large and clear enough to be read by the everyone in the audience.

- How much money do you need? How much money has been invested in the venture already? Be specific about this. Your cash-flow projections should tell you exactly how much you need. There is no need to show the cash flows at this time. If the deal is interesting, the investor can look at them in detail later.

A Professional Presentation Is a Performance

The business plan, the oral presentation, and the writing of letters to potential investors – these things all go together.[2] First, you need the business plan, which we just discussed. Then, you're ready to develop your presentation which you will present to a group or an individual investor. And last (the last section in this chapter), you'll have follow up activity with a potential investor, which will usually be by written letter. Preparing each of these forms of communication will probably result in improving your other two complimentary forms of communication.

For your presentation, your goal is that it leads to a follow-up by a potential investor to seek additional information and this hopefully leads to a cash investment. An easy analogy for you to imagine how to prepare and present your presentation is to think of it as a theater play. You've all been to plays. Plays are organized into a series of acts (think slides) and each act will be in some particular setting and have some group of actors communicating among themselves,

---

[2] Dr. Rick Evans, Director of Engineering Communications, Cornell University, contributed to this section.

but the actors are really communicating with their audience. The acts making up the play each have some particular purpose that lead the audience to the next act. And all plays have a climax that the audience has been prepared for by the actors' actions up to that point. At the end of the play, the audience applauds; there are curtain calls; there are meetings with the actors in the lobby afterward, signings of programs, and so on. You have an expectation of how the play will be conducted and the rules that surround this event, both for you and for the actors. A presentation to a group of potential investors is really the same thing. We'll give you some detail and guidelines to help you in constructing your performance, but keep the theatrical play in mind as you do.

Guidelines for Oral Presentations

The world might be an easier place if there were but one right and one wrong way to do something. Unfortunately (or fortunately depending on your perspective, sometimes labeled "philosophy of life"), there are many right ways and many wrong ways to do something. So pay attention to the fact that we've labeled the following directions as *guidelines*, and they should not be considered absolute by any means. The guidelines given are largely based on directions I received to make a presentation to a venture capital fair that was assembled by Investor's Circle (Brookline, MA).[3] One of the organizers of this venture fair was strongly considering making an investment in my fish start-up company (which he did, by the way).

Giving a presentation is not just presenting information. You are building a story so listeners want to read your business plan, and this will lead them to make an investment. In reality, you are preparing to give a performance intended to solicit a response by a possible investor. This performance will have several distinctive characteristics defined by the expected genre of this type of event:

- Special Time. Your presentation will be a specific time, scheduled well in advance, which gives you plenty of time to prepare. The audience as a result expects a polished performance. A venture capital fair is such an event and will be sponsored typically by a venture fund, a university, or a local government group.
- Programmed. The sponsor or host will publish rules on how your performance is to be conducted, for example, time allotted, are questions allowed, and number of people allowed to present.
- Participatory. The role of the audience will be scripted, for example, how they interact with you before, during and after. As you give your presentation, try to talk to your audience and not at them. You should create a reason for why the audience will want to talk to you after your performance; for example, you mention additional information that you can discuss after

---

[3] Investors' Circle, 320 Washington Street, Brookline, MA 02445; email inbox@investorscircle.net.

the performance ("Please see me at the lunch sponsored after the presenta-tions"). So design something into your presentation that will help them to approach you afterward, that is, their final participatory response.
- Reflexive. The audience will generate an impression of the performers giv-ing your company's performance. Your goal is to create an impression of high competence in your members. Your potential investors want to believe that you really "know your stuff." Your hope is that the reflexive response should be that your performance has demonstrated to the audience your internal competence as a collective team.

AUDIENCE

Target your presentation to the investors in the audience who you expect to be interested in your company; realize that not all of them may be interested. Ask yourself the following questions:

- What do they know or not know about my project?
- What do they want to hear?
- What do they need to hear?
- How will I address their knowledge, wants, and needs?

You should have a good idea of the type of people who are making up your audience and of what their background knowledge levels are. The investors' objective is to decide whether they want to meet with you during a follow-up time slot (usually a venture fair will have some specific opportunity for investors to meet with you as part of the program, e.g., at a "tradeshow" portion of the fair or over a lunch, dinner, or reception).

Because audiences vary, you may need to adjust your presentation's content based upon whom you expect to be attending and what their expectations might be. Typically, many in your audience will have heard many presentations in the past, and they are experienced at quickly judging their degree of interest in any particular proposal. This means that the audience has an expectation of what they will see, that is, rethink the theatrical play analogy. On the other hand, many of the investors in the audience may not have highly technical backgrounds or knowledge of your particular marketplace. It is critical, therefore, that you pres-ent your company and its financing requirements as clearly, concisely, and force-fully as possible.

LENGTH OF THE PRESENTATION

The audience will have an expectation for how long your presentation will last, just as you have an expectation as to how long a movie or a television commercial will last. When you present at a venture fair, the organizers will define how much time is allocated for your presentation. Ten minutes (or some other specific time limit will be given) is a typical length allocated for individual presentations. *Your responsibility* is to make this allotted time to be more than adequate to present the important concepts associated with your business and why someone would want to make an investment. In a timed-presentation format, an audience will *not* have the opportunity to ask questions during this presentation. If you exceed the ten-minute period, you will be interrupted and forced to stop, which *will make you and your team look like complete fools and no one invests in a fool.*

> No one invests in a fool!

SLIDE INFORMATION

The real issue with slides is how you (the presenter and the audience) will interact with the slides that make up your presentation. You are using the slides to facilitate an interaction between you and the audience. When you change to a new slide (think of the curtain going up or a new act begins in a play), the audience will immediately focus on the slide and not on you. So, they won't hear anything you're saying until they finish reading the slide themselves. They can only focus on one thing at a time! The audience will focus where you lead or direct them; for example, you say, "Audience, please look at the handout I have placed at each of your tables." Guess what – the audience will now look at the handout. Use the slides to help convey your ideas. Remember that the slides are not meant to be a linear shortened version of your entire business plan. You won't have much time, so you have to be careful in what you present.

In general, fifteen slides seem to be the optimal number for a ten-minute presentation. Most presenters will find that at least ten slides are needed to effectively communicate their story with the presentation appearing dull when fewer are used and rushed when closer to twenty slides are used. The first slide in each presentation should include the name of the company; the company logo, if any; and the name of the presenters. The very last slide should include the name of the company and the company logo or whatever you want the last thing to be on their mind when you finish speaking.

Typically, you go into a question-and-answer session immediately following the formal presentation segment. I cannot tell you how many times the most frequently asked questions relate to the last slide presented (or nearly the last)! So take advantage of this when you can.

SLIDE MECHANICS

Every slide should have the company logo included on it. All slides should be presented horizontally. The size of the text should be no smaller than that which would fill a slide top to bottom with twelve lines single spaced to ensure that the text is large enough to be read at a distance. The problem with slides that cannot be read easily is fundamentally that it hinders interaction with the audience. *The use of small text size is a major complaint at venture fairs!* You might have a slide with a table on it that no one can read, but the intent here is that they not read the content but can appreciate some other message you are trying to convey; of course, then the key here is that you tell the audience what your point is.

On slides that feature charts and/or graphs, make sure that the axis labels feature the same type size (or larger) as is used for the text. Each slide should contain no more than eight lines of text. Information should be provided in short, concise statements.

There is a wide variety of opinions on how "glitzy" your slides should be. Glitzy slides will not make bad content better! The absolute rule is that the audience needs to be able to read your slides easily under the lighting conditions that you present your slides. (For additional discussion and guidelines, see Additional Resources at the end of the chapter and www.edwardtufte.com, written and created by Dr. Edward R. Tufte.)

> Glitzy slides will not make bad content better!

FINAL COMMENTS

The success of your presentation will largely depend on the quality, relevance, and opportunity described by content of the presentation. The best way to create a better presentation is by having better content. However, you can still destroy superb content by having a poor oral performance, in which your delivery and overall impression you create destroy any positive impressions that the audience should have made. Performance will be critical.

When you attended your last theatrical performance, you probably didn't hear any of the players flubbing their lines. This is because they practiced and practiced. You must do the same if you expect to pull off a polished performance. When you are presenting as a team, consider the overall choreography of the performance:

- How to divide responsibilities: no gaps, overlaps
- Plan transitions: verbal, visual, physical
- Consider crucial timing concerns
- Strive for consistency: pace, delivery issues, and visuals
- Avoid (typically) any jokes (these are extremely difficult to execute successfully and can incur high risk; remember that you are not a professional comedian).
- Attire: your team should look like a team in dress code, for example, your costumes or uniforms convey a desired impression.

Everyone experiences stage fright. This is a positive because you will have some adrenaline-driven energy, and this should allow you to perform at your highest level. To properly channel your energy, however, you will need to rehearse and practice your presentation. As you practice giving your presentation, keep track of the following relative to your actually delivery of your presentation:

- Vocal: projection, pace, enunciation, inflection
- Body language: mannerisms, eye contact, stance
- Word choice: jargon, slang, grammar, "fillers"

- Demonstrations: handling, explaining
- Polish: poise, smoothness, practiced

Finally, the day will come when you will give your presentation. Don't let all that hard work be compromised by some last-minute snafu! Consider the following logistical issues and use the following helpful hints to ensure your success:

- Make sure you have clear directions of how to get to the site (you don't want to get lost and be late getting there; minimize your stress levels as much as possible).
- Arrive early: arrange the speaking area and become familiar with the audio visual equipment; bring your own laptop (two preferably) and multiple thumb drives with your electronic presentation.
- Conquer your nerves: breathe; drink; focus; if you must, use small note cards to help you keep on track; this is better than getting lost during your presentation, which is a total disaster.
- Consider planting a question in the audience to bring out important additional information (but make sure that this is *not* obvious or done in a manner that could create any level of resentment).
- Provide refreshments – always popular (if the venue is appropriate, using your product as part of the presentation if appropriate).
- Look professional. Appearance counts. Play and look the part you have in this event.

And last, you are probably doing better than you think you are. Relax, and enjoy the moment. Many of the people listening want you to do well and are looking to make an investment in your company. Good luck!

8.7 Writing an Effective Investor Letter

OK, we've just about completed the three-part puzzle, the first two parts being the business plan and the presentation. Now, the last part of the puzzle is constructing the letter you'll write to a potential investor and you'll do that in the form of a business letter. Writing this letter requires technique and a command of letter writing etiquette. Again, as we discussed in the previous chapter section, there are expectations from your audience members as to what they will receive in a letter. Avoid embarrassment for yourself or your company (and probably a loss of a potential investor) by demonstrating appropriate etiquette in such a letter, for example, addressing them by their proper surname. In the early stages of your start-up, one of your most likely investor types will be an angel investor. We discuss letter writing in this context.

The Angel Investor Letter

Writing the letter to an angel investor may be your most important step in moving your company forward, and almost assuredly, it will be a step you must complete to secure an investment. I use the word 'letter' generically here to include emails. Today, the majority of our communication is by email. There is nothing wrong with contacting a potential angel through email, but be sure to follow the same formality you would use in a conventional letter that is put in the USPS.

You'll be working with angel investors much before you will ever have a venture capital firm interested in your business. As a guideline, VCs are looking for companies generally that have at least $10 million in annual sales, compared to angel investors who will be interested in much smaller investments, for example, $50,000 or $100,000. If you are successful in your first phase of your start-up's business strategy, then the same angel investor might be willing to invest a much larger amount and/or link your company to other angel investors or VCs.

Remember that the angel investor is human and naturally wants to "look good" to his or her friends for having invested in your start-up company. Appeal to those "ego" interests, whatever they might be. You have to secure the angel investor's interest by first getting a foot in the door; if you get your foot in, then the rest of you has a chance to follow.

COMPONENTS OF ANGEL INVESTOR LETTER

This will be one of the most important letters you ever write, because it is the means by which you will secure capital to expand your business. No capital means no business! So, put a lot of effort into this letter. An effective angel investor letter should include the following information and in the suggested sequence, the first point – your connection to the angel investor – being critical:

1. Your personal connection to the angel investor (you may have already interacted with them at a venture fair)
2. That you are credible (you are not a nut or just a dreamer)
3. It should be a fun experience for them (some problem they might be particularly interested in)
4. Gist of company and its upside value potential, for example, size of market
5. Funds invested to date and principals' investment to date
6. Quick terms, for example, pre-value, funds sought, timing
7. Definitive message that allows you to execute a follow-up interaction of some type (meeting, business plan, site visit).

The item 7 is critical. Make sure your letter ends with a proactive response on your part, for example, "I will contact you in a few days as a follow up to this letter." This will keep your letter on the angel investor's desk and have him or her anticipating your call, which will tend to keep your letter in their mind. Then, hopefully, when you do make the call, the response won't be "Who?"

It takes skill to construct an angel investor letter, because some of the previously listed components have to be expressed in a sometimes less-than-direct manner, for example, pre-value. Revealing too much information may compromise your ability in future negotiations with the angel investor, and not giving enough information may result in the angel investor not being interested enough to take your follow-up call. However, you need to provide some sense of what you think your company's value is, which could be done by indicating the funds invested to date or sales generated. If you think your prevalue is a large multiple of these numbers, though, you will probably need to provide some indication that you are thinking this way. Stating a clear pre-value may be counterproductive to later negotiations.

In stating your connection to the angel investor letter, it is not enough to say that so-and-so suggested that you contact the angel investor unless the angel investor knows so-and-so. You must establish the connection between you and the angel investor, no matter how indirect, for example, "My uncle and you were fraternity brothers at the University of Wisconsin. You may have already met them at a venture fair, but you need to remind them of this in your letter.

EXAMPLES OF INVESTOR/ANGEL INVESTOR LETTERS AND BUSINESS FORMAT FOR LETTERS

The following is a series of examples using actual letters written to an angel investor who also had a small venture capital fund. Names have been changed to protect the innocent. After the angel investor letters, we also provide some suggested standard formatting for a business letter for your future reference.

EXAMPLE 1 (AS WRITTEN)

Letterhead for National Toner, LLC
Date June 1, 2012

Dear Ms. Lotsofdoe:

Fred Sanford of Regional Outcasting suggested I contact you concerning investment in our venture.

National Toner LLC is a social purpose business venture designed to help economically disadvantaged Arizonians enter the workforce. We will provide transitional employment for program participants; our affiliated nonprofit agency, Welchol, will provide National's employees with training and supportive services.

We have raised $50,000 in seed financing, founders have committed $100,000, and we have leased a manufacturing facility in Tempe, AZ. We now seek $750,000 in capital to commence operations.

National will remanufacture laser printer toner cartridges and distribute them in the Tempe area. We have identified an underserved market for high quality remanufactured cartridges, which we will sell for 20% less than new. We will use our social mission as a selling tool and target firms with an interest in community development and public service, such as banks, law firms, and institutional nonprofits. We will also take advantage of our strategic location close to Phoenix to provide expedited delivery.

We will provide an exit for investors through a debt-financed recapitalization, and we project delivering a 27% IRR after five years.

We have developed a comprehensive business plan, which we would be pleased to provide on request.

If our venture is of interest, please contact me. Also, I understand that you are knowledgeable about angel investors in the Tempe area, and I would be grateful for any suggestions of potential investors who might be interested in our venture. Thank you for your consideration.

Richard Toner, President
National Toner, LLC
Street Address
Tempe, AZ 85312
Telephone: (555) 123–4546

EXAMPLE 1 (WITH COMMENTS)

Letterhead
Date June 1, 2012

Dear Ms. Lotsofdoe:

Fred Sanford of Regional Outcasting suggested I contact you concerning investment in our venture.

National Toner LLC is a social purpose business venture designed to help economically disadvantaged Arizonians enter the workforce. We will provide transitional employment for program participants; our affiliated nonprofit agency, Welchol, will provide National's employees with training and supportive services.

We have raised $50,000 in seed financing, founders have committed $100,000, and we have leased a manufacturing facility in Tempe, AZ. We now seek $750,000 in capital to commence operations.

National will remanufacture laser printer toner cartridges and distribute them in the Tempe area. We have identified an underserved market for high quality remanufactured cartridges, which we will sell for 20% less than new. We will use our social mission as a selling tool and target firms with an interest in community development and public service, such as banks, law firms, and institutional nonprofits. We will also take advantage of our strategic location close to Phoenix to provide expedited delivery.

We will provide an exit for investors through a debt-financed recapitalization, and we project delivering a 27% IRR after five years.

We have developed a comprehensive business plan, which we would be pleased to provide on request.

If our venture is of interest, please contact me. Also, I understand that you are knowledgeable about angel investors in the Tempe area, and I would be grateful for any

Comment callouts:

> Should have complete name, title, and address of person here. Then you do the salutation.

> Good intro, but missing a connection between Fred Sanford, angel, and writer.

> Excellent way to give an indication of prevalue and specifically states funding request. Good.

> This letter even gives their exit strategy! Good.

suggestions of potential investors who might be interested in our venture. Thank you for your consideration.

Richard Toner, President
National Toner, LLC
Street Address
Tempe, AZ 85312
Telephone: (555) 123–4546

> There is a major error in this letter. Do you know what it is? (does not provide the proactive follow-up from the writer.)

EXAMPLE 2

From: Gene Smith
To: Richard Jones
Subject: Investment Opportunity
Date: Wednesday, February 16, 20xx 3:13 PM

> improper format; memo format used

I hope everything is going well. Thanks again for the San Francisco real estate advice. The search continues. We are actually focusing on Fremont more now, but still keeping our eye on San Fran. We'll see what happens.

> Apparently, Gene knows Richard and is making him aware of an opportunity. So the less than formal style here is more acceptable.

I came across an interesting investment opportunity today that I think you may be interested in. I had lunch with an old college fraternity buddy who is an associate at Raven Products and his partner who is an associate at MYOU. Both of them are leaving their respective firms effective March 1 to start an Internet company, which they have been developing over the last year. The premise of the Company is that it will provide professional service firms and other companies with a web-enabled application that expands the dining options for their employees and automate the purchase, billing and payment process of meal ordering.

> Note how the connection is made, the basics of the opportunity etc. are all made effectively.

They currently have a private placement memorandum and are attempting to raise $1M, with a minimum investment of $25,000. If you're interested in learning more, I can put you directly in touch with the founder.

Supposedly they already have several other commitments. My take is that they could probably benefit substantially from the business expertise and experience of someone like you.

> Other than a nonpositive follow up here – this is a good one; and no mention of pre-value (a shortcoming, but could be addressed in next meeting; may have been intentionally left off as a strategy to accelerate discussion.)

Hope to hear from you soon.

EXAMPLE 3 (GOOD FORMAT IN TERMS OF STYLE)

February 16, 20xx

VIA-FEDERAL EXPRESS

Harvey Smith
Principal
555 Maple Dr.
San Francisco, CA 92222

> *Bad* example, *no* personal connection or linkages between letter writer and angel investor.

Re: Proposed Investment in Hillcrest

Dear Mr. Smith:

We are pleased to offer you the opportunity to invest in Hillcrest (the "Company"). Our Company will provide professional services firms and other employees with a Web-enabled applica- tion to eliminate certain inefficiencies in their administrative processes. The Company's private placement memorandum in connection with its sale of 2,000,000 shares of Series A Convertible Preferred Stock (representing 20% of the Company's capital stock) (the "Offering") enclosed herewith more fully describes our business model and the terms of the Offering.

> Good, a document with more detail is attached; might have waited for a phone discussion to suggest this.

We are confident that an investment in the Company will be a rewarding and profitable chance to participate in the tremendous growth of the Internet. To that end, you have the Company's unequivocal guarantee that all of its resources will be devoted to enhancing shareholder value.

We will be sending you additional documents shortly that will govern an investor's relationship with the Company and related mat- ters, including (1) a Subscription Agreement, (2) a Co-Sale Agreement, (3) a Registration Rights Agreement, (4) an Amended and Restated Certificate of Incorporation, and (5) the Employment Agreements of Messrs. Fiddle and Palin. We encourage you to have your attorney review these documents. If you or your attorney has any questions or comments, please contact either George Sanford or M.B. Free at the Company at (555) 755–1222.

> Good, because it puts follow up in their hands

We look forward to a mutually rewarding relationship and welcome the opportunity to discuss this Offering with you.

> Could leave Harvey with a more definitive follow up action on their part

Sincerely yours,

Business Form Letters: Style

Modified block format

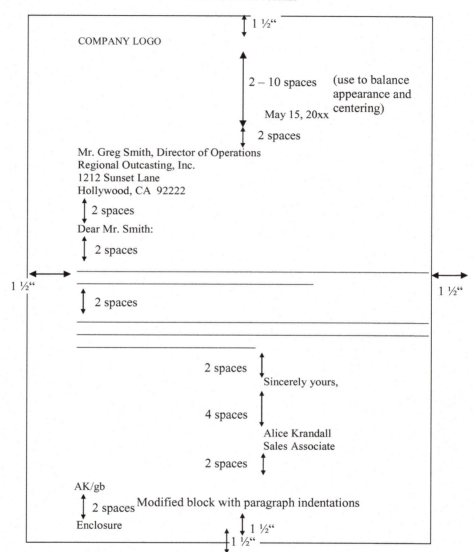

COMPANY LOGO

2 – 10 spaces (use to balance appearance and centering)

May 15, 20xx

2 spaces

Mr. Greg Smith, Director of Operations
Regional Outcasting, Inc.
1212 Sunset Lane
Hollywood, CA 92222

2 spaces

Dear Mr. Smith:

2 spaces

1 ½"

1 ½"

2 spaces

2 spaces

Sincerely yours,

4 spaces

Alice Krandall
Sales Associate

2 spaces

AK/gb

2 spaces Modified block with paragraph indentations

Enclosure

1 ½"

1 ½"

1 ½"

Full block style – everything aligned along the left margin

1 ½"

2–10 spaces

COMPANY LOGO

May 15, 2000

2 spaces

Mr. Greg Smith, Director of Operations
Regional Outcasting, Inc.
1212 Sunset Lane
Hollywood, CA 92222

2 spaces

Dear Mr. Smith:

2 spaces

2 spaces

2 spaces

Sincerely yours,

4 spaces

Alice Krandall
Sales Associate

2 spaces

AK/gb

2 spaces

Enclosure

1 ½"

1 ½"

1 ½"

8.8 Additional Resources

There are many resources to help you in writing business plans and assisting your start-up venture in general. A partial list follows.

PRESENTATIONS

See www.edwardtufte.com for books and discussion written by Dr. Edward R. Tufte on developing effective presentations.

EFFECTIVE WRITING

Barabas, Christine. *Technical Writing in a Corporate Culture: A Study of the Nature of Information.* City: Ablex Publishing, 1990.

BUSINESS PLANS: WEBSITE SUPPORT AND SOURCES

www.Cambridge.Org/Timmons or www.cornell.bee489.com These sites provide support for the entire text. Links within this page will direct you to specific topic areas such as Chapter 8: "Writing Business Plans." (Note to reader: link needs to be created; for Cornell community, go to http://blackboard.cornell.edu/; course name BEE4890)

BPLANS.COM

You can find a dozen examples of "standard outlines" for business plans at this site. This site is sponsored by Business Plan Pro (software provider). http://www.inreach.com/sbdc/book/bizplan.html

"How to Start a Business: Business Plan," Small Business Development Center

AMERICAN EXPRESS, HTTP://WWW.AMERICANEXPRESS.COM/SMALLBUSINESS/

Advice given by the American Express Small Business Exchange "Creating an Effective Business Plan"

SMALL BUSINESS ADMINISTRATION, HTTP://WWW.SBA.GOV

This is one of the best sources for free advice on small-business planning. Included are many free spreadsheets, worksheets, and fact sheets covering all aspects of starting a business. Although the materials, worksheets and resources are not specifically focused on agriculture, the tools found on this site have been widely used by entrepreneurs and are sound in terms of the focus and approach (note that this is not true of everything you will find on the Internet about business planning!). Look for The Business Plan – Road to Success: A Tutorial and Self- Paced Activity, see http://www.sba.gov/starting/indexbusplans.html.

BIZPLANIT, HTTP://BIZPLANIT.COM

The virtual bizplan on this site is an excellent guide to what to include in every section of your business plan. The straightforward explanations are accompanied by lists of common mistakes to avoid and section-specific tips. The site includes an article from a free monthly newsletter related to business planning.

AGRICULTURE INNOVATION CENTER, HTTP://WWW.AGINNOVATIONCENTER.ORG

Created by the Missouri State Department of Agriculture, this website has a wide variety of resources, including lists of possible value-added products. For each type of product, a description is provided along with an extensive list of resources related specifically to that particular market niche.

AG INNOVATION NEWS HTTP://WWW.AURI.ORG/NEWS/AGINNEWS.HTM

Ag Innovation News is a newsletter published by Minnesota's Agricultural Utilization Research Institute (AURI) to inform the food, agriculture and business communities and the general public about developments in new ag-based products.

BUSINESS PLANS: TEXTBOOK RESOURCES

Burton, E. James. *Total Business Planning: A Step-by-Step Guide with Forms.* (New York: Wiley, 1999.

Covello, Joseph, and Brian Hazelgren. *The Complete Book of Business Plans: Simple Steps to Writing a Powerful Business Plan.* Naperville, IL: Sourcebooks, 1994.

Gorman, Robert T. *Online Business Planning: How to Create a Better Business Plan Using the Internet, Including a Complete, Up-to-Date Resource Guide.* Franklin Lakes, NJ: Career Press, 1999.

Gumpert, David E. *How to Really Create a Successful Business Plan – Step by Step.* 4th rev. ed. Lauson Publishing Company, Needham, MA 02494, 2003.

McKeever, Mike. *How to Write a Business Plan.* Berkeley, CA: Nolo Press, 1994.

McLaughlin, Harold. *The Entrepreneur's Guide to Building a Better Business Plan: A Step by Step Approach.* New York: Wiley, 1992.

Nesheim, John L. *The Power of Unfair Advantage.* New York: Free Press, 2005. See Appendix A: Business Plan Outline.

Packsack, Susan M., ed. *Business Plans that Work.* Chicago: CCH, 1998.

Pinson, Linda, and Jeny Jinnett. *Anatomy of a Business Plan.* Fullerton, CA: Marketplace Press, 1999.

Sahiman, William. "How to Write a Great Business Plan." *Harvard Business Review* (July 1997): 98–108.

BUSINESS PLAN SOFTWARE

There are software packages (e.g., Business Plan Pro, see http://www.paloalto.com/ps/bp/) that can help you write your business plan. These packages can be helpful for creating nicely formatted presentations of data that you enter. Many software packages step you through a series of questions intended to help you think through the planning process. The investment community easily recognizes these types of plans and your credibility suffers; that is, if you can't write your own business plan, what else can't you do? My opinion is that you need to invest the time necessary to learn how to write a good, attractive business plan. Take the time and then you do it.

8.9 Video Clips

Directions for viewing clips on the Prendismo Collection (formerly from www.eclips.cornell.edu):

- Visit http://prendismo.com/collection/. (You must subscribe with the site for full access to all features.)
- Click on the **Subscribe** at the top of the site. Students receive reduced rates.
- Once you have subscribed, type in the name or title of the clip in the **Search** at the top right corner of the screen.

- Click on the clip you wish to view.
- Click on the play icon to view the clip. If you have not become a subscriber, you will only view the first 20 seconds of the clip before being prompted to log in.

There is also a "landing page" – where an outline for a business planning class is presented and a single e-clip message is associated with the various topic titles: http://eclipsco.org/businessplanning.htm.

1. Mark Brandt (Notiva)

- Graduate of Cornell University
- Worked at Andersen Consulting and Cargill Inc.
- Working at Notiva, a retail financial software company
- Founder of The Maple Fund, which invests in emerging technology
- Title of video clip: "Mark Brandt Discusses Having A Short Business Plan"

2. Anita Stephens (Opportunity Capital Partners)

- BA Economics from Cornell University and MBA from Golden State University
- A general partner of Opportunity Capital Partners
- Capital Partners is a private equity firm providing equity capital to early-stage companies with leading edge technologies.
- Title of video clip: "Anita Stephens Discusses What a Good Business Plan Should Look Like"
- Title of video clip: "Anita Stephens Shares Common Mistakes Seen in Bad Business Plans"

8.10 References and Bibliography

Dumaine, Deborah, Write to the top: Writing for corporate success, 3rd revision, New York, NY: Random House Trade Paperbacks, 2004

Bauman, Richard. *Verbal Art as Performance.* Long Grove, IL: Waveland Press, Inc., 1977.

Gumpert, David E. *Burn Your Business Plan.* Needham, MA: Lauson Publishing Co., 2002.

From National Venture Capital Association 655 North Fort Myer Drive, Suite 850, Arlington, VA 22209 (XX – this is the reference to the table I added on number of IPO's)(retrieved June 2013).

8.11 Problems

1. Why are there so many ways to write a business plan?
2. Who are you really writing to when you write a business plan?

3. Construct an easy to view pro forma statement for your business plan.

4. Develop a management chart for your company.

5. Write a complete business plan for your start-up (this is probably a several-week assignment).

6. Develop a presentation for a venture fair–type audience (assume a ten-minute presentation time limit). This is to complement your business plan assignment.

7. Assume you have made a presentation at a venture fair and several people have given you their business cards and ask that you follow up with them in the next week or so. Write an angel investor letter to one of these individuals who will take you to a next step. (This is done after the business plan and presentation are completed or in draft form)

9 Fund-Raising

Start small, but think tall!

Robert H. Schuller[1]

9.0 Entrepreneur's Diary

Raising money is hard work. You have to be prepared. Anytime a potential investor senses that you have not properly prepared for a meeting, you instantly will have lost any chance of landing this person as an investor. When I was raising some stage 2 financing for Fingerlakes Aquaculture, I had a meeting with a group of potential investors in Boston, friends and acquaintances of my angel investor, Peter (I think I've mentioned him before). We met in the high-rent financial district. Big buildings … marble hallways … all that type of stuff. I had spent many hours refining my presentation down to about twenty minutes to cover selected details so they would ask for more information that I could tell them was covered in the full business plan and that I had copies with me for them if interested. Part of my presentation was to actually prepare some tilapia fillets from my farm so the guests could really "get a taste of what it was all about." (This was back in 1999, when most people had no idea what a tilapia fillet tasted like or even what a tilapia was!) The meeting was to start at 11:00 a.m. I was getting ready to start my presentation, including the cooking arrangements for the fillets. The invitees (there were five in total) got there about ten to fifteen minutes early and started chatting with my angel investor Peter. It didn't take long before they became so curious about the product and what it tasted like that I had no choice (in my opinion) but to go ahead with that segment of the presentation, even though it was supposed to be at the end of my planned presentation right about lunch time (I thought this would be perfect timing). They all loved the product. Two of the people in attendance became investors in my company. I never did get to show them my PowerPoint presentation. But if I had, it was a good one. I was prepared.

---

[1] Robert H. Schuller, "Infinite Possibilities in Little Beginnings," in *Hour of Power*, October 15, 2008 (television show).

9.1 Financing the New Venture

At some point in your entrepreneurial adventure, you will probably find yourself concluding that your future is constrained by a current lack of capital. Assuming you've been reading this book pretty closely, you should understand that taking your company beyond the "I've got a great idea" to a real company will require funding beyond what you and family and friends can or are willing to do.

One thing that I have learned, which should be obvious but seems to be ignored by many beginning entrepreneurs, is that no matter with whom you are dealing, no one likes to lose money. Even if one of your investors has a net worth of $500 million, that $50,000 investment they made in your business start-up is very important to him or her. Using the ill logic that this person has lots of money so what is $50,000 to them is a mind-set that will lead quickly to disaster. This concept strikes at the heart of fiscal responsibility and you must have it. You must respect every investor's capital as if it were your own. Every dollar is important.

The Importance of Financing

You will need money to get your business going. A growing business consumes cash. As you build sales, your cash flow in will be temporarily exceeded by cash flow out because you incur your cost-of-goods expenses before revenues are collected for these same sales. This lag between the cash going out to build product and the cash coming in from sales can cause an imbalance that can be disastrous! This lag time results from the time required to build the product and create inventory and the additional time between billing customers and receiving their payments. You have to have the cash on hand to pay the bills incurred.

Memorize: cash flow is king.

In contrast, a successful mature business will have positive cash flow and has no need for financing, because it will have already weathered the lag problem and now have adequate inventory and revenues being received into the company bank account on a consistent basis. A mature business such as this is sometimes called a cash cow. The phrase that should be memorized for all start-ups is "cash flow is king," because your survival will depend on it. If you run out of money at any stage, you can assume that you will also lose much of your ownership position, often to a level of insignificance.

Types of Financing

There are two types of financing: debt and equity. Debt means you have obtained cash by borrowed money, which means the company has not given up any of its company ownership in the process (although you may have pledged company

Table 9.1. Comparing Characteristics of Debt and Equity Funding

Characteristic	Debt	Equity
Repayment terms	Fixed periodic repayment with interest	Repayment in future; set by negotiation; no repayment in case of failure (hence higher risk)
Rate of return expected	Typically set in relationship to the prime lending rate	• Family – highly negotiable • Community venture funds (20%–50%) • Venture capitalists – Double digit returns (50%–150%)
Frequency of payment	Typically monthly	Negotiable – often deferred; sometimes not required, simply treated as an investment in future potential
Shown on balance sheet	Liability	Shareholder's equity
Cost to entrepreneur	Comes from on-going cash flow	Entrepreneur gives up ownership and therefore a portion of long term income growth prospects
Participation in business	Minimal as long as payments are met	Variable – high in the case of angels and venture capitalists
Risk to source	Legal obligation to make payment	Risk shared by investor
Ownership implications	All ownership retained	Some portion of ownership is given in exchange

assets as collateral to secure the debt). Raising money by selling equity (or stock) results in your exchanging some fraction of the company for the cash infusion. In a corporation, the claims of the debtors come first before the investors. That is, if the company is ever dissolved or sold, the debtors are paid first from sales of assets, and any remaining funds can then be distributed to shareholders (preferred shareholders first and then common stock shareholders).

Debt is not necessarily "bad"; however, it can be, if servicing the debt (monthly payments) results in your company having a negative cash flow. Negative cash flow is sometimes called a ***burn rate***, with the message being that you have to return to positive cash flow before you've burned all your cash! Refusing to borrow money often constrains the growth of a firm. Imagine if people had to get jobs and work after high school until they had the $200,000 needed for a four-year college degree, before starting your first class. Table 9.1 presents some of the important differences between debt and equity.

One thing that clearly emerges from looking at the table is that because equity investors are absorbing more risk, they expect higher returns and more participation in the business than banks. In contrast, using debt to start a business is dependent on whether your cash flow can support the payments, especially during the early months and years of the business. Bank debt is not a realistic avenue for early stage start-ups, because there is simply too much risk for a bank to be interested. You might seek a loan through a Small Business Administration (SBA)–backed loan, but this still requires you to have a local sponsoring bank to administer the debt. Once your company has reached positive cash flow, then banks become a viable option. Even here, however, be prepared for someone to personally guarantee the bank loan in addition to whatever assets are pledged from the company. A personal guarantee to the bank is very serious business, because this means that the bank will hold you personally accountable for any debt incurred by your company that the company is incapable of paying back; this is true even when the company is an LLC or any other corporation. Borrowing money from an existing business or from friends and family can be an option, but these type loans often end with hard feelings between the parties if you can't repay the loan. In such cases, it is crucial to clearly define the terms and the lender's expectations and/or how to structure the debt to avoid problems.

Funding Strategies and Sources

Your choice of funding source depends on your mission and business strategy. If you want to retain 100% ownership of the company, then you will have to suffer the consequences of typically being constrained by a lack of capital and not being able to grow the business as fast as it could be grown. This has its own risk by allowing competitors to emerge and take over market share. In addition, your business will not generate the rapid income growth, which will compromise your ability to attract outside investors that are looking for high-growth opportunities. By contrast, high-growth businesses that are aimed at large markets must be funded adequately for scaling up the business.

It is important to understand the steps in financing your business, the sources and trade-offs for where to obtain growth capital and the basics of venture-backed funding. As you start down this path, think about funding in concentric circles. You start with sources closest to you and work your way outward. The farther you go from the center, the more control you give up, the more formal you plan is expected to be, and the higher the rate of return that is expected. Figure 9.1 illustrates the sources you might use for financing and how they differ in terms of ties to the founder and expectations regarding returns. Remember these points as you go forward from your first round of seed capital to later rounds of venture capital.

Finding capital to expand your start-up will probably be your most difficult task to date, because you probably have not raised any money previously, which means you have no experience or track record in this regard. Every entrepreneurial venture will require capital. Some will require a lot more than will others, but all will require some. Your initial investors are assuming the first risk of a start-up, which is – can the product or idea be developed into a sellable item (see Chapter 10, "Rules for Investing," which discusses the five risks of a start-up company).

The progressions of raising capital will somewhat go from the center of the circle (Figure 9.1) and then outward as larger sums of capital are needed. Generally, most of us will start out small. You have an idea or an invention that can be self-

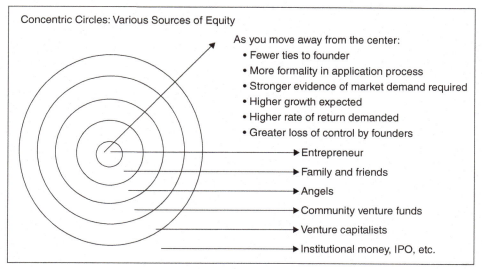

Concentric Circles: Various Sources of Equity

As you move away from the center:
- Fewer ties to founder
- More formality in application process
- Stronger evidence of market demand required
- Higher growth expected
- Higher rate of return demanded
- Greater loss of control by founders

→ Entrepreneur
→ Family and friends
→ Angels
→ Community venture funds
→ Venture capitalists
→ Institutional money, IPO, etc.

Figure 9.1 Characteristics of various sources of equity.

financed, and you have made some initial steps toward commercializing the product. Generally your product is not ready for general market introduction; you are still refining and testing. At this point, you may have spent $10,000 to $50,000. Things are looking pretty good, and we are now ready to expand. Depending on a whole host of circumstances, for example, how much potential is there for the product and what is your current cash flow and market penetration, you may go to any of the financial resources listed earlier. Iterating between yourself and family friends is common, but eventually to make a real business, you will have to go beyond your immediate (closest to the center source) circle.

The reality is that most start-ups go through several stages of funding. At each stage, you need to consider the amount of funding needed, whether you are eligible for borrowing (positive cash flow requires collateral assets), whether you have achieved significant milestones, and how much ownership you are willing to give up. Table 9.2 provides a summary of likely funding sources related to your current status or stage of business development.

Equity Investments

Equity financing should be directly related to a series of milestones achieved by your company, that is, developing product, acquiring customers, obtaining revenues, and achieving growth targets. The type and amount of equity funding you are seeking should be coordinated closely with these milestones. For example, a venture capitalist will be interested in your company only after you have achieved all of the previously listed milestones.

Deciding where and how to fund your business is intricately linked to your business strategy. If you have a business that is a cash cow from the beginning, you may be able to borrow money rather than give up equity. On the other hand, if

Table 9.2. Stages of Business and Appropriate Sources of Financing

Stage	Description of Business and Milestones Achieved	Sources of Financing
1	Start-up stage; You have your first business plan written	Informal investors, such as • friends or relatives, partners
2	You have a refined business plan and product samples but no revenues	Same as stage 1
3	You are starting to execute your business strategy and have pilot programs in place.	Angel investors;
4	You have been in operation for some time and have documented revenues and expenses; generating revenues in excess of $1 million	Same as three; can start to look to small VCs depending upon revenues and their potential.

your company requires significant cash investments before you attain profitability, giving you no cash to make loan repayments, you will have to pursue equity investors. This period is also called your ***pre-revenue*** period, which defines your company as a pre-revenue company at this point.

One very difficult area for new businesses owners seeking investment is determining for a given equity investment how much of the company ownership is shared. This subject is discussed in detail in Chapter 10, "Rules for Investing." Additional information on financing is provided at the end of the chapter under Additional Resources.

9.2 Financing Sequence

Personal Funds

The first place to look for funding is always in your own pocket. What percentage of total funding should come from your own pocket? The degree that you fund the business from your own resources depends on how you feel about sharing ownership and distributing risk. Most start-ups don't survive, so remember that when deciding how much of your own savings to invest.

Your credit card, if you haven't maxed it out yet, is an instant source of cash to start your business. You can borrow against your credit card, but high interest rates are imposed. Credit cards are a very expensive source of funds for a new business. Using them only makes sense if there is a relatively short time between selling your product/service and collecting the money and you are borrowing to cover a small gap in cash flow. Even so, remember that in most cases, you will be paying double-digit interest rates.

If you have a home, you probably have some significant equity built up if you're already halfway or more through the term of the original loan. You can borrow against this asset if your home partner agrees. Banks will only loan against a portion of your equity.

When you can start a company with your own funds, we generally call this a ***bootstrap*** company. Such companies can become quite successful, with Dell

Computer being one of the most famous examples. Michael Dell started Dell Computer in his dorm room at the University of Texas. He didn't rent an office, hire five people, and spend lots of money on office furnishings. Bootstraps typically start out slowly and gain momentum. The successful bootstrapper has a strong self-belief, has a good product, and avoids activities or strategies that require significant capital.

Friends and Family

Friends and family are the most common sources of funds to get a business going. To keep your friends and family, it is imperative to construct a formal agreement (simple, but *formal*) so that expectations are clear on both sides of the table. There are way too many stories about families that didn't realize that if the business went under, they would not be able to recover any of the principal they had contributed to the company. If you don't want your uncle, brother, or friend telling you how to run your business just because they invested money in your company, you best get the terms down on paper so that there are no surprises when things go well or poorly.

You may want your friends or family to loan you money (as opposed to investing), but this means that you have to pay these people back with some interest. Make sure the terms of the loan are well defined and whether your debt obligation is voided if your company goes bankrupt. The upside scenario for you is that the terms of your written agreement would allow you to convert their debt to equity or to repay the loan at your discretion if the company becomes successful. There are no rules here other than what you negotiate with the lender, but remember that you are dealing with friends or family and that you should try to maintain a mutually satisfying relationship (following the golden rule in this case would be a good guideline).

Angel Investors

Angel investors are people that have significant capital for investing, are interested primarily in companies in their community, and have an interest in being involved with a start-up company. Angel investors are willing to participate in early or "seed" funding rounds before a company has a proven record of sales. Angel investors are qualified investors, meaning that minimally they have a million dollars of net worth or exceed $200,000 in annual income requirements. How many angel investors are there? Well, there's around 50,000 recorded angel investment deals made each year, and the average investment deal will be several hundred thousand dollars.[2] Remember, however, that there are approximately 600,000 new businesses started each year. The competition is stiff.

---

[2] See details on angel investment at Center for Venture Research, University of New Hampshire, http://wsbe.unh.edu/cvr.

Angel investors must be very attracted to the idea about which you have started your company (VCs only care about financial returns). Angel investors are primarily interested in the *concept* and the fun associated with your potential venture. They are much more flexible than are venture firms in terms of evaluating the suitability of your company as a sound financial risk. Angel investors don't demand as high a return as venture capitalists do, but they still expect large upside opportunities for being an early-stage investor and for taking this high risk. Thus, securing angel investment funds requires a completely *different strategy* than does obtaining a personally guaranteed bank loan (although deep down, the same fundamentals must be there).

The key to finding an angel investor is networking, mostly so that you both know that the other party is not a crook. Your best networking linkages to an angel investor is by a personal connection from a relative, a business associate, and/or a university connection or something that allows you to make a logical initial connection. Most people like to help someone they know. Although you really can't go to Yellow Pages under "Angel" to find your godsend, in this regard, the Internet is becoming more useful for seeking out potential angel investors who may have formed into an angel investment group. For example, Milwaukee, Wisconsin–based Silicon Pastures (www.SiliconPastures.com) is a group of angel investors who assess business plan proposals and share background research on the company and its principals (called ***due diligence***) for possible investments. This angel investment group hosts monthly investment presentations at upscale Milwaukee-area restaurants. Each presentation event consists of one or more pre-screened investment opportunities. After the entrepreneurs make their pitches, members of Silicon Pastures meet privately to discuss the viability of the investment. Each member makes an individual decision whether to invest. The key here is that you use the Internet to locate this potential group of angel investors. So, if you were located in the greater Milwaukee area, then one of these persons could be a potential angel investor, particularly if you somehow have a linkage from your own network of contacts to this person. The investment review process used by Silicon Pastures is fairly typical of what you'll go through to secure an angel investor.

After you've identified a likely angel, then you must convince the angel that you have the following:

- You have credibility.
- Your company has upside value (and will be fun for the angel investor to be involved in).
- You have a reasonable chance of success (and, in doing so, will make the angel investor look good to his or her friends; you will have given something back to society to make the world a better place).

Avoid thinking of an angel investor as someone who will give you money just because they like you or your cause. Although an angel investor can be a godsend, a bad fit can be the death of your enterprise. Here are six things to consider before taking money from an angel investor:

1. **Who is your angel investor?** Make certain that you really know your angel investor. Understand his or her motivation and expectations for exit strategy and return on investment. Your angels will probably have created wealth by building and running a business of their own (read this as this person thinks they know how to do it). Invest time and energy in interviewing and doing a background check on your potential angel investor. Is their money legitimate, or was it obtained by illegal means or in a manner that would offend you (drug money)?

2. **What are their real qualifications?** Prospective angel investors will want to review your résumé and understand your business credentials. You have every right to ask them about their business experiences and philosophy on how to manage a start-up company. Be careful about angel investors who can't admit their own shortcomings or areas of ignorance, particularly as related to your business market. Your angel investor may become a key management person in your start-up and may fill voids that are currently lacking in your bare-bones staff. Your angel investor is probably motivated by wanting to help you succeed and can bring tremendous energy to the table for your benefit.

3. **Disclose fund-raising goals.** Clearly describe your intentions to raise additional capital to meet your business goals (this should be laid out in your business plan). Angel investors are often placed on your board of directors. Now, if you end up with a board that is unwilling to raise money when the company needs it (e.g., for expansion), then your goals and the company's goals are not the same, and you will be severely constrained in moving your company aggressively forward.

4. **Exit strategy.** Have a realistic discussion with your prospective angel investor about an exit strategy for your company. Most companies will never approach the criteria necessary for an initial public offering (IPO); more details on IPOs are given in a later section (see Public Offerings later in this chapter and Chapter 10). A more reasonable target is to build your company in order to sell it at a strategic point. If your angel investor needs their principal back in twelve to twenty-four months, then it is unlikely that person is a match for you, because it generally takes much longer to build a successful company, for example, five to ten years.

5. **Potential conflicts of interest.** Be on the lookout for potential conflicts of interest. You do not want an angel investor who is heavily involved with a competitor's company. Another potential conflict can arise from angel investors who indicate that they will network your company with the others with which they are involved in. As long as it's all done at arm's length, and your company isn't unduly pressured to do business with one of your

angel investor's other investment partners, such networking can work to your advantage.

6. **Follow protocols for business.** Even though your company is private, and you have a small four or five member board, follow formal protocol during board meetings. Take detailed notes and carefully review the minutes for accuracy and completeness. Let an outside firm handle the administration of your stock. Even if only a handful of shareholders exist, avoid passing this critical responsibility on to your corporate secretary. Careless or amateurish handling of your company's stock could kill a later deal. If you ever hope to sell your company to a publicly traded firm, you don't want regularity problems to pop up during due diligence. Sales of your securities to unaccredited investors at prices (see Rules and Regulations later in Chapter 10) that are arbitrary and can't be supported by your financials will send investment bankers and public companies running in the opposite direction.

Banks

If your company is moving forward successfully and already has positive cash flow (almost an absolute criterion), you may want to maintain your current ownership percentage and obtain capital from a bank. Banks stay in business by avoiding risk. If your company has any financial weakness that might compromise your ability to make those monthly payments to the banking lender, you will not be able to obtain a loan. Even in cases where your company has a current positive cash flow and no debt, do not be surprised when the bank asks for collateral well beyond the assets in the company to more than cover the total loan request. Part of this is related to an expected steep discount on any assets you pledge as collateral if the bank has to try to sell them to cover your loan balance.

In some cases, the bank will ask you to *perfect* the loan personally. This means that you will put assets in the name of the bank that will cover the principal of the loan. The bank will also discount the value of nonliquid assets; for example, stock might be valued at 75% of current selling price. Bankers do not favor loans that are used to cover operating capital needs, but exceptions with strong justification can be made. A banker will make a loan based upon the **Six Cs**:

1. Character (of you and your track record on paying back debt),
2. Credit (your credit history and the company's),
3. Capacity (your ability to repay, both your business and you personally),
4. Capital (what your company has at risk),
5. Condition (current climate for loans and viability in the sector area of your business; for example, dot-com or "smart-phone app" type business start-ups are out of favor), and
6. Collateral (additional being pledged, compliments capital borrowed).

> If the banker requires you to "perfect" the loan, think twice (at least)

As you can see, working with people at a local bank is an advantage in terms of some of the Cs, because they will tend to be more familiar with you and probably the market being addressed. A rule of thumb, however, is never borrow short term to cover long-term obligation (building assets) and never borrow long term to cover short-term obligations (salaries, costs of goods). Bankers favor using borrowed funds to purchase capital assets and for "cost of goods" usage (parts, inventory) because these assets can be salvaged, compared to using borrowed funds for other operating costs.

> **Rule of Thumb:**
> Never borrow short term to cover long-term obligation (building assets), and never borrow long term to cover short-term obligations.

Generally, for an entrepreneur, the whole context of your proposal to a banker should focus on the costs associated with scaling up your business. You need to convince the banker that the loan makes good sense for you and the bank. In your written proposal to a banker, you will need to describe:

- the amount of loan request (appear in letter and probably will be identified in your cash flow statement),
- the uses of the loan proceeds,
- a three-year pro forma statement and balance sheet,
- a one-year cash flow analysis in monthly increments, and/or
- the perfected collateral you will provide.

Remember that the bank has many more people asking to borrow money than it has money to lend. It is looking for good customers, and your cash flow statement showing a clear ability to repay your loan is a good start at identifying you as a possible good customer. You need to work hard to cultivate a strong relationship with a commercial banker. Think of your banker as becoming one of you development team members as your company progresses.

Banks may also be an avenue to obtain an SBA (Small Business Administration) loan. The SBA doesn't actually loan you the money; instead, it guarantees up to 90% of the loan value being made by the actual lender (your local bank). Although the rules are outlined on the SBA website (http://www.sba.gov/financing/fr7aloan.html#general), to understand whether your business is eligible for an SBA-backed loan, it makes sense to visit with your local bank or credit union. An SBA loan will also require personal guarantees and will require at least one additional bank fee in the loan process. You will probably be better off just obtaining a bank loan directly from your local bank and ignoring the SBA option.

COMMON ERRORS WHEN BORROWERS APPROACH BANKS

According to bankers, the following mistakes are ones that business owners make when approaching them for a loan:

- Are not prepared even with a simple written business plan with financials that have been reviewed by a qualified accountant
- Don't realize that the management team and repayment ability are the top considerations
- Mix up short-term and long-term debt – most loans are asset based, and the life of the loan must coincide with a reasonable lifespan of the asset.
- Have no collateral or are unaware they will be asked to make a personal guarantee
- Think they can borrow 100% of value of asset (usually 60%–75%)

Venture Capitalists

We are targeting our advice in this book towards early stage companies, which means for you that a VC is an unlikely investor. VCs and other major investors (representing multiple millions of dollars) are not interested in your company unless you already have a product, sales, and a history of successes to date. VCs do not invest in early-stage start-ups, no matter how good you think your idea is! Venture capitalists are looking for investment opportunities of $5 million and higher and for companies that predict potential for fast growth and large market potential (approximately $100 million in sales in year 5). This means that your business must be scalable to a national or global marketplace.

You should understand VC investment criteria so that if and when you meet the preceding criteria, you can approach them when you need capital for expansion. Even when you are just starting out, you should try to think big, or as Robert H. Schuller (noted evangelist with the Hour of Power television ministry) likes to say, "Start small, but think tall!"

The first step to obtaining an investment from a VC is being able to talk to one. Although you can find venture capital firms by searching the Internet, your mostly likely avenue is relying on one of your angel investors to network an introduction for you, that is, open the door for you. If you don't have such angel investors already, you'll still need some type of personal connection to a venture firm to have any realistic chance of obtaining investment funding. It's generally recognized that VC firms will see thousands of business proposals each year and fund only a handful. The only way to break through the mathematical odds is with the personal connection. This will give you some credibility from the very beginning.

Before you talk to the VC, try to have a basic understanding of its criteria used to evaluate a potential investment. Also, research the firm to estimate the size of investment it typically makes. Partners in VC firms can manage five to ten company investments at any one time. Figure out how many partners are in a

VC, and you can guess about how many deals are ongoing. Take this number of companies being managed and divide it into the fund's total assets, and you'll get an estimate of the size of investments being made by this particular VC.

The number one consideration for VC investors is the quality of the management team. Angel investors are less interested in this aspect, because they can play (and like to play) this role when they think management is lacking. This surprises many start-up entrepreneurs, who expect investors to be more focused on the idea; true for angels, but not for VCs. Even though the management term criteria is number one on the list, during the negotiations, some VCs will pound into your head that your team lacks some management experience, but not to worry, after they invest, they will supply missing management skills via their designated CEO. They will have a stash of such executives available whose salary will be $15,000 to $25,000 per month (a lot to an engineer, but just enough to keep people in these circles out of poverty). The new CEO then replaces your team with their cronies. This scenario is probably more typical with early-stage start-ups that required modest VC funding.

The second consideration regarding VCs is the size of the potential market. Generally, VCs are looking for market potentials in the $500 million and larger size. Small niche markets are not on their radar screens. Third, VC investors are expecting extremely high rates of return on their investment. This is so the small percentage of winners, probably about one out of three investments or fewer, will compensate for the losers and provide the firm an average high overall return. The farther out from the center of the funding circle you move (center being self, see Figure 9.1), the higher the rates of return are expected. This is actually fair, because equity investors are assuming much of the financial risk of the business, and accordingly, they expect and demand very high returns on their investments.

> In valuing your company, VCs consider your management team as their number one consideration

This book is targeted toward engineering innovators trying to execute a start-up business. Unfortunately for us, VCs know how to deal with engineers, but engineers don't know how to deal with VCs. VCs take advantage of this situation to maximize their equity positions. Engineers often end up being short-changed. Engineers who better understand how VCs operate can negotiate more equitable solutions.

VCs are generally finance oriented and not technically grounded. Like all people, VCs dismiss what they don't understand – your ideas – and focus on what they do know – financial lingo and rules. Further, VCs assume that if you can bring this product to market, then there must be plenty of other engineering types who could do it just as well and probably at a cheaper price.

> VCs assume that if you could do this, then there are plenty of others who could do it just as well and probably at a cheaper price.

At the beginning, engineers tend to have an awe-struck view of the VC, who has the money and the power. Engineering type entrepreneurs tend to be totally honest and frank in their discussions with VCs in terms of how long it will take and how much money to achieve some well-defined set of milestones. Engineers will be 100% behind their proposal and will firmly believe that they can do what they say they are going to do. Often, we are naïve in this respect. The following is the *2–2–2 Rule* you should understand and I have added my own additional "2" rule, making this the 2–2–2–2 rule: twice as long, twice as hard, twice as costly, and half as much revenue.

2–2–2–2 Rule for Business Start ups
1. Twice as long
2. Twice as hard
3. Twice as costly
4. Half as much revenue

What we thought might take one year to reach, might take two or even three or more years. We will experience difficulties none of us expected (even though we are smart engineers with lots of experience), the project will cost twice as much as we predicted, and the price we end up getting for our product/service is half what we had in our business plan (where did that competition come from?)

Prior to making an investment, the VC will send at least one "expert" to evaluate your company and product. This expert may have very little background on what you are doing, but the VC has designated them as the expert. Hence, the VC's opinion counts most to him or her. Don't expect the expert to understand what you are doing as well as you understand your processes. In fact, the expert's distorted understanding of what you are doing may actually hurt you tremendously. So, view this as a positive opportunity to make sure the expert has a very clear understanding of what you are doing and your strengths and your value opportunity. The expert can become your interpreter and emissary back to the VC. *View the expert not as an adversary, but as an ally.*

Once you've secured the VCs investment, life will not be one happy love affair for ever after. Often, your company will run into financial problems. At this point, the VC will nuzzle into his or her position of power and will start to explain the facts of life to you (read you are about to get taken to the cleaners). This will be a classic example of the golden rule: whoever controls the gold makes the rules! Often this leads to your team members being given their walking papers or being diluted to a trivial level of ownership (see Chapter 10, "Rules of Investing"). Be prepared and be realistic.

Public Offerings

There are several methods of *soliciting investments* from the public, the most well-known being an *Initial Public Offering* (IPO). As mentioned, start-up companies are not anywhere near being ready for such an undertaking. Less well known and more recently available to small companies are direct public offerings (DPOs) and are covered by a series of Securities and Exchange Commission (SEC) regulations, for example, *Regulation D* and *Regulation A*. Sometimes the DPOs are called Internet offerings and allow companies to raise as much as $1 million with minimal paperwork or oversight. Before you pursue any of these methods, consult competent legal help before proceeding. Violating security regulations can end up with you serving mandatory jail sentencing, so this is nothing to fool around with on a casual basis.

Government

SMALL BUSINESS INVESTMENT COMPANY (TAKEN FROM *HTTP://WWW.SBA.GOV/*)

Congress created the Small Business Investment Company (SBIC) Program in 1958 to fill the gap between the availability of venture capital and the needs of small businesses in start-up and growth situations. SBICs, licensed and regulated by the SBA, are privately owned and managed investment firms that use their own capital plus funds borrowed at favorable rates with an SBA guarantee to make venture capital investments in small businesses. There have been several changes in these programs over the last several years, so be sure you have up-to-date information before pursuing these funding sources. SBIC financing is not appropriate for all types of businesses and financing needs. The SBA offers a wide variety of financial assistance programs designed to suit the varied needs of small businesses in the United States. To learn more about other financing options available through the SBA, refer to the website at www.sba.gov or call 1–800–UASK–SBA (1–800–827–5722).

SBIRS (SMALL BUSINESS INNOVATION RESEARCH PROGRAM)

If you are already an up-and-running company, the SBIR program can be an excellent avenue to fund short-term research on product development. The SBIR is a highly competitive program that encourages small business to explore their technological potential and provides the incentive to profit from its commercialization. By including qualified small businesses in the nation's R&D (research and development) arena, high-tech innovation is stimulated and the United States gains entrepreneurial spirit as it meets its specific research and development needs. More detail can be found on the government SBIR site: http://www.ed.gov/programs/sbir/index.html. Generally, the SBIR program defines a small business as having fewer than 500 employees, but there are other definitions that broaden this classification by industry type and class. These programs generally consist of a phase I short-term grant (six months, funds of about $100,000) and then successful phase I grantees can submit a phase II grant (two years, funds of approximately $500,000 and up). Competition for phase I grants is extremely competitive (about one out of twenty proposals are funded), whereas successful Phase I awardees have almost an even chance of securing phase II funding.

There are eleven federal agencies which participate in this program, including the Departments of Education (ED), Agriculture (USDA), Commerce (DOC), Defense (DOD), Energy (DOE), Health and Human Services (DHHS), Homeland Security (DHS), Transportation (DOT), the Environmental Protection Agency (EPA), the National Aeronautics and Space Administration (NASA), and the National Science Foundation (NSF). The program is administered similarly by each of these departments.

ENTERPRISE ZONES

Since the mid-1980s, more than 1,000 enterprise zones have been created across the country in depressed economic areas. These districts revitalize working communities where big employers have shuttered their plants, but which already offer enormous potential for new companies: strong workforce, good site for logistics, and centered in growing markets. All they need is the trigger of incentives, and an extraordinary package of handsome benefits, usually in the form of tax credits and tax refunds based on investment and job creation, has been developed by many states. Contact your own states government website to see what might be available in your geographic area of interest.

9.3 The Venture Fair Process

Investor funds come in a wide range of flavors, colors, and magnitude. Remember that this book is focused on angel investors or extensions of the family and friends network. Venture capitalists (VC's) are generally focused on companies that are farther along in the start-up process and are seeking investments in the $5 million to $10 million or larger category (see Chapter 10, "Rules for Investing," for more details). However, there are venture groups that are made up of angel investors that use the venture company as their filter to find attractive opportunities (see the earlier discussion of Silicon Pastures, an angel investor group). The key is that your company profile and objectives must match up with theirs. So review their individual websites to gather information on the size of equity placements being made and the type of companies being chosen for equity investment. Strongly avoid blanket mailings to a who's who list of venture firms. This will damage your credibility. Just as investors will qualify you, you must qualify the potential venture firm as being an appropriate match to your company. You can waste enormous amounts of time working with a potential investor that turns out not to be a good fit and as a result, not investment is obtained.

You need to have an inside link to someone in the venture fund, or your chances of obtaining funding are next to zero. Even with such a connection, you will find this a very challenging way of finding equity financing. In my case, my contact was my angel investor, Peter, who had a good friend, Willy, who was a member of Investor's Circle (Investors' Circle, Brookline, MA 02445, www.investorscircle.net). Willy guided me through the process, and I was able to secure an investment as a result of presenting at the Investor's Circle venture fair.

The basic process was to submit an application and Investor's Circle circulated the application among its members. If selected as one of ten or fifteen groups, they then gave you more instructions on how to proceed. See the advice given in Chapter 8, "Business Plans, Presentations, and Letters," on how to prepare and present an oral presentation. All venture fairs seem to follow a similar format, e.g., submit an application with a brief description of your company and what is your linkage to their particular fair. A very exact set of guidelines will follow if you are accepted for presentation. Do not deviate from their directions.

Our venture fair was at a Four Seasons Hotel in Boston. As part of my presentation, however, not as part of the ten minutes I was given to present, I provided tilapia fillets to the Four Seasons chef, and they were served as part of the luncheon for all guests. This made a great conversation topic among potential investors as they approached me for more details on the Fingerlakes Aquaculture LLC investment opportunity. More details on this particular venture fair is provided in the website that support the text.

9.4 Additional Resources

Community Development Venture Capital Alliance (CDVCA)

http://www.cdvca.org/

An example of a fund with a community development spin, the CDVCA site features resources for entrepreneurs and the ability to join the virtual community or "shop" your business.

Grow New York's Enterprise Program

http://www.agriculture.ny.gov/GNYRFP.html This site tells about a special project for "agribusinesses in New York State" and businesses that are eligible for funding under certain conditions.

Rural Development

http://www.rurdev.usda.gov/rbs/coops/vadg.htm

This USDA site describes special funding for value-added businesses.

9.5 Video Clips

Directions for viewing clips on the Prendismo Collection (formerly from www.eclips.cornell.edu):

- Visit http://prendismo.com/collection/. (You must subscribe with the site for full access to all features.)

- Click on the **Subscribe** at the top of the site. Students receive reduced rates.
- Once you have subscribed, type in the name or title of the clip in the **Search** at the top right corner of the screen.
- Click on the clip you wish to view.
- Click on the play icon to view the clip. If you have not become a subscriber, you will only view the first 20 seconds of the clip before being prompted to log in.

1. Rob Ryan (Ascend Communications, Inc. and Entrepreneur-America)

- Founder of Ascend Communications, which was sold in 1995 to Lucent Technologies
- Founder of Entrepreneur America
- Discusses history of Ascend Communications
- Discusses intellectual property, technology transfer, networking, and mentoring
- Undergraduate degree from Cornell University
- Title of video clip: "Rob Ryan Discusses Final Round Meeting with Venture Capitalist Investors"

2. Mark Brandt (Notiva)

- Graduate of Cornell University
- Worked at Andersen Consulting and Cargill Inc.
- Working at Notiva, a retail financial software company
- Founder of The Maple Fund, which invests in emerging technology
- Title of video clip: "Mark Brandt Discusses Importance of Developing a Network and References to Gain Funding"

9.6 References and Bibliography

Arkebauer, James B., and Ron Schultz. *Going Public: Everything You Need to Know to Take Your Company Public Including Internet Direct Public Offerings (DPOs)*. Chicago: Dearborn Financial Publishing, 1998.

9.7 Classroom Exercises

1. How do you find an angel investor?
2. Why is incurring a large debt load in lieu of equity financing a dangerous thing to do?
3. What are the biggest handicaps for an engineer in the fund-raising game?

9.8 Engineering Economics

Present Worth Analysis

Investors may evaluate your business opportunity by calculating its net present worth or value (NPW or NPV) as related to your pro forma analysis. Using NPW is also a convenient comparative economic analysis technique and is probably one of the more important skills that an engineer must possess to function in the real world. Engineering designs are based not only on their physical feasibility but also heavily on their economic feasibility. A bridge made out of solid titanium may be very strong, but a bridge with the same specifications could be made much more cheaply with steel or concrete. The actual specifications for models, such as the dimensions of a device or a tank, may also heavily depend on the cost and availability of that product. It is an engineer's prerogative to be efficient in his or her design and therefore be efficient in cost as well.

For the pro forma case, the P, or present cost, signifies the *initial cost* of the project or investment made to launch the current phase of the business. The F, or *future worth*, which could be the projected value of the company (as in the previous section example) or it could be the *salvage value* of some particular capital expense incurred that was implemented to increase revenues for the company. For example, my fish farm purchased an automatic feeding system for $100,000 to reduce labor costs and to improve feed conversion efficiency (you can imagine your own examples, e.g., a circuit board press to increase the number and quality of boards produced per day). Salvage values are often zero because there is no likely customer for a used "such and such." The salvage value is dependent on the condition of the equipment, the inflation, and its technological value. The blue book value of a car for a bank loan after five years and 75,000 miles is a good example of the salvage value of that car at that time. Loan value is what the bank thinks it could get for the car at wholesale, thus the car's residual or salvage value.

For a piece of equipment there is usually an *annual cost* or operating-maintenance-repair (**OMR**) cost, which represents the cost of operation, maintenance, energy, repair costs, and so on that it takes to run the equipment and produce a product. The sale of the products that the equipment creates produces *annual revenue* (**AR**), which can be either positive or negative. If the analysis is done on an annual basis, we determine the *net annual worth* (**NAW**), which can be uniform and/or a gradient depending on the process and is defined as annual revenue less annual cost (*AC*) or

$$NAW = AR - AC. \tag{9.1}$$

The interest rate for a cash flow analysis could be any of the following:

- Inflation rate
- Interest rate that you would expect from a savings account
- Bank loan rate available to the project
- Expected rates of return on invested capital

The last bullet point can result in a very high interest rate value. VC-type investors would probably use their firms expected rates of return on capital investments for the interest rate to do an analysis, which will be in the 30% to 40% range.

CALCULATING NET PRESENT WORTH

The term *net present worth*, or *net present value*, (**NPW** or **NPV**) applies to the equivalent present worth of a cash flow, which is the "net" difference of the present worth of the benefits (revenue) and the costs.

$$NPW = PWB - PWC, \tag{9.2}$$

where

PWB = present worth of the benefits and

PWC = present worth of costs

The P designator indicates this is a current or present value for beginning amount of money, or the worth of money at time zero before any interest or inflation factors have been applied. Obviously, if the benefits outweigh the costs, then the net present worth is positive and vice versa. The benefits in a cash flow are defined as all of the positive flows, and the costs are all of the negative flows.

Net present worth is useful for analysis because you can easily compare the net present worth's of two or more options in order to quickly decide which option is most desirable. Net present worth also represents the money needed at the beginning of a project (if NPW is negative) in order to carry out the project. The *disadvantage* of the NPW calculation in comparing alternatives is that the two alternatives may have different useful lives. When this is the case, you usually are better served to calculate an equivalent *uniform annual cost* (EUAC) to compare the different alternatives, discussed in a later section.

It is possible to determine the present worth of a future payment (P/F), a uniform annual payment (P/A), an arithmetic gradient payment schedule (P/G), or a geometric gradient schedule. Any payment made before interest begins to take effect (before year 1, or the end of the first interest period) becomes added to or subtracted from P directly.

Example 9.1

Assume the cost of your new technology for your production plant will cost you $75,000, and you need the board to provide this money by an additional round of investment, because your company is already cash-strapped at this point. You expect (at least you tell your board) that net revenues will increase by $25,000 per year for the next five years. Will the board approve this request? What interest rate do you use?

Solution. The missing information is what do you use for the interest rate? You have inside information that the board members achieve an annual rate of return of 30% on their investments. Thus, you must use this interest rate to evaluate the suggested improvement.

Year	Benefit
0	−$75,000
1	25,000
2	25,000
3	25,000
4	25,000

$PWB = 25,000(P/A, 30\%, 4) = 25,000(2.166) = \$54,150$

$PWC = \$75,000$

$NPW = PWB - PWC = \$54,150 - \$75,000$

$\quad\quad = -\$20,850$

So, the board turns down your proposal because the NPW of the benefits is less than the NPW of the cost. But, they respond by asking you to tell them what the **internal rate of return (IRR)** on this investment would be, and if it is more than 15%, they might reconsider. You just happen to have your laptop computer handy with Excel so you instantly tell them to give you a second and you'll tell them.

Invest	Year 1	Year 2	Year 3	Year 4
−75,000	25,000	25,000	25,000	25,000
12.6%				

The syntax on the Excel function is IRR(values, guess). Note that the investment is negative and the benefits are positive. This function could, of course, be used to calculate a very complicated series of cash flows and benefits and costs. So, back to our story in this example, your board would once again turn you down unless you could convince the members that a 12.6% return on this company improvement should be done for other reasons as well, for example, staff morale? Also, note that in year 4, you are only including the additional revenues ($25,000) and you have not included any salvage value here for the initial capital investment ($75,000). *If* the salvage value were, say, $7,000 (so year 4's cell increases by $7,000 to $32,000), then

Now, maybe you could convince them?

Invest	Year 1	Year 2	Year 3	Year 4
−75,000	25,000	25,000	25,000	32,000
15.2%				

Payback Period

Although not used typically by investors to determine whether to make an investment in a new company, you might use ***payback period*** as a way to determine whether to implement some in-house improvement to your existing company operations. It can also be useful in determining how much you can charge for your product while remembering the definition of marketing, which is to solve customer's problems profitably. You could price your product based upon the customer realizing a payback period of some specific target.

Payback period is the time required for the benefits (profits) from an investment to equal the cost of the investment. In other words, the payback period is the analysis time (n) that is required in order to get NPW = 0 (and assumes that there is no salvage value in the investment). The payback period can be considered the time it takes for a venture or project to become profitable. Under the most simple definition, the calculation is done by assuming that interest has no impact on the benefits; that is, future money has the same value as today's money. Although this approach is obviously flawed, it is still the most commonly used. For this case,

$$Payback\ period = \frac{Initial\ Cost}{Uniform\ Annual\ Benefit} = \frac{P}{A}$$

This equation not only ignores interest effects, but it also only works if the annual benefit is a constant. If we are working with complicated cash flows that do not work with this method, we can rearrange the above equation to a form that includes interest,

$$P = A\ (P/A, i, n)$$

where $n = Payback\ period$.

Example 9.2

You are the persistent company manager, and you are still not satisfied with the board turning down your request to implement the new improvements to your production plant ($75,000), so you suggest another alternative that would accomplish the same thing but would cost less. You will need to change vendors of your oxygen supply company, and this switch will save you $10,000 per year but will cost $50,000 to change some equipment around to accept the new vendor's oxygen. You know that this investment will not satisfy the board's outside investment criteria (30% per year), but you suspect that the board members might respond to an attractive payback period. What is the payback period and what is the internal rate of return on this investment over a ten-year period?

Solution.

$P = \$50,000 \quad A = \$10,000 \quad IRR = \% \ n = ?$

$Payback\ Period = P/A = 50,000/10,000 = 5\ years$

$NPW = -50,000 + 10,000(P/A, i\%, 10) = 0$

$(P/A, i\%, 10) = 50,000/10,000 = 5$; solve by iteration, $i\% = 15\%$

(or use the Excel IRR function: IRR(−50,000:10,000 for 10 cells)

With enthusiasm, tell the board that the payback period is five years and has a rate of return of 15% over a ten-year period, and they might go for it!

The other wrinkle to calculating simple payback periods is when the payback period is some fractional number of years, for example, 5.3 years. Because companies function on a yearly basis for profitability (tax year), the payback period would be six years in this case, because the company would have to operate more than five years to fully "payback" the item. If you are preparing to take the national Fundamentals of Engineering (FE) Exam, read the question and the possible given (multiple-choice) answers very carefully. Try to determine if the question is asking for a decimal value or the number of years (integer value that is a roundup of the fractional value). Also, if the company profits are basically constant each month, then calculating a fractional year makes sense; that is, you don't have to round "up" to the next integer year.

Replacement Analysis

In this chapter, we have discussed the different sources of capital to expand your business. Investors will want to know what you are going to do with their money and why. When some of the capital will be used to replace an existing line of equipment or to add some mechanical capacity to increase product output, you will need to provide a solid economic analysis of why you are intending to do this. Engineering economic analysis is appropriate in these cases, particularly when you are borrowing money from a banker; justifying the expense is a valuable part of your presentation.

Some things we use wear out or become too costly to continue to use, or they become obsolete or out of fashion compared to newer items. Most of you reading this book will know this from your experience of keeping your computer up to date and being able to run all the programs that you want to run. The question you may have asked yourself is, "Should I replace my computer or my car?" If the cost is relatively small, you probably don't spend too much time thinking about it. For high-cost items, however, some analysis may be beneficial and lead you to a more sound financial decision.

The usual goal of all such replacement analyses is to identify whether it is economically beneficial to replace at the current time an item you now have with another one you can buy. This analysis can be performed each year or each decision period, each time updating your estimates of future costs and benefits. So, suppose you are thinking of buying a new car to replace your current car. We call the current car the defender. It would rather have you as its owner than someone else, the challenger. It is defending its desire to remain being your car. The potential new car is challenging your defender, hoping it will replace your current car and become your new car. This defender-challenger terminology, of course, applies to any item that might be replaced.

Buying a new item, such as a car, will obviously cost much more than operating your current car another year, but over several years, it may be the better financial decision. Our task is to pick the lowest cost alternative. But the costs we need to compare are not the purchase price compared to the current car's operating costs, but instead the minimum equivalent **annual** costs of both choices. The number of years at which the minimum equivalent annual cost occurs is called the economic life of the item. Comparisons are made by calculating the equivalent uniform annual cost (EUAC) for the different options. This type of analysis is further described at the book's website supplementary materials (see Chapter 9: Economic Life).

Annual Cash-Flow Analysis

Annual cash-flow analysis compares the *annual cost* and benefits of the cash flows of two or more alternatives. Your choice can be affected both by the uniform annual costs and the annual benefits being generated as a result of making an investment. To do this analysis, we break the analysis into costs, *equivalent uniform annual cost* (**EUAC**), and the benefits generated, *equivalent uniform annual benefit* (**EUAB**), both expressed on an annual basis. Using these definitions, the net annual worth or value (NAW) is therefore defined as

$$NAW = EUAB - EUAC. \tag{9.3}$$

Analysis using EUAB and EUAC is performed by converting cash flows into either an EUAB or an EUAC using the previously presented equations. The alternative with the highest NAW should be the one that you will choose. Remember that you can convert a net present value (NPV) into a NAW by using the A/P function given in the interest tables.

The advantage of using an annual cash-flow analysis is that you do not have to choose a common analysis period for two different options, say, one with a lifetime of five years and the other with ten years. If you calculate the NAW for an alternative with a lifetime of five years and the NAW for another alternative with a lifetime of ten years, then the NAW's are being compared on an annual basis in both cases.

Example 9.3

Two options are being considered for a construction project. If interest is 8%,

	A	B
Initial Cost	20,000	30,000
Annual Tax Benefit	1,000	2,000
Salvage Value	5,000	10,000
Useful life	5	10

which option is the better deal?
Solution.

$NAW_A = \$1,000 - 20,000(A/P, 8\%, 5) + 5,000(A/F, 8\%, 5)$

$$= \$1,000 - 20,000(0.250) + 5,000(0.170)$$
$$= -\$3,156$$
$$NAW_B = \$2,000 - 30,000(A/P, 8\%, 10) + 10,000(A/F, 8\%, 10)$$
$$= \$2,000 - 30,000(0.149) + 10,000(0.069)$$
$$= -\$1,780$$

Option B has the lower cost and is therefore the better choice

Capital Recovery and Sinking Fund Methods

Closely related to the various calculations for Annualized cost are two commonly used terms: **Capital Recover Method** (CRM) and **Sinking Fund Method** (SFM). Often you need to calculate how long it would take to recover the **initial cost** of some item purchased, for example, a car. You buy a car at a cost of P, and the car dealer finances you the car by having you pay equal payments at some interest rate (per period) for a given number of periods (typically five years, broken into monthly payment schedule). These annual cost recovery calculations are also called the *Capital Recovery Method*; the symbol terminology for this method is

$P =$ amount that needs to be recovered (paid back),

$A =$ annual payment value (by period, with period being consistent with the interest rate),

$i =$ interest rate for the period used, and

$n =$ number of periods, for example, years.

The opposite of recovering capital is to save cash for some future event or reach a targeted "nest egg" that you plan to use for some activity (like retirement; buying a vacation home). This is called a *Sinking Fund Method* or *Calculation*; the terminology here is

$F =$ the future value, F;

$A =$ amount invested each period; and

i and $n =$ same as before.

These uniform payment series factors are given in the compound interest tables in the Appendix.

9.9 Problems

1. Ronald works for a company that has allotted him the equivalent of $10,000 per year for his project. Ronald can decide between two different ways to do the project, and one method will finish his project much faster but has

	Method I	Method II
Initial Cost	50,000	100,000
Salvage Value	10,000	5,000
Lifetime	7 years	5 years

a higher cost. The economic factors for each method are shown in the following table:

a. Which method should Ronald choose with an internal ROR of 8%?

b. What is the present worth of the money left over (out of $10,000 per year) if the method from (a) is used?

2. Mariah wants to purchase a new car. She will either purchase the car and "run it into the ground" by driving it for ten years, at which point she considers its value $0, or she will keep the car for five years and then assume she can sell it on eBay for $5,000. For a car whose initial cost is $25,000, which option is better? The market interest rate is 6% annually.

3. You are looking at two alternative technology bids from two vendors trying to convince you to purchase their solution to your problem. Alternative A has an expected life of three years, and Alternative B has an expected life of five years. What method (NPW or NAW) should you use to determine what is in your best interests and why?

	Bid A	Bid B
Initial Cost	$70,000	$100,000
Salvage Value	$10,000	$0
Lifetime	3 years	5 years

4. Alan Adams is a college senior who is considering doing a two-year master's program after he graduates in June. He believes that having a master's degree will increase his future earnings from employment. The graduate school that he has selected charges an equivalent present worth of $50,000 in total for the two-year program. However, he predicts that once he has his master's, he can earn $5,000 more per year for the first ten years of his job, and afterwards his salary will increase by $10,000 every ten years for forty years of employment (i.e., $15,000 higher in years 11–20, $25,000 higher in years 21–30, $35,000 higher in years 31–40). Assuming he retires after forty years of employment, what rate of return has Alan received from his master's education?

Year	Cash Flow
0	−$300
1	−100
2	+100
3	+200
4	+300
5	+400

5. Solve the following cash flow (profit) for the rate of return to within a 1/2%.

6. A man is considering installing a new drinking-water treatment system in his home so that he can make the current water more potable and taste

better. Currently, he buys water in jugs at the local grocery store for $1.50 a gallon. He predicts that his entire household consumes 5 gallons of water per week or 260 gallons per year. His new treatment system would cost $10,000 to install and would draw from his own well so that he would then be paying nothing for water. What rate of return will this man receive on the water treatment system if the system is considered to last

a. forever?

b. 100 years?

c. 50 years?

d. Would you recommend that the treatment system be installed? Explain.

7. A small strip mall complex in Jim's neighborhood is being sold for $200,000. There are currently five stores in the complex, and each store has a long-term lease that expires in two years. The rent paid is $800 per month for each space. Maintenance and other expenses for the mall owner are $5,000 each year. Several new housing developments are being built in the area, and it is expected that the complex can be sold for $250,000 after two years.

a. What is the internal rate of return for this investment (assuming he can sell the mall for $250,000 as described)?

b. Should Jim buy this complex if his MARR is 10%?

8. A restaurant wants to buy a new energy efficient oven for their kitchen. It can be purchased today for $10,000. The annual energy savings from the oven is calculated to be $5,000 and its effective lifetime is ten years. Expect the maintenance costs to be $1,000 for the first year and increase $500 per year after that. The salvage value of the oven will be $1,000. What is the rate of return on your investment to buy the new oven?

9. You have been leasing space since you started your repair store to service the hydroponic systems you've been selling. Now, you think you have enough volume to justify building you own small repair shop. Your contractor's bid to build the shop is $100,000. You have been paying $12,000 per year in rent (paid monthly in $1,000 increments) for similar space but it is time to reconsider signing a new lease. You can borrow the $100,000 from your local bank using a ten-year loan at a 6% interest rate to pay for the new shop. Your plan is to invest the $12,000 you save in no rent payments annually into your brother's hedge fund that has been producing annual rates of return of 18% per year for the last several years. Can you expect the funds generated by the hedge fund to be sufficient to cover your loan payments?

10 Rules of Investing

It's only business. Nothing personal.

Michael Corleone (*The Godfather*)[1]

10.0 Entrepreneur's Diary

I recently was involved in raising capital for starting an indoor shrimp farm. Just about everyone likes to eat shrimp. Interestingly, just about all of our shrimp (90%) is imported, creating a fairly obvious market opportunity. So, several years back, we put some seed money together and started working on the constraints that prevented someone from successfully raising shrimp indoors and making a profit. Well, after three years of research and demonstrating a prototype production system, we were ready for our next round of financing (we needed about $500,000). We prepared a presentation and invited a group of high-net-worth individuals. We presented in one of the high-rent office buildings in the financial district of Atlanta. You know, people do like shrimp, but they also like to have a sense of confidence about the people they are dealing with. My cousin John, a businessman in the Atlanta area, brought that credibility to the table. Our challenge was to present the opportunity to the individuals in the room in a form they could understand. We had all the necessary legal documents with us. We knew the rules and followed them. We were successful that night and raised most of the equity capital we needed. The investors all seemed to like and respect my cousin. And they all ate a lot of shrimp that night, too!

10.1 Overview

At some point in your entrepreneurial adventure, you will probably find yourself concluding that your future is constrained by a lack of capital. Assuming you've been reading the book pretty closely, you should understand there are a lot of steps to take you beyond the "I've got a great idea" stage to actually putting a company in operation, which usually starts as a very small one with minimal overhead. A

---

[1] *The Godfather*, dir. Francis Ford Coppola (Hollywood, CA: Paramount Pictures, 1972).

normal sequence is that you've used "family and friends" capital to demonstrate your prototype and verify that customers will use your product or service, that they like it, and that they will pay you more for it than it costs to deliver and manufacture. Now, you are ready to expand your business and, hence, the need for more capital. Knowing this, you have crafted a well-written crisp business plan (Chapter 8) and are now ready to hit the streets with it so to speak. Well, raising money is a slippery and treacherous path, but hopefully this chapter will give you some basic insight into the general process and what to watch out for along the way. We cover the basic investment rules that you must follow and understand to win an investment partner.

In this book, we are primarily focusing on smaller companies and start-ups. We are not trying to address efforts that secure tens of millions of dollars, but more likely the phase after friends and family (self-financing). So right now, you probably should be thinking in the hundreds of thousands of dollars as a target for your fund-raising goal. Surprisingly, whether you are negotiating with a small investor for a $10,000 placement or a venture capital firm for $10 million – the expectations and effort from the investor's perspective are basically the same. The smaller (angel) investor has probably been involved in or had experience in major investment deals and will bring their experience to the table along with associated expectations on what you must do. You must meet these expectations, or you will not secure their investment. This also probably explains some of the often heard investment banter that it is easier to raise $10 million than $10,000. Why, because the expectations and the required diligence on behalf of both sides of the equation are about the same in either case. The risks might be higher financially for the $10 million investment, but the effort involved in securing either amount of funding are not that much different. My advice to you is that the principles are essentially the same for any level of fund-raising. Be prepared and know the rules.

10.2 Investment Capital Principles

Investing in start-up ventures is like a pinball game: you have to keep the ball in play. When you run out of balls to play, you are out of the game. Substitute cash for balls in the previous sentence, and you can understand start-up financing. Thinking like the potential investor in your start-up company can be invaluable as you try to raise equity capital. Reviewing some of the basic reasons "why or why not" a company will succeed or fail can be useful, regardless of whether you are targeting angels or a much more serious level of investment with venture capitalists. Risks or failures of a start-up can generally be classified into the five following categories (see Figure 10.1[2]):

1. The Development Risk: Can you develop the product?

---

[2] A. D. Silver, *Venture Capital: The Complete Guide for Investors* (City: John Wiley & Sons, 1985).

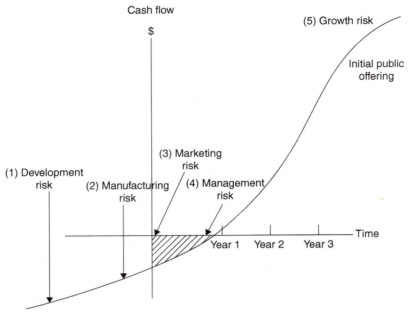

Figure 10.1 The five risks in a start-up company.
Source: From Silver, *Venture Capital.*

2. The Manufacturing Risk: If you can develop it, can you produce it?
3. The Marketing Risk: If you can make it, can you sell it?
4. The Management Risk: If you can sell it, can you sell it at a profit?
5. The Growth Risk: If you can manage the company, can you grow it?

Start-up companies have the first two risks to overcome. This means there are still three more risks to overcome. Positive cash flow is not seen to occur until sometime during the end of the third risk phase and into the fourth. This negative cash flow during the first series of risks is called the ***pre-revenue*** stage making you a pre-revenue company. The capital during these stages has to be supplied by investors (sometimes you) who are willing to take high risk, so the idea had better be very appealing. Larger investments will occur during the last two risk phases and are generally provided by the venture capital firms. Keep these stages of risk in mind when you are approaching the investment community and what type of investor you should be targeting.

Investment Laws

Although we have focused primarily for start-ups and angel-type investment opportunities when writing this book, it is good to have an understanding of some of the basic principles that venture capitalists use in evaluating investment opportunities. In this arena, venture capitalists have developed their own laws to guide their investment decisions, which can also be used as good guidelines for any investor considering a start-up venture.

LAW 1

Accept no more than two risks per investment (see the five risks in a start-up in Figure 10.1). This then tells you at what stage a venture capitalist is interested in your company; no need to approach them before that (marketing risk stage or later).

LAW 2

Invest in value opportunities, where value is quantified by the **Value Equation**,[3] which is the product of three factors:

$$V = P \cdot S \cdot M, \tag{10.1}$$

where

V = valuation,
P = the size of the problem,
S = the elegance of the solution, and
M = the quality of the entrepreneurial team.

Venture capitalists will place a relative value on each of the three terms making up the Value Equation, with each term having the same potential maximum value. For example, each term might be bracketed by a maximum value of three, so the maximum score that could be achieved would be 27. The point is that all three terms have to score high, if the investor is to be interested.

The P value is evaluated from your analysis of the market that you are targeting, which has been discussed at greater length in Chapter 4, "Developing Your Business Concept." The P term is maximized when all four of the following marketing factors are present:

1. Large numbers of potential customers
2. Existence of qualified customers (can afford to buy product)
3. Homogeneity of customers ("I'll make a Model T in any color you want as long as it's black!!" – Henry Ford)
4. Word of mouth is principal advertising form

The second term in the value equation is the S term, which is the elegance of the solution. The S term is composed actually of two factors:

$$S = B \cdot T, \tag{10.2}$$

where

B = the quality of the business plan or the solution-delivery mechanism and
T = the existence of low-priced technology.

Again, a subjective judgment is being made here to arrive at a value for S, particularly because it is made up of two terms, B and T. The T or technology component of the S factor is important in achieving an elegant S. New technologies

[3] Introduced in A. D. Silver, *Venture Capital*.

can be protected via basic patents, process patents, or lead time, and these three kinds of protection are of great importance to the investor. The technology factor might be considered strong for example, particularly if also protected by patents, if it involved micro-mechanicals (MEMS), cloning biotechnology techniques, or some other new technology coming into the market place. The value of T might drop the overall value of S if it involved relatively expensive products to implement a solution or the government could block or delay product entry.

Successful venture capitalists look for an "elegant S." This means that if the solution is a product, it should be proprietary and if it is a service, it should have a non-duplicable distribution system. The P and M factors would have to be high to cause an experienced venture capitalist to invest in a non-elegant S company.

There is an axiom in the business-launching arena: anything worth doing is worth copying by others! This means that if an investment opportunity is attractive to one group of investors or to one entrepreneurial team, it will surely be attractive to several others. A better-faster-cheaper system or device, no matter how thoroughly protected by patents or some non-duplicable delivery system – will very shortly be copied, duplicated, and simulated. Success attracts significant competition, quickly. Natural monopolies simply do not last more than three years, unless protected by the government in some way.

In the case of non-proprietary products, such as one frequently finds in the service industries, venture capitalists tend to prefer delivery systems that are either non-duplicable or very expensive for the second or third company in the industry to copy or emulate. Perhaps the best example is Federal Express, which identified an opportunity to deliver small, time-sensitive packages overnight on a guaranteed basis. Their idea was not unique, but their superior execution is what has made it a very successful company. They still attracted competitors, for example, UPS and DHL.

LAW 3

Invest in big P companies, because the investor community will accord to them unreasonably high Vs, irrespective of S and whether or not you are entering an emerging market (the E term is defined below). The P term must have certain attributes for a venture capitalist to be interested:

- $50 million to $500 million by year 5 in revenues
- Probable market penetration target of 10% with 15% being optimistic
- Market minimums of $0.5 to $5 billion

It takes a pretty major success for a company to reach a $50M or more in revenues. Angel investors are attracted to companies with much smaller revenue projections, for example, $1 million to $2 million, but are certainly hopeful that your company will grow into such revenue streams.

THE OPPORTUNITY EQUATION

Investors in start-up ventures are even more interested in what I like to describe by an equation I developed called the Opportunity Equation:

$$O = E \cdot T \cdot C, \tag{10.3}$$

where

O is the perceived value of the opportunity,
E is whether the market is emerging,
T is the timing of your entry, and
C is your competitive advantage.

This equation is closely related to the **Value Equation**, but includes what are probably the two key factors for why an investor will invest:

- the current level of service to a market opportunity and
- your competitive advantage.

The **Opportunity Equation** also includes the key questions that anyone should ask about a market before trying to enter it: is the market definable, reachable, and can it be profitably served? (See Chapter 4, Section 4.2 Market Analysis.) The three factors of the Opportunity Equation basically determine if you can serve your defined market in a

profitable manner for both yourself and your customer. Similar to the Value Equation, each factor in the Opportunity Equation is assigned a value between 0 to 3. All three factors are assumed to have the same overall value in terms of contributing to evaluation of a business opportunity. You can see some similarity between the Value and Opportunity equations. However, the Opportunity Equation is probably more important in evaluating the potential success of a start-up company because it emphasizes the timing of entry, with management expertise assumed or built into the C factor, competitive advantage.

> Key Questions: Is the market definable, reachable, and can it be profitably served?

It is very expensive to capture market share once it is already being served by other providers. Hence timing of market entry is critical. We all recognize right now that China represents an **emerging market** for many products and services. But, the key question for the entrepreneur is whether or not any of these emerging markets are currently underserved; that is, is your timing right? The more a market is already being addressed, the less likely it is a quality opportunity for a start-up. Start-ups do better by identifying those opportunities that no one else has currently "seen" or at least has yet to attract multiple providers.

> The more a market is already being addressed, the less likely it is a quality opportunity for a start-up.

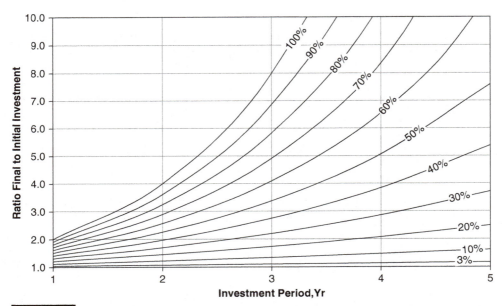

Figure 10.2   Target rates of return for venture capitalists based on years of investment and target return ratios.

Venture Capital Expected Rates of Return

A venture capital firm will have target returns of

- $5x$ investment in three years.
- $10x$ investment in five years.

Figure 10.2 demonstrates target rates of return for a venture capital (VC) firm based on years of investment and target return ratios. Remember that the target returns are average returns for the firm. For every ten investments made by a VC firm, the success of the start-up firms over the next three to five years will generally result in a success ratio as follows:

- three outright failures
- five that are hanging on (near cash flow positive)
- one that has significant positive cash flow
- one winner (exceeded projections)

Although VCs don't like to publicize these expectations, they are a reality nevertheless. As a result, this is why the VC is actually looking for the big-winner types from the onset of the investment decision process. These winners provide the types of returns that allow the successful VC's to obtain average overall returns around 25%. You need to remember this as you decide your company's strategy and what in fact will attract venture capitalists. A kiss of death for fund-raising is if your company has had bad publicity related to some aspect of your product; this news spreads quickly essentially killing any possible deal with a VC. Avoid the temptation of talking to newspapers and trade magazines until you are 100%

sure that your product is ready for market. An early failure of a product will kill chances of obtaining future funding. Also, that first wave of publicity you receive from the press will lead to people calling for product. Be sure you are ready to deliver.

> **Advice:**
> Avoid all publicity until the product works and is ready for delivery.

Returning to a mathematical approach to evaluate investor expectations, we can calculate the target value for an investment in a start-up company for a VC to be interested:

$$F = P(1+i)^n, \tag{10.4}$$

where

n = number of years investment made and

i = average rate of return targeted by investment group.

If the investor targets a 50% annual return on their investment (ROI) for five years, then your company's value future value (F) must increase from its present value (P) by the following ratio:

$$F = P(1+0.5)^5 \; or \; \frac{F}{P} \cong 8.$$

Basically, this is a "wow" result. The VC-type investor is looking for opportunities that over five years would give them back nearly eight times their original investment level. As you might suspect, these types of opportunities do not come along easily. This should also help you understand what investor class you should be targeting when you are seeking investment capital. Uncle Joe's investment criteria might be somewhat less demanding! Figure 10.2 depicts in graphical form the relationship between investment return, years of investment, and rate of return. Additional example calculations are demonstrated in the Engineering Economics section at the end of this chapter on how to calculate rates of return (ROR) from pro forma income statements.

10.3 Due-Diligence Process

Once you have attracted investors' interest in your plan, their next step will be to perform a due-diligence review on you and everything associated with your company. Typically a due-diligence process will consist of five audits prior to an investment being made and usually in the following order:

1. Business plan (what got them interested)
2. Market size

3. People
4. Financial
5. Legal

In one investment deal I went through, the investor (Gene) started his due diligence with a legal review of our company. Now Gene had a law degree, so this partly explains his approach, but Gene's logic was that if our company had some fatal legal flaw, then there was no reason to execute the other due diligence reviews. This makes sense to me. Why waste time and money if there is a fatal flaw legally? This also emphasizes the point that every stage of your company must be on solid legal footings. Follow all the rules.

A deal that survives the due diligence process will generally receive an investment proposal. A potential investor will take great pains to ensure that these five steps are effectively and thoroughly completed, because this increases their odds for selecting successful investment opportunities. You would think that the odds of success for your company would be extremely high once an investment has been made after a complete due-diligence process, but often more times than not, an investor sponsored company will fail. Some of the reasons would include the following (so don't do these):

- Faulty business plan
- Incompetent management
- Market failed to develop
- Bad luck (not really under your control)
- Overpayment for the investment

The business plan is the primary instrument that conveys much of the information needed during the due diligence process, so make it as easy as possible for the reviewer to find the information they are looking for. Your business plan must answer six basic questions (see Chapter 8 for details on how to write an effective business plan):

1. What problem is being solved?
2. Is your solution proprietary or unique?
3. What stage is your company currently in?
4. Has your company or product been endorsed by a highly regarded person or entity?
5. What are the strengths of your management team including your board of directors?
6. How much capital is required?

Serious investors like to monitor your progress and interact heavily with the company managers, in other words, keeping a tap on the financial sanctity of their investment. Thus, an investor or a firm will want your principal place of business to be located within a one or two-hour drive or flight time from the investor's office. You should start to get the idea that you are going to have an intimate relationship with your investor, whether they are an angel or a venture capitalist. This should also tell you where to look for potential investors, that is, nearby.

> An investor will want your principal place of business to be located within a one or two-hour drive or flight time from the investor's office.

No-Deal Routine

If you make it successfully through the due diligence process, you should start to expect a deal to be offered, but before success, there is often disappointment. You will face your share of rejections. For example, after three months of interaction with your potential investor, they determine that the problem or market you have identified is relatively small and that other companies have begun developing product solutions for the target market as well, although slightly different from yours. The investor will then say something in a letter or on the phone to the following effect:

> We think the market you have identified is smaller than our typical target range, and there are several other companies geared up to sell into that market which makes your potential share even smaller than described. We will have to continue to monitor your progress.

In other words, the investor is saying no deal; they have given you a turndown.

Another reason that you as an inexperienced young entrepreneur might be turned down is that you have not taken the time or have been unable to find an experienced partner as part of your paid team or board of directors or advisors. The venture capitalist will say something similar to this:

> We would prefer to see your management team strengthened by a strong local board of advisors and perhaps an experienced marketing manager to assist you in launching your new company.

And in the end, you may have done everything just right – and you still get turned down by a slew of investors. Experienced investors have a variety of ways of saying no without really saying so directly, in a way that they try to keep the door open for future involvement. Don't hold your breath, however. Learn to recognize these signs of noninterest. Table 10.1 summarizes the most common ones with each having several variations that mean the same thing: no deal!

10.4 Company Valuation

Raising capital by selling equity in your company is what enables you to implement your business strategy. The amount of equity you are able to retain will depend to a large degree on how well you present your value proposition and then how well you negotiate with the new investors to convince them of this value.

Table 10.1. Commonly Used Vague No-Deal Expressions

1. We cannot review your deal at this time due to unusual time constraints on our end.
2. We are not investing in start-ups (or first-stage companies, etc.) at this time.
3. We are not investing in your industry at this time.
4. We are not investing in companies located in your region of the country at this time.
5. Although one or two of us liked the concept, we were unable to obtain unanimous consent at the investment committee level.
6. We are not investing in companies that require less than $10 million (or some specific amount) at this time.
7. We are not investing in companies that lack x at this time [fill in x].

> Determining the pre-value of your company is the most important point of your negotiation.

The negotiated valuation of your company with your new investors is probably the most important part of the whole investment process, because it determines what percentage of your company you will be able to retain as you exchange equity for capital. New investors will have a pretty clear idea of what they are willing to invest and it will come down to them being convinced that they have are being given a fair exchange of equity in your company for their cash infusion.

Your ownership retention and the investor's equity percentage ultimately comes down to what pre-value is being placed on your company prior to outside money being accepted for equity. This value is largely a matter of negotiation. If you can present the investor with your own calculations estimating your company's value, perhaps you can influence the investor to revise their figure. In a friendly negotiation with angel-type investors, they may be willing to agree to a higher valuation if you can provide them with reasonable justification.

Advice

Your "best" offer from an investor will not necessarily be from the one giving you the highest valuation. Remember, a good investor is more than just a cash source, but a person who should become a vital, contributing member of your team.

In negotiating a valuation, be realistic on your company's value particularly with your first investors. This will empower them to be your financial partner and then let them pitch your company to future investors, where they are the ones pushing the value of your company, not you. Your first early stage investors are taking a very high risk. Reward them for their vision and support. If you do not understand this, you will have a uninterested smorgasbord of individual investors. Your chances of long-term financial success will increase if you start to think of your investors as partners for the future. Investors are actually pretty simple creatures; they simply want to make lots of money by helping your company to be very successful. Toward this end, an angel investor and particularly a venture capital investor will be very helpful in the following ways:

- Providing post investment monitoring and advice
- Providing added value to your company's portfolio
- Bringing in other sources of capital
- Rescuing your company when you enter a troubled period (financially or strategically)
- Helping you to liquefy investment (sell company)

Remember, your investors are your partners in success.

10.5 Company Valuation Methods

Company valuation should be an open and full-disclosure process. There is nothing wrong with, and it is to your benefit to show a potential investor how you calculated your company's pre-value. Once you can agree on pre-value, then there are even several ways to calculate investor ownership percentage that will be given for cash invested. Your method of determining company value is not intended to be a secret process nor should it be; sometimes it is even a mutual effort in which you and your angel investor are trying to find a fair value. I'll review several techniques, and you can use the one that demonstrates your highest value. Hopefully you're getting the idea that valuation is ultimately a *negotiated* value between you and the investor.

Pre-Value Determination Methods

THE VENTURE CAPITAL METHOD

When a company is in the start-up phase, it often has negative cash flows, negative earnings, and a highly uncertain future. The Venture Capital method values the company in the future when it is projected to have positive cash flows and/or earnings.[4] The calculated future value is then discounted back to a present value using a high discount rate, usually somewhere between 35% and 150%. The present value is found by discounting the future value using the familiar formula:

$$PV = \frac{FV}{(1+i)^n},$$ (10.5)

where

PV = present value;
FV = future value;

---

[4] For extensive details about this method, see John L. Nesheim, *High Tech Start Up: The Complete Handbook for Creating Successful New High Tech Companies*, rev. and updated ed. (City: Simon & Schuster, 2000).

i = interest rate, discount value; and

n = number of years

The future value, FV, is based on the pro forma analysis as some multiple of predicted earnings or sales, depending on the convention (if there is one) for the particular industry. The PV is obviously very subjective to what discount value is used. The Cayuga Venture Fund[5] is happy with projected returns of 35% to 60% (but require you to keep your company in the upstate New York region). High-profile venture capitalists will often insist on something much more toward the top of the mentioned range, for example, 75 to 100%. Using these high discount values allows a firm to obtain average returns of 25% to 50% on all its investments by compensating for their investments collectively in companies that failed or had minimal value increases.

Although the Venture Capital Method is popular with some investor groups, there are objections to using it. Your major criticism (being the founder) is the use of the very large discount rates. The venture capitalists justify these high discounting rates for several reasons:

- Many of their investments in start-up companies end up with little to no value.
- Venture capitalists provide valuable services to the company being funded and the large discount rate merely helps provide them with compensation for their time and effort.
- High discount rates are used to compensate for the illiquidity of private firms; that is, stock in a privately held company can be very difficult to sell or cash in.
- Projections made in business plans are notoriously optimistic, so the high discount rate is a means of bringing some realism to the valuation.

Accuracy in predicting future values of companies may be addressed by using more complex methods of evaluation. For example, probabilities can be placed on the occurrence of each event and values and payoffs can be assigned using decision trees, Bayesian analysis, and utility theory.[6] You can pretty much make valuation methods as complicated as you want to, so be careful on what your intent is. You should not try to confuse the investor; if you confuse them, they will walk away, no deal. The more complicated analysis tools are more often used on established companies with documented cash flows for several years. Newer-stage companies (the focus of this book) will generally use very simple methods to determine their pre-value, for example, using some form of equation 10.5, or even what might seem an arbitrary value "pulled out of a hat!" The use of an arbitrary value arises when investors have a preconceived idea about what percentage ownership they want for making their cash investment.

[5] The Cayuga Venture Fund is a small venture capital group in Ithaca, New York. In general, members have a strong connection to Cornell University and/or local development interests.

[6] Barry Render Ralph M. Stair, Jr., and Hanna, Michael E., 2012, *Quantitative Analysis for Management* (Upper Saddle River, NJ: Prentice Hall).

THE NET PRESENT VALUE METHOD

This is the classic method taught at business schools for valuing corporations (usually large, publicly traded ones). The basic premise is that a successful business is a cash-generating machine, and the value of the company depends on how much cash is generated over time and then discounted by inflation or some other interest rate, for example, market interest, opportunity cost on a company's money, or targeted internal rates of return. Sophisticated cash-flow analysis can be used to include the effects of tax shields due to borrowing and the related effects on value of the company.

In this method, a net present value (NPV, which becomes your company's pre-value estimate) is calculated based upon a series of cash flows for some distinct period of analysis, for example, five years, and a terminal value for the cash flow in the last year; the yearly cash flows would include the effects of depreciation, interest effects, and so on. So there are two steps: first, determine your yearly cash flows, and then discount the cash flows back to a present value.

Cash flow by year can be taken from your Proforma Statements shown with your business plan:

$$CF = EBIT \cdot (1 - \tau) + D - CapEx - \Delta NWC, \tag{10.6}$$

where

CF = cash flow,

$EBIT$ = earnings before interest and taxes,

τ = tax rate,

D = depreciation expense (for tax purposes, a noncash expense),

$CapEx$ = capital expenditures (which use up cash and result in depreciable assets), and

ΔNWC = changes in net working capital = current assets − current liabilities,

where

Current assets = *cash* + *inventory* + *accounts receivable* and

Current liabilities = *accounts payable and short-term debt.*

Net working capital typically grows with the sales of the firm.

Next, the *NPV* of the company is calculated by discounting the future cash flows back to their present day value,

$$NPV = \frac{CF_1}{(1+r)} + \frac{CF_2}{(1+r)^2} + \frac{CF_3}{(1+r)^3} + \cdots + \frac{CF_n + TV_n}{(1+r)^n}, \tag{10.7}$$

where

r = targeted rate of return, decimal and

TV = terminal value (see equation 10.8 for calculation).

The *r*-value used in equation (10.7) is much less than a venture capitalist uses in the earlier venture capital method (interest rate of 35% to 150%), and might be a company's average targeted rates of return or a generous market rate of interest within the current banking community. If you go out a few years (at least five), the terminal value (*TV*) or salvage value of the company will have a minor effect on *NPV* because it will be heavily discounted by time, if you have decent cash flows in the few years before the terminal value year. However, few start-up firms have this characteristic, so this method might not be a great choice for valuating your start-up company.

You can estimate the terminal value in year n, using

$$TV_n = \frac{CF_n(1+g)}{r-g},$$ (10.8)

where

g = the growth rate of the company in year n.

If you extend the analysis to where your start-up is a mature firm, you can use a target value of g of 8% to 10% or so, if you cannot find values from other mature firms in your industry.

OTHER METHODS – GROSS SALES

In low-margin products such as food, the value of the company is often equated to yearly gross sales or something close, for example, use a 1.2 multiplier. In my fish company, we used the "*F*" approach and a value of 1.2 for some premium in determining our pre-value.

$$Prevalue = F * Gross\ Sales,$$ (10.9)

where $F \cong 1.0$ to 1.2.

The nice thing about this approach is that it is hard to argue against because establishing your company's gross sales is already a matter of record (your tax return). Once again, it all comes down to negotiating your company's value, but don't expect much more than gross sales in many cases (hence the need to generate sales, which then creates value).

More sophisticated methods are discussed in the supplemental materials given on the book's webpage, see the supplemental materials for Chapter 10, Valuation. There, we review the Black-Scholes option pricing model, the capital asset pricing model (CAPM), and Monte Carlo simulation analysis. These methods would rarely if ever be used for a start-up company's valuation.

Methods to Calculate Ownership Percentage

RETURN ON INVESTMENT (ROI) METHOD (THE HOCKEY STICK METHOD)

If you attend a venture fair or talk to some investor types, you will probably hear something about the "Hockey Stick Method" of evaluation. Remember what a hockey stick looks like: flat for a little while and then, wooomphhh, a rapid upward slope. There is never an argument on the flat portion of the graph, that is, when your company is not making much money or, more typically, losing money

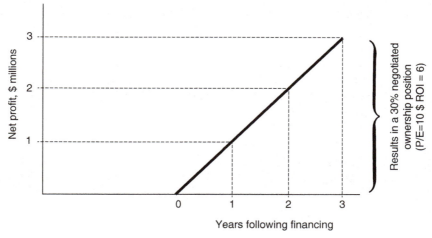

Figure 10.3 The hockey stick method of determining VC ownership.

(negative cash flow). But the rate of slope of the hockey stick can become contentious. Using the Hockey Stick Method, investors first decide what return on investment (*ROI*) they want when they cash in on their investment, usually after only three or four years. Next, investors and the entrepreneur must agree on a price earnings (P/E) ratio or the multiplier at the point where the hockey stick begins its rapid upward slope (profitability).

So, for example, assume that an investor wants to invest $1.0 million and wants a ROI of 6 or (6 * $1.0 million = $6 million). Assume that your company's projected net earnings beginning the third year are $2.0 million, and you agree on a P/E (price to earnings) ratio of 10. After this negotiation, the calculation is shown below that determines the investor's ownership position after investment is 30%:

$$Ownership\% = \frac{\$Invested \ \cdot \ ROI}{Net\,Profit\,Year\,"n" \ \cdot \ (P\,/\,E)} * 100\% \qquad (10.10)$$

Substituting our values from the example into equation (10.10), we obtain

$$\frac{\$1.0\,million \times 6}{\$2.0\,million \times 10} * 100\% = 30\%.$$

See Figure 10.3, which graphically demonstrates the preceding calculation.

DILUTION FACTOR METHOD

Assume that you and your potential investor have agreed that your company currently has a net worth or pre-value of $100,000 and the new investor is going to invest $25,000 (you can add zeros to make this more exciting if you want to). The percentage of the new capital as part of the combined post-value gives a term called the dividend. Using the Dilution Factor Method, you then divide this dividend by the dilution factor to determine the percentage ownership of the new investor.

$$Dividend = \frac{New\,Capital}{Net\,Worth + New\,Capital}.$$

$$Ownership = \frac{Dividend}{Dilution\,Factor} * 100\% \qquad (10.11)$$

For example, if the dilution factor were 2, then the investor in the current example will own only 10% of your company after investing instead of 20% if no dilution factor were used:

$$\frac{\$25,000}{\$100,00 + \$25,000} \cdot \frac{1}{Dilution\,Factor} = \frac{0.20}{2} * 100\% = 10\%.$$

Using this method, the argument with the new investor will center around both the dilution factor and the original net worth of the company being used. So, this creates two points to argue about! You should assume that the investor doesn't really care about anything more than their ending percentage of ownership for the cash they will invest. My advice is to just go with a dilution factor of 1.0 and forget about arguing what the value of the dilution factor is, because it then drops out of the preceding calculation, and ownership is calculated based just on pre-value and the investment being made, that is, the simple pre-value method shown next.

SIMPLE PRE-VALUE METHOD

I prefer to keep the negotiation process with the potential investor as simple as possible when determining the investor's percentage ownership for the cash they are about to invest. The investor pretty much has a preconceived idea of what they want, so keep the math simple and calculate their percentage of ownership as follows:

$$Ownership = \frac{New\,Capital}{Pre\text{-}value + New\,Capital} * 100\%. \qquad (10.12)$$

This approach reduces the negotiation to only one item: agreeing on what the pre-value of your company is.

Example 10.1

Let's assume that your company needs a $1.5 million capital infusion, and the investor makes an offer for 25% post ownership. This means the investor has defined the post value at $6 million. This can be seen by rearranging equation (10.12):

$$Pre\text{-}value = \frac{New\,Capital}{Ownership\%} - New\,Capital$$
$$= \frac{\$1.5M}{0.25} - \$1.5M = \$6M - 1.5M = \$4.5M$$

If you accept her offer, then you have agreed on a pre-value of $4.5 million and the dilution factor by definition is 1.0. If the investor were to say that they want 50% for their $1.5 million, this means that you are both agreeing to a post-value of $3 million (or a pre-value of $1.5 million). This is the "shortcut" approach of doing the Simple Pre-Value Method.

SHARE PRICE METHOD

If you have founded your company, then you have already issued shares of stock (corporation) or member units (LLC). These shares of stock will have been issued in exchange for company assets, for example, intellectual property contributed to the company by the founders or for cash to outside investors. Selling shares to outside investors has been executed at some specific share price. As a result, a company's pre-value can be calculated directly by the product of last share price used and the number of outstanding shares:

$$Pre\text{-}value = Shares\,Outstanding * Share\,Price. \qquad (10.13)$$

Note that this calculation does not address the impact of unissued shares remaining from the original number of **authorized shares** or additional shares that might be authorized in the future (dilution is addressed in a later chapter section). If you feel that your company has a much higher value than is indicated by using this share-price method, then use an alternative method. This would be the case when your company has had some significant achievements or gained important assets or capabilities since the last shares were sold at what you now consider an undervalued price.

MULTIPLE SHARE PRICE METHOD

Finally, there is the sublime method of calculating ownership percentage related to human nature! A sophisticated venture capital fund manager actually told me this: investors will not want to pay more than four times (the factor in equation 10.14) whatever the last stock was sold for.

$$Share\,Price_{New} = Factor * Share\,Price_{Old}, \qquad (10.14)$$

where
   $Factor$ = premium allowed on the last share price of record.

Logic is something along the lines that the stock just couldn't have appreciated more than four times the last value that had been negotiated by others only two or three years earlier.

10.6 Ownership Positions

Ownership, Majority, and Valuation

Company ownership is broken into some specific number of shares of stock for a corporation (member units in an LLC) at some specific point in time. When new shares are sold, all outstanding shares, those shares that have been issued to date, are considered to have the same new value. But as your company goes forward, value usually changes, hopefully higher, so the next round will once again re-establish the new share value for all outstanding shares. Part of the reason that share price may change is that the number of shares can be increased, almost without limit as determined by the company's bylaws and some majority of shareholders agreeing to do so.

Generally a supermajority (defined as being greater than two-thirds of the outstanding shares, usually interpreted as 70% by lawyers) is required to do something that affects shareholders' material interests, whereas a simple majority (defined as greater than 50%) of shareholders votes can decide minor issues, for example, voting on officer positions. Thus, one of your "ideal" goals is to maintain a supermajority of your company's shares (you and your team, but preferably just you) so that you can make major decisions that might not be in the best material interests of some minor shareholders. Your next threshold goal of ownership is to try to maintain a simple majority during a negotiation of exchanging equity for capital. If you drop below 50%, then it really doesn't matter too much what your ownership percentage is and your only consideration is pre- and post-value of your total shares. In this case after a round of financing, your obvious goal is that your stock value (number of shares × share price) should go up even though your percentage ownership went down, which sometimes prevails over a goal of maintaining some majority percentage position. The concept here would be that 25% of a $10 million company is better than 51% of a $2 million company.

The investor will generally have some pre-determined ownership position that they are comfortable with. Your job is to figure this out and make sure it is compatible with what you are trying to achieve. There are several ways for the seed-round investor to state their intentions; be familiar with all these:

- "I'll put in $300,000, based on a $200,000 pre-money valuation."
- "It's worth $500,000 post-money, and I'll take 75% based upon full dilution."
- "I'll take 6 million shares, with a $500,000 post-money cap."

Pre- and Post-Ownership Positions

At the time of incorporation, a company is required to define the number of shares authorized for issue, called authorized shares. Although companies generally issue either common or preferred classes of stock, we will assume for the next example where valuation is illustrated that the corporation only issues common stock. Most venture capitals will want their initial stock issued as preferred stock that is convertible 1:1 with common (at a future time or event occurrence of their choosing), so the distinction is transparent in our examples.

Example 10.2

Consider a start-up company, High Tech, Inc., looking for seed-round financing. There are several founders, and they have issued themselves a total of 2 million shares of common stock from an authorized number of 10 million shares. The CEO owns 1.1 million, and the others own a total of 900,000 shares. They approach a seed-round (angel) investor. The seed-round investor says that she will invest $300,000 and that she *pre-values* the company at $200,000. The High Tech founders agree. There are no reserved option shares. Please answer the following questions:

a) How many new shares of stock will be issued?
Answer: 3 million (pre-value is $0.10/share, from $200,000/2 million shares)

b) How many shares of the stock will there be outstanding after the funding?
Answer: 2,000,000 (existing) + 3,000,000 (new) = 5,000,000 shares

c) How much (as a percentage) of the company will the CEO have after the seed round funding?
Answer: 22% (1.1 million shares divide by 5 million shares)

d) How much cash will the seed round investor hope to walk away with after a successful sale of the company in five years (assume venture capital–type return of 100% per year compounded)?
Answer: Company sold for $(1+100\%\text{ yearly return})^5 = 32$, so 32 × investment of $300,000 = $9.7 million (60% of company selling price).

Solutions
a) 3 million (pre-value is $0.10/share, from $200,000/2 million shares)
b) 2,000,000 (existing) + 3,000,000 (new) = 5,000,000 shares
c) 22% (1.1 million shares divide by 5 million shares)
d) Company sold for $(1+100\%\text{ yearly return})^5 = 32$, so 32 × investment of $300,000 = $9.7 million (60% of company selling price).

Example 10.3

The founders of New Tech are a little more savvy and decide to reserve 2 million shares for stock options for employees (like themselves). The other details are the same as Example 10.2. The seed-round investor again says that she will offer $300,000, with a pre-money valuation of $200,000 including the reserved shares for stock options (another way to say this would be $200,000 pre-money, fully diluted, referring to the 2,000,000 shares held in reserve, but not any additional balance of authorized shares after the issuance of the shares to the new investor for their $300,000; see the further description of this term later in this section). Please answer the following questions:

a) How many new shares of stock will be issued?
b) How many shares of the stock will be outstanding after the funding?
c) How much (as a percentage) of the company will the CEO have after the seed-round funding?

Solutions:

a) 6,000,000 (pre-value is $0.05/share, from $200,000/4 million shares, where number of shares considered is the outstanding plus the reserve shares, or existing 2 million + 2 million in reserve = 4 million shares)
b) 6 million (new shares = $300,000/$0.05/share) + 2 million (existing) (does not count the reserve) = 8 million shares outstanding
c) 13.75 % (1.1 million divided by 8 million)

> **Fully Diluted:** This term may have several interpretations. Be careful.

Meanings of Full Dilution

Fully diluted is a phrase that is used by investors with more than one meaning to determine how they are computing the valuation of a company in which they are proposing to invest. This term should appear in investors' term sheet where it is defined how the investor places a valuation on the company for investment purposes (see previous section for methods to determine company valuation).

Fully diluted refers to the number of shares a company will be considered as having issued for the limited purpose of determining the percentage of shares an investor will acquire for their investment in the current round. In the extreme case, this could be the total number of shares that have been authorized.

For example, an investor might offer that they will purchase $1 million of company stock at a post money valuation of $10 million on a "fully diluted" basis. Just what does this mean? *You must understand* that the more shares investors can include in their definition of "fully diluted" for the same defined post-value used to calculate share price, the more shares investors will receive for their cash investment. Conversely, it is in your (the founder or current owners of stock) interest to reduce the number of shares defined as being part of the "fully diluted" definition (see list at end of this section), for example, reserved shares and/or any non-issued shares are not included in determining share price. This maximizes share price for an agreed-to company value. Then if you sell your shares as a part of a contract buyout (often you will be forced out against your wishes), your payout is maximized.

> Your goal should be to minimize the number of shares considered as part of the fully diluted lot, because this maximizes share price for an agreed-to company value.

Generally, any currently defined or reserved share options are included in the calculations of value based on a fully diluted basis. This means that these reserved shares will suffer dilution when the options vest, whereas an investor who has negotiated a percentage of shares based on full dilution that included these options will not. If you can negotiate this point, go ahead, but most of the time, options are considered as being subject to dilution when more shares are issued by the company. A company's option-eligible employees should place value appropriately on these options (usually awarded in lieu of cash salary), for example, if there is a large percentage of authorized shares that have not yet been allocated, you as an option holder should assume that eventually the shares will be issued, which almost always causes a devaluation in share price and what the value of your option was. Over time, undoubtedly it will be in someone's best interest that these unallocated but authorized shares are issued. Generally, it will happen at the worst possible time for you! The manner in which "fully diluted" is defined directly effects share price and therein how many shares a company must sell to raise a targeted amount of capital and the degree of dilution imposed on current shareholders. Shares included in a definition of "fully diluted" are (listed by descending frequency of appearance):

- shares of a company's capital stock that have been issued to (including the current round) and are still held by the company's shareholders;
- shares of a company's capital stock that have not yet been issued but can be purchased under outstanding options or warrants that have been issued by the company;
- shares of a company's capital stock that have not yet been issued but can be acquired on conversion of existing rights to convert debt or other rights, as in shares a lender can acquire under a contract that lets it convert some or all of its debt into stock;
- shares of a company's capital stock that have not yet been issued but have been reserved (and are therefore available) for issuance to company employees under a company stock plan, but which have not been issued or committed to employees by the issuance of options, warrants or other stock plan grants;
- shares of a company's capital stock that have not yet been issued but are reserved (and are therefore available) for issuance to identified future business partners on completion of some defined milestone or event;

- shares of a company's capital stock that have not yet been issued but are expected to be used to complete business deals in the future; and
- any other shares that might be issued in the future (this is the most extreme case in that all remaining *authorized* shares are assumed to be issued).

As shown in the preceding list, fully diluted can be calculated in different ways, so be sure that it is explicitly defined for your own negotiation. The sooner this term is defined, the better, because it could be a source of major disagreement as to what the investor is expecting versus what you are willing to give. Note that all these issuances are limited by the quantity of authorized shares, but this number can be increased subject to constraints of the bylaws (corporation), operating agreement (LLC). If a majority of the shareholders (particularly if it is a supermajority of greater than two-thirds of outstanding shares) vote to authorize more shares, then this will happen. So, in effect, your position that you thought was fixed because it was based on the number of authorized shares when you made your investment, is just more smoke and mirrors because you could be diluted to a trivial percentage as more shares are authorized.

Example 10.4

Assume that in the previous New Tech example, another year has passed and that the reserve shares have now been issued, so the company now has 9 million shares of outstanding stock, with 1 million shares remaining from the initial authorized number of 10 million. A new investor is willing to invest $1 million to own 10% of the company fully diluted. Without any additional shares being authorized, this means the investor would own one-tenth of the company's issued stock (1 million of 10 million total) and that the stock is being valued at $1/share, in order that the investor receive the 1 million shares requested. Because there are only 10 million shares authorized at this point, the company will have sold all of its authorized shares if it were to agree to the deal. So, before the sale of stock is performed (making passage of the motion to authorize additional shares less likely), the company votes in a special meeting to authorize an additional 90 million shares to the currently authorized 10 million shares.

Now, how does this new authorization of shares affect the calculation for how many shares the new investor receives? If fully diluted now included all 100 million shares authorized, then the shares are valued post round at $0.10/share ($10 million/100 million shares), so the investor would receive 10 million shares ($1M/[$0.10/share]). WOW, what a difference based on how fully diluted is defined. In both scenarios, the investor still winds up with 10% of the post-round value of the company, but in the second case, the investor winds up with 10/19ths of the outstanding shares after the current round is completed (note also that this gives the new investor more than 50% of the outstanding shares). This investor's initial controlling position (53%) could allow them to pass some motion they would propose that requires only a simple majority but could be dramatically in their own self-interest!

Maximum Dilution: The Cram Down

Founders of a start-up can implement means to gain additional equity ownership by awarding options in relation to milestone achievements. Conversely, outside investors will continually be diluted unless they continue to make further investments in each succeeding round. Once your start-up team loses majority control of the company, your team is at the mercy of the outside investment group. Even if an investor or a group of investors are not in a majority ownership position, they can quickly assert control if a cash-flow crunch occurs. When this happens, a minority investor who has cash to invest can force your team to accept a "deal" that – other than providing the cash you need to keep the company running and have no other means of obtaining – you find completely repugnant.

> Your alternative to accepting a "cram-down" offer will probably be Chapter 7 bankruptcy.

In the preceding cash-flow crunch, get ready for what is called the "cram down." A cram down is literally just what it sounds like. If your company runs out of money, an investor with ready cash will be in a position to force terms on you that you will probably have to agree to, for example, accepting a share value of 10% of what was used for the preceding round. This results in your ownership position being diluted to about 10% of the ownership position you had before you ran out of money. While this may seem harsh, it is probably a better alternative than declaring chapter 7 bankruptcy and losing all that you have. Remember that cash flow is king and the king makes the rules. *Don't run out of money.*

10.7 Types of Stock and Value Protection

By this point, you may have discerned that I'm not the biggest supporter of going to your lawyer every time you have any decision to make. However, selling stock for equity is not one of those times. Once you start to take anyone's money except your own, you need to proceed on solid legal ground. So, before you collect any capital, visit your lawyer and have an in depth conversation of your capital-raising goals for both the near and long term. I cannot emphasize enough that you need to be prudent, cautious, and thorough and to exercise full disclosure with your lawyer in this regard. Having a securities fraud on your record is not where you want to end up.

Selling equity in your company means you are selling ownership in the form of stock shares or member units (in an LLC; LLC member units are also commonly called shares). There are two kinds of stock, common and preferred. Common stock comes last in terms of liquidation preference. If a company is liquidated and sells all of its assets, the debtors are paid off first, then the preferred shareholders, and then finally (if there is any money left) the common shareholders. In a bankruptcy proceeding, sometimes the common shareholders negotiate for a small amount, but they often get nothing at all. Preferred stock is used much more often in a non–publicly traded company. You can essentially think of preferred stock as being almost similar to debt, because the defined preferences generally include how stockholders are reimbursed for their original investment before common stock shareholders. Preferences are detailed further later in this section.

Stock Value Protection Methods and Antidilution

When a company issues additional shares of stock in a new financing round, investors in previous rounds find themselves with ownership of a smaller percentage of the company, unless they continue to participate by buying additional shares. As a result, investors will often insist on some antidilution provisions be added as a condition of having purchased their initial shares, making these shares preferred by this mandated condition. This means that these shareholders will be issued additional shares without cost so that they own the same percentage of the company after each subsequent round of financing is completed.

Antidilution clauses may end your company if you need to raise more money. How? For example, a current investor has an antidilution clause and declines to make any additional investments in the company, but your company needs cash. New investors say they will not invest unless they own a certain percentage of the company, which is not compatible with your other non-investing members with the antidilution clause. This stalemate can end your company because of a lack of cash, and while this makes absolutely no sense, some investors will do this to you. Be careful who your investors are.

In any financing involving preferred stock, the preferred shares convert to common as follows:

$$Number\ Shares\ Common = \frac{Original\ \$\ invested}{Conversion\ Price}. \qquad (10.15)$$

For example, $3 million is invested at a conversion price of $2; the preferred shares will convert to 1.5 million shares of common stock. The investors will purchase 1.5 million shares of preferred at $2 per share, and the preferred will convert to common at a one-to-one basis. This ratio may change, however, if price protection antidilution measures are in place.

In a successful company, each round of stock will usually sell for more than the previous round. If a company runs into trouble, yet must raise cash with a stock offering, the stock price may be less than the previous round. This is known as ***down-round financing***, which is the dreaded scenario for current investors who no longer choose to participate in investing additional cash to protect their current ownership position.

The down round goes something like this. You needed cash, so you previously had some outside investors take equity in exchange for giving you cash. You're happy because you (founding team) still own 90% of the company. A year passes and now you are short of cash for a whole list of reasons and unexpected circumstances. So, you go back to the investors with this problem. The investors say, "Fine, I'll give you so many dollars at a share price of 10% of X," where X was the share price for the last round. This is the classic down-round scenario. This is a favorite technique of some investors to take control of your company.

> **Down Round:** Current investors will ask if this will be a "good money after bad" deal and may simply write off their previous investment as a bad decision.

So, what's the rule? Don't run out of money. When you do, you are at the mercy of your investors and generally they have no mercy. It can be even worse, for example, when merciless investors say that because the new share price is so much lower than what they paid previously, it is only logical that they paid too much in the last round and they want an immediate adjustment by having the shares reissued at the new price for the earlier investment. If you don't agree, then they say, "Fine, keep your company, and I'll pick it up for nothing at bankruptcy." You may have thought the investors were your friends, and maybe they are, but when you get squeezed by the investors with these types of alternatives, they like to use Michael Corleone's line from *The Godfather*: "It's nothing personal; it's just business."

Your only defense to the down-round attack is to have other interested investors. But, eventually, you will have to lower the stock price until some investors are willing to invest. An extreme down round is a way to transfer control of the company to only those who are willing to continue to invest cash; for example, a new share price might be $1/100$ or $1/1,000$ of the previous round's share price. When this happens, current investors will bring out the adage to guide their decision as to whether this will be a "throwing good money after bad" deal and simply write off their previous investment.

PRICE PROTECTION

The following example demonstrates the need for price protection for an investor in an early-stage company. The company has the following structure: 1 million shares of common stock exist followed by Series A preferred shareholders who purchased 1 million shares at $1 per share. The conversion price is $1 per share. Expressing the Series A preferred in an as-converted-to-common basis:

Common	1,000,000	shares	50%
Series A Preferred	1,000,000	shares	50%

At the end of the above round of financing, the company has a value of $2 million. Now, the company runs into trouble and must find financing to meet payroll and other pressing obligations of $50,000. In this extreme example, the company is able to sell 500,000 shares for only $0.10 per share, but you desperately needed the $50,000, so you do the deal. The previous Series A investor does *not* invest further.

WITHOUT PRICE PROTECTION

Common	1,000,000	shares	40%
Series A	1,000,000	shares	40%
Series B	500,000	shares	20%

Without price protection, the Series A investor finds him- or herself with 40% (1 million shares of a total issued of 2.5 million) of a company now only worth $250,000. His or her investment has decreased in value from $1 million to $100,000 and, hence, the term *down round*. (Note that the new investor with the new Series B shares might be the same owner of the Series A stock.)

FULL RATCHET PROTECTION

Under full ratchet protection, the conversion price of the previous round of preferred stock is altered using the formula:

$$NCP = OCP \cdot \frac{NOP}{IOP},\tag{10.16}$$

where
  NCP = new conversion price,
  OCP = old conversion price,
  NOP = new offering price, and
  IOP = initial offering price.

Assume the same situation as our previous example, except consider a full ratchet price protection on Series A stock. In this example, we have

$$NCP = \frac{\$1.00}{share} \cdot \frac{\$0.10/share}{\$1.00/share} = \frac{\$0.10}{share}.$$

Hence, the Series A shares will convert to 10 million shares of common stock (to make the original value of stock to be identical at the lower price, or $1 million), and the post-financing situation is

Common	1,000,000	shares	8.7%
Series A	10,000,000	shares	87%
Series B	500,000	shares	4.3%

You can see the extreme effect on the founders, who hold the common stock. After the full ratchet protection has kicked in, they own less than 10% (8.7%

in fact) of the company. The founders have been "diluted out." Now, seeing the potential impact of the full ratchet protection on the number of issued shares may eliminate any interest the new investor has in purchasing shares. Note that if merely one share of stock had been issued at the low price, the full ratchet protection would have kicked in. Full ratchet financing is reserved for very risky ventures in which the founders have not yet proved their ability to move a company to success. You should strive to have some time limit on the full ratchet provision, for example, to end after twelve months or to end after the occurrence of a significant event, such as the issuance of a patent or some other significant milestone, which has ostensibly removed some identified risk factor.

WEIGHTED AVERAGE PROTECTION

This is the most common type of price protection. The adjustment to the conversion price is weighted to reflect the number of shares before the down-money financing and the number of shares outstanding after. The formula for the conversion price is

$$NCP = OCP \left[\frac{OB + MI/OCP}{OB + SI} \right], \tag{10.17}$$

where
NCP = new conversion price, \$/share;
OCP = old conversion price, \$/share;
MI = money invested, \$;
OB = total number of shares outstanding (including common) before this round; and
SI = number of shares issued in new round.

Let us assume that Series A preferred was subject to weighted average price protection. In this case, having issued 500,000 new shares at some lower price, we find that

$$NCP = \frac{\$1}{share} \cdot \left[\frac{2,000,000 + \$50,000/\$1}{2,000,000 + 50,000} \right] = \frac{\$0.82}{share}.$$

Then, Series A will convert to \$1,000,000/\$0.82/share or 1,219,512 shares, common share value is *not* affected, giving

Common	1,000,000	shares	36.77%
Series A	1,219,512	shares	44.84%
Series B	500,000	shares	18.39%

So, the Series A investors have seen some protection, at the expense of the common shareholders, but the effect is not as severe as in the full ratchet case.

STRUCTURAL ANTIDILUTION

In the event of a stock split, reverse split, or stock dividend, the preferred share-holders or common can preserve their ownership positions by requiring structural antidilution as part of their preferences or as a general condition on common stock. Structural antidilution is (and should be) included in all equity purchase agreements. For example, a stock split might be necessary if the pre-valuation revealed that the stock was worth $1,000 per share. Because most companies like to sell shares in the range of $10 to $20 per share (particularly if this were an IPO event), the company might do a 100:1 stock split prior to the stock offering to make the share price in the previous example $10/share, meaning current share-holders would receive 100 times the number of shares they currently possess. No one loses value in this process, so no one should argue against it. The key here is the shareholders after the splits or reverse splits still own the same percentage of outstanding shares; that is, ownership percentages did not change.

Preferred Stocks

Preferred shareholders are just that, their stocks are preferred over owners of common stock by assigning some set of defined privileges or so-called prefer-ences. Venture capitalists almost always demand preferred shares. Angel inves-tors are much less likely to demand preferred stock.

There is no common formula for defining the preferences attached to a round of preferred stock, except the round will have some designator attached to it. Once you issue a round of preferred stock, then each subsequent round will almost always give birth to another series of preferred stock. This series of stock issuances may be listed as Series A Preferred, Series B Preferred, and so on. Others like to use the letter *A* followed by a number, Series A1, Series A2, and so on. In any event, each series of preferred stock can and often does have different rights, features, and provisions associated with it. Also, you can assume that new investors will try to negotiate that their preferences must be fulfilled before any previously granted preferences are met. These negotiations can be extremely dif-ficult, because some previously granted preferences may make it impossible to grant any additional preferences.

Holders of the preferred stock will usually require that their stock be converted from preferred to common in the event of any of the following:

- IPO
- Sale of the company
- At the option of the holder or when some condition is met as defined by the holder or the company

Some venture capitalists prefer to keep the preferred stock preferences clean and simple, for example a preference is given that the current round of investors will receive their investment dollars back before and profits are distributed to common shareholders in any fashion. They will argue that elaborate downside protection schemes complicates the deal unnecessarily, which can *compromise* the capability of securing future rounds of financing if needed. I agree! Try to avoid issuing preferred stock whenever possible.

Other venture capitalists will insist on a full range of downside protections and antidilution features. They claim that they are putting up the bulk of the cash money for the business (the founders have typically contributed much less than they are about to commit), and they deserve to have protection against losing all of their cash due to managerial incompetence (referring to you of course!).

Venture capitalists like to hold senior preferred stock to "monitor" their companies in a fiduciary manner. The senior adjective means that their stock will always overrule any other preferred stock privileges from other stock issuances. *Monitor* is a complex word. It means closely watch and influence the decisions of management. Without senior securities, it is difficult for the investors to maintain their board seats or have other privileges not shared by common stockholders. And if venture capitalists are unable to exert leverage on their company managers, many of the advantages inherent to venture capital investing would be eliminated. If a venture capitalist cannot make or influence management changes when things go wrong, then venture capital investing loses its leverage and control. Entrepreneurs (our side) like to call this whole thing "micro-managing" – and we don't like it.

Defining Preferred Stock Preferences

LIQUIDATION PREFERENCE

Liquidation is usually defined very broadly as the sale of the company or substantially all of the company's assets. **Liquidation preference** generally means that investors receive their original investment back upon liquidation of the company, nothing more. If Series A Preferred was sold to investors for $0.50 per share, it will liquidate at $0.50 per share. That is, if the Series A Preferred shareholders invested $2 million for 40% of the company and there were no other series of preferred stock issued, the holders of the Series A Preferred would receive the first $2 million in cash distributed to the shareholders, in preference to the common shareholders.

If the company were sold for more than $5 million (after taking any debt into account), it would make sense for the Preferred Series A shareholders to convert to common stock (because 40% of $5 million plus is greater than $2 million). For example, if the company were sold for $9 million, the Series A Preferred shareholders would convert to common and share in the distribution of the cash to all common shareholders, in this case receiving 40% of $9 million, or $3.6 million. Remember, the preferences can be *written in any manner* that is negotiated between the company owners and the new investor.

PARTICIPATING PREFERRED STOCK

When a company is sold at a modest price, the deal is referred to as a "sideways" deal. Such deals do not make a venture capitalist very happy (remember that they need high returns), and they try to figure out a way to prevent them from happening. For example, going back to the previous example, assume that the Series A Preferred investors have put in $2 million for 40% of the company, and the company is sold five years later for $7 million. The Series A investors will convert to common shares and receive $2.8 million, representing a minimal return of only 7% per year. In contrast, assume that the founding team had contributed a total of $50,000 in original cash investment. They would get $4.2 million (60% of $7 million) or a return of more than 142%. In this case, you (the founders) say, "Great," and say that this is only reasonable for the thousands of hours of sweat equity you have poured into the company.

The previous scenario drives venture capitalists crazy, since they are of the opinion that they created the success and therefore they should receive the majority of any financial upside. So, *participating preferred stock* was designed to close this loophole in the event of a *sideways deal*. Here, the holders of the participating preferred stock are paid their original investment first from the sale proceeds. Then, they get to participate in the common stock distribution as well, by converting to common stock. The amount of common stock they get is the same as in the simple case; that is, all preferred shares convert on a one-to-one basis to common shares. This is also called "getting your bait back."

In this example, the Series A Preferred shareholders are given the $2 million cash of their original investment returned to them (the preferred part) from the $7 million paid for the company, and their preferred shares are also then converted to common (the participating part) shares. As a result, they receive 40% of the remaining $5 million distribution as well, for a total payback of $4 million and a ROI of 32%. Not an astronomical return, but not too bad, and much better than the 7% return in the first part of this example.

Sometimes it is written that in an exit deal above a certain level, the preferred shares just convert to participating shares, and there is no return of the initial investment. For example, preferred shares would be considered as participating if the company was worth less than $30 million (or some other arbitrary value; investors get their initial investment back and the conversion to common shares), but the preferred provision would be inactive if the company was worth more than that amount (converts to common shares without the initial investment being returned before distribution of proceeds to common shareholders).

PARTICIPATING PREFERRED PLUS DIVIDEND

Sometimes there is a cumulative dividend on the preferred stock, where the dividend is typically 6% to 8% per year (defined in the negotiation when the investors purchased equity in the company). This dividend accumulates and is not paid off until the company is sold or successfully executes some other exit strategy, for example, an IPO. Then, the original investment is returned to the preferred shareholders along with the accumulated dividend. This is used to ensure that the investment earns at least a small rate of return.

REDEMPTION RIGHTS

Redemption rights forces a company to repay the investors their initial investment price or some other agreed-to value at some time in the future, sometimes in equal increments over two or three years (seemingly to make the requirement appear to be more reasonable). This innocent little provision can "end" your company! For example, an investor has negotiated a preference that requires you will pay them 25% of their original investment at the end of two years. If you don't have it, the investor can take your company or force other very unpleasant recourses. Advice here is as I have stated elsewhere, *do not run out of money,* which could be caused by some mandatory redemption rights that forces you to be out of money. You can't blame potential investors for asking for redemption rights. It may be their only means *of ensuring* a way of retrieving their investment with a return.

One problem with redemption rights is with investors in future rounds. They might argue with justification that the company will simply take the proceeds of the new financing round to redeem the shares of the investor from a previous round. And, once redemption rights have been given to one series of stock, all subsequent investors will want redemption rights for their stocks as well.

The company founders should try to avoid including redemption terms. If the investors insist upon them, you should try to push out the redemption date well into the future, five to seven years, if possible. Also, you should try to spread out the cash payments to such an investor over two to three years, which may enable you to finance the redemption out of ordinary cash flow. Finally, redemption rights should terminate in sale of the company or an IPO, because they otherwise cause lingering problems for the company's new owners (in the event of a sale) or put downward pressure on the valuation. And consider adding some milestone that when achieved would terminate the redemption right.

> Consider adding some milestone that when achieved would terminate the redemption right.

Right of First Refusal or Preemptive Rights

Right of first refusal or preemptive rights are preferences given to preferred shareholders to share in future offerings to maintain their ownership percentage. For example, consider a company with one preferred investor holding 4 million shares, which is 40% of the company's outstanding stock shares (10 million shares outstanding). A second round of funding, selling 2 million shares for $1 per share is pursued. If the original investor holds the right of first refusal and decides to exercise it, the ownership will look like the following table after the new round has been completed (the series A holder purchases 40% of the new shares):

Common	6,000,000	shares
Series A	4,000,000	shares
Series B purchased by original Series A investor	800,000	shares
Series B, purchased by new investors	1,200,000	shares

Note that the Series A investor purchased 800,000 shares for $800,000. The other investors have purchased 1.2 million shares for $1.2 million. The Series A investor still has 40% of the company. This is often fine, but sometimes, the new investor may have a specific minimum investment threshold and exercising the Series A preemptive rights may dip them below that target threshold. If the new investor wanted to purchase a minimum of $1.5 million worth of shares, the preemptive rights in this scenario of the Series A holder would reduce the available shares for purchase so that the new investor may no longer be interested (as in the original Series A holder's ability to control 40% of the company may prevent the new Series B investors of obtaining some minimum required level of ownership).

Preemptive rights take time during the deal, as the original investors often have fifteen to thirty days to consider whether to participate. These rights can delay any future deals being closed, so company management, in the interest of streamlining future financing rounds, should try to avoid giving preemptive rights if possible. These conditions should all be spelled out in the company's operating agreement or bylaws.

10.8 Term Sheet Definitions

Eventually you hope that your negotiations will bring in capital. The first sign of this is when the prospective investor presents you with a term sheet. The term sheet will outline the specifics of the deal, so you must make sure that you understand this term sheet fully. We give you some of the basic jargon involved and the definitions for commonly used terms. Stock preferences may be part of the term sheet, which were previously discussed and are repeated here. A good site for looking up legal terms is http://www.investorwords.com/.

> Go to your lawyer <u>after</u> you have discussed and understood the terms of the proposed deal, but **before** you agree to anything.

The investor and you (company founders) should take the time to review the term sheet very carefully until you fully understand the terms and the investor's reasons for including them. In this manner, possible misunderstandings may be avoided. After you fully understand (at least you think you do) the term sheet, then go to your lawyer. It's generally not a good rule to start the negotiations using your lawyer, because lawyers tend to overprotect you to the point that the other side may view your conditional terms as being so far from acceptable that the possible investor just gives up before they even start. Angel investors are not looking to spend their time arguing with lawyers over term sheets, and there are a multitude of start-up opportunities that they can choose from.

Two example term sheets are presented in the supplemental materials given on the book's webpage: www.Cambridge.org/Timmons (see Chapter 10 Supplemental Materials – Term sheets). A listing of commonly used phrases and terms seen in Term Sheets are listed in the next section.

Additional Term Sheet Definitions

ADDITIONAL PROVISIONS
Methods of modifying the agreements, means of notifying one another, who will pay legal fees, and other primarily legal items.

AMOUNT OF THE INVESTMENT
A short, succinct description that clearly outlines the amount of capital being invested. Needs to be on every term sheet so that there is no confusion about how much money is being raised in the current round.

CONVERSION PRICE AND TERMS OF THE PREFERRED STOCK
This is where valuation is established as well as the dividend (interest) rate, if any; the date the money is repayable if the company has not done well enough to encourage conversion into common stock; and any other preferences being requested are listed here, if not defined elsewhere. See '*Debenture*'.

CO-SALE AND "TAKE ME ALONG"
This section deals with offers by outsiders to buy the stock of management and with how the entrepreneur must have the offer made to the investors as well. Should they receive a similar offer, they must take the entrepreneurial team along. This prevents either group from bailing out without the other. This is a form of 'piggy-back' rights,

COVENANTS – AFFIRMATIVE
"Dos" are a variety of actions that the entrepreneur agrees to do regularly so long as any of the preferred stock (debentures) is outstanding. These include making dividend (interest) and principal payments in a timely manner, tax payments, and insurance payments when due, as well as submission of financial statements, unaudited and audited, calling board meetings regularly, maintain properties, books, records in good order, and meeting other reasonable requests.

COVENANTS – NEGATIVE

"Don'ts" are a variety of actions that the entrepreneur will not take or permit to occur without the consent of some or all of the preferred stockholders (debenture holders, and it depends upon the wording of the debenture holder's wording). This list can be relatively long or relatively short, depending on several things, including the mutual trust of the parties, the financial health of the company, the number of investors involved, and the desires of the investors' counsel. A standard listing consists of the following "don'ts":

- Change the type of business engaged in
- Pay any common stock dividends
- Incur liens on the company's assets
- Make any loan or guarantee any loan to any person or entity above a minimum dollar amount; same for investment
- Acquire or merge with any other company
- Change the company's capital structure
- Increase officers' salaries above agreed limits
- Incur lease liabilities above some maximum
- Pay employees more than agreed amounts
- Restrict dealings with insiders, such as an officer or director selling a service to the company.

DEBENTURE

A term that refers to unsecured debt or an obligation that has no collateral other than what was defined by the terms of the preferred stock (see Conversion Price definition).

DEFAULT

In the event that the company fails to comply with one or more affirmative or negative covenants, the company is in default and the amount invested by the venture capitalists becomes immediately due and payable. However, there is usually a "cure" period of 30 to 120 days, which can be extended by mutual agreement, to permit the company to take whatever steps are necessary to cure the default.

DEMAND RIGHTS

If you see a term sheet that includes terms called Demand Rights (S-1 and S-3 being the most common), this can require the company to register with the SEC some fraction of the issued shares, subject to defined restrictions and timing of events. Be sure to understand their meaning and ramifications (who pays for the registration) meaning you should seek legal advice in this regard. This can be a fairly expensive process (legal and filing fees). If you are at all unclear on this point in a term sheet, consult your attorney before signing anything.

The investor's threat to invoke their registration rights is the only legal means available to compel a company to go public or to accept some alternative strategy that meets the minority shareholder's interests. As such, the registration right is a very valuable right for the investor and one that founders try to restrict because of the time and cost involved if those rights are invoked – it can easily cost $200,000 or more to file a registration statement. Rarely will registration be implemented against the desires of management, because management is an intimate part of presenting the offering to investment bankers and brokers who will be buying the stock.

PARI PASSU

If there are several rounds of capital funding, there will likely be a number of series of preferred stock. The new money has the leverage, and they may insist on preferential treatment with respect to the current holders of preferred shares. However, for good relations with your existing investors, as well as setting a precedent for further financing rounds, the new investors may consent to treating all preferred shareholders equally, which is known as ***pari passu*** ("of equal step" in Latin), or more particularly defining that one series of equity will have the same rights and privileges as another series of equity.

PIGGYBACK RIGHTS

A piggyback right is the right to participate in an offering initiated by the company. These rights are appropriate for both small and large companies. Piggyback rights guarantee that the holders can participate equally in any type of deal that is being made for the company. This is particularly important to any shareholders who want to liquidate their shares to an outside party during a company stock offering. Also, see definition of 'co-sale'.

PREFERRED STOCK

Will define the preferred terms for this round, if any (previously discussed).

PREDEMPTIVE RIGHTS

Predemptive rights are preferences given to preferred shareholders to share in future offerings to maintain their ownership percentage.

REGISTRATION RIGHTS *(see Demand Rights)*

REPRESENTATIONS AND WARRANTIES

These terms require the entrepreneur to prepare a list of documents, normally exhibits to the contracts, which prove various statements that they put into the business plan or made to the investors during the due-diligence process. These include evidences of incorporation, trademarks, patents, indebtedness, leases, contracts, capital equipment, and other items, evidence of which might give the investors greater comfort. The entrepreneur, on the other hand, may ask the investors to represent and warrant that they are able to provide the time of their senior people, an additional round of financing if necessary, or other items that might give the entrepreneur greater comfort.

SECURITIES

Generally, convertible preferred stock or common stock in the case of venture capital funds or subordinated convertible debentures in the case of SBICs.[7]

10.9 Raising Capital through an IPO

Raising money through an *initial public offering* (IPO) is not relevant to a start-up company, and is not a relevant exit strategy for newer companies. The math is just not in your favor. For example, in 2008, there were only twenty-nine IPOs executed, and a typical year will see less than 100 IPOs executed.[8] In contrast, in recent years the USA has seen around 600,000 new small businesses created each year.[9] However, if you think your company is at a stage at which an IPO merits consideration, then your first step is to contact legal counsel to review the process, because the legalities of this procedure are well beyond the intended scope of this book. Seriously consider all the ramifications of going public, and ask yourself why some public companies actually choose to buy back their stock and return to being a private company. Advantages of going public include stronger capital base, increased financing prospects, executive compensation, owner diversification, and increased company and personal prestige. Disadvantages of going public include trading restrictions, short-term growth pressure, disclosure and lack of confidentiality, costs (both initial and ongoing), and loss of personal benefits.

WHAT IS THE REGISTRATION PROCESS?

Going public requires an S-1 Registration Statement, a carefully crafted document that is prepared by your attorneys and accountants. It requires detailed discussions pertaining to the following:

- Business product, service or markets
- Company information
- Risk factors
- Proceeds use (how are you going to use the money)
- Officers and directors
- Related party transactions

---

[7] Small Business Investment Companies or SBICs are lending and investment firms that are licensed and regulated by the Small Business Administration. The licensing enables them to borrow from the federal government to supplement the private funds of their investors. SBICs prefer investments between $100,000 to $250,000 and typically have much more generous underwriting guidelines than a venture capital firm.

[8] From National Venture Capital Association, 655 North Fort Myer Drive, Suite 850, Arlington, VA 22209 this is reference to the IPO statistics.

[9] From **SBA.gov**, the official website for the U.S. **Small Business Administration**

- Identification of your principal shareholders
- Audited financials

After your registration statement is prepared, it is submitted to the Securities and Exchange Commission (SEC) and various other regulatory bodies for their detailed review. When this process is completed, the management team will do a "road show" to present the company to the stockbrokers who will then sell your stock to the public investors.

How much does going public cost? Lots. The following figures are considered minimums, and many larger offerings will have costs that greatly exceed these numbers.

- Legal – $50,000 to $150,000
- Accounting – $20,000 to $75,000
- Audit – $30,000 to $200,000
- Printing – $20,000 to $80,000
- Fees – $10,000 to $30,000

There are also additional costs for underwriter commissions and expenses as well as numerous expenses on the part of the company. Notice that the company usually pays the expenses; the investors (if allowed to participate) are only subject to the underwriter discounts.

10.10 Additional Resources

Avenmore Developments

http://www.avonmoredevelopments.com/aboutba.htm

This site features a concise and useful discussion of angel investors.

Community Development Venture Capital Alliance (CDVCA)

http://www.cdvca.org/

A good example of a fund with a community development spin, the CDVCA site features resources for entrepreneurs and the ability to join the virtual community or "shop" your business.

Site for Looking Up Legal Terms

http://www.investorwords.com/

10.11 Video Clips

Directions for viewing clips on the Prendismo Collection (formerly from www.eclips.cornell.edu):

- Visit http://prendismo.com/collection/. (You must subscribe with the site for full access to all features.)

- Click on the **Subscribe** at the top of the site. Students receive reduced rates.
- Once you have subscribed, type in the name or title of the clip in the **Search** at the top right corner of the screen.
- Click on the clip you wish to view.
- Click on the play icon to view the clip. If you have not become a subscriber, you will only view the first 20 seconds of the clip before being prompted to log in.

1. Laurie Linn (Communique Design & Marketing)

   - Fifteen years of experience in marketing consultation for major financial institutions
   - Has worked with Chase, Bank One, and smaller financial institutions around the world
   - Discusses challenges of having her husband as her business partner
   - Title of video clip: "Laurie Linn Discusses First Business Plan, Finding Funding and Creating a Relationship with a Financial Team"

2. Kevin McGovern (McGovern and Associates)

   - Chairman and CEO of McGovern and Associates
   - Discusses professional and education background and decision to attend law school
   - Discusses international business, entrepreneurship, and specific examples from SoBe beverage company start-up
   - Undergraduate degree from Cornell University and law degree from St. John's University
   - Title of video clip: "Kevin McGovern Discusses Sources of Funding"

3. Dan Miller (The Roda Group)

   - Cofounder and managing director of the Roda Group – a venture capital company
   - Discusses founding of the Roda Group and professional experience prior to starting venture capital firm
   - Discusses networking, marketing, and factors for success in business and entrepreneurship
   - Undergraduate degree from Cornell; master's degree from Stanford
   - Title of video clip: "Dan Miller Shares Thoughts on Venture Capital Firms"

10.12 References and Bibliography

Arkebauer, James B. and Ron Schultz. *Going Public: Everything You Need to Know to Take Your Company Public Including Internet Direct Public Offerings (DPOs)*. Chicago: Dearborn Financial Publishing, 1998.

Black, Fisher, and Myran Scholes. "The Pricing of Options and Corporate Liabilities. *Journal of Political Economy* 81, no. 3 (1973): 684–654.

Nesheim, John L. *High Tech Start Up: The Complete Handbook for Creating Successful New High Tech Companies*. Rev. and updated ed. New York: Free Press, a Division of Simon & Schuster, 2000.

Render, Barry, and R. Stair. *Quantitative Analysis for Management*. Upper Saddle River, NJ : Prentice Hall,, 1977. See chapter 3.

Silver, A.D. *Venture Capital: The Complete Guide for Investors*. Hoboken, NJ: John Wiley & Sons, 1985.

10.13 Classroom Exercises

1. Why is so much material presented on valuation methods?
2. What can go wrong when you provide an antidilution restriction to a new investor (now or in the future)?
3. When might it be a good idea to reduce your ownership from 51% to, say, 25%?
4. When is an IPO a realistic exit strategy?

10.14 Engineering Economics

Internal Rates of Return (IRR)

Translating your pro forma statements into a single statistic represented by rate of return takes a few steps to arrive at the answer. There are two ways most generally used to analyze this: first is using the internal rate of return analysis, or IRR (there is an Excel function called IRR to do this for you). **IRR** is defined as being the interest rate that would result in a net present value (NPV) of zero for today's investment over the period that the money was invested, in the current sample, five years. This means the value of the revenues generated by the investment is balanced by the expenses of the investment including interest costs (time value of money of the investment).

Closely related to IRR is **XIRR**, which is still by definition the interest rate that results in a zero NPV. Fortunately, Excel has functions that calculate both IRR and XIRR. IRRs are reserved for cases in which the payment periods are constant periods such as end of the year, whereas XIRR is a function that returns the internal rate of return where the schedule of cash flows is not necessarily periodic.

The other way a project is analyzed is by comparing the rate of return (ROR) to some internally defined minimal rate of return that your company expects on all investments, called **Minimum Attractive Rate of Return** (MARR). Using your company's established or targeted MARR, you then calculate the net present worth or value (NPW or NPV) of the project and if the NPW is positive, then you can deem the project worthy of your company's investment. A negative NPW means you'd reject the project unless there is some other strategic reason that is not solely financial.

Examples: How To Calculate Rate of Returns (ROR) on an Investment

EXAMPLE 10.5

Assume that our high-tech fish farm Fingerlakes Aquaculture has presented an investment opportunity to potential investors that shows the company will be worth five times its current valuation at the end of five years. The current round of investment will result in 1 million shares of stock being outstanding and the share price currently asked is $3 per share. This means the book value of the company at the end of this round would be $3 million. So, in five years, the pro forma analysis indicates the company is projected to be worth $15 million. So, if an investor had purchased preferred shares that required the company to redeem their shares for five times his investment, what rate of return would the investor realize? So, in equation form we have

$$P = \frac{A_0}{(1+i)^0} + \frac{A_1}{(1+i)^1} + \cdots + \frac{A_5}{(1+i)^5}. \qquad (10.18)$$

The investor is not to receive any payments until the fifth year, so equation (10.18) reduces to

$$P = \frac{A_s}{(1+i)^s} \quad or \quad 3,000,000 = \frac{15,000,000}{(1+i)^s} \quad or \quad i = (5)\frac{1}{5} - 1.0.$$

Solving for i, we find

$$i = (5)^{\frac{1}{5}} - 1 \quad or \quad i = 1.38 - 1.00 = 0.38 \quad or \quad 38\%.$$

Thus, in the preceding example, the investor would have received an IRR or Rate of Return (ROR) on their investment of 38%.

EXAMPLE 10.6

Your engineering team suggests that you install a new plastic moulding line so you do not have to outsource this particular component. The cost of the machine is $500,000. It has a four-year life and no salvage value. The yearly maintenance is expected to be $50,000. The moulding line will give you a net savings of $200,000 a year over the four-year period, not considering the maintenance costs. Your company's MARR is 20%. Should you buy the moulding line?

Answer:

Set this up in the following Excel format:

	Year 0	Year 1	Year 2	Year 3	Year 4
Cash Flows	$ (500,000)				
Net Savings on replacing out-source		$	200,000 $	2 00,000 $	200,000 $ 200,000
Additional Maintenance		$	50,000 $	50,000 $	50,000 $ 50,000
Net Cash Flow	$	(500,000) $	150,000 $	1 50,000 $	150,000 $ 150,000
MARR		20%			
		($93,074.85) NPW of cash flow			

Because the NPW is negative, you'd reject this project. By calculator, you'd discount each year back to time zero and then sum the results to obtain the NPW; for example,

$$NPW = -500,000 + 150,000 \ (P/F, 20\%, 1 \text{ year})$$
$$+ 150,000 \ (P/F, 20\%, 1 \text{ year}) + 150,000 \ (P/F, 20\%, 2 \text{ years})$$
$$+ 150,000 \ (P/F, 20\%, 3 \text{ years}) + 150,000 \ (P/F, 20\%, 4 \text{ years}).$$

10.15 Problems

1. Your company was formed with 1 million authorized shares. To date, you have sold 400,000 common shares for $1/share. In the next round, you are going to sell 100,000 common shares at $10/share. After this round is complete, what is the post-value of the company?

2. Continuing with the question 1, in which 1 million shares were authorized shares, after an investor has purchased 100,000 common shares, what is the smallest percentage of the company the investor can own in the future in a real world situation assuming there are not any nondilution clauses in their purchase of shares?

3. You had previously sold 1 million shares to George, an investor, for $1/share, and he had you include a full ratchet clause. Your company needs some more cash, so you sell 1 million shares at $0.20/share (company had dramatically decreased in value). What are the ramifications of this event on George's position?

4. Name two of the three rules for a venture capital investor.

5. You are trying to raise $1 million in a new round of financing to do your company launch after your initial two-year period (start-up) of refining the product, developing some customers, and so on. Justify some pre-value based upon the following information.

	Start-up years	Year 1	Year 2	Year 3
Sales	$25,000	$100,000	$500,000	$2 million
EBITDA	-$175,000	$10,000	$100,000	$500,000

6. You were a super-successful entrepreneur having launched a company right out of college and after 5 years you sold off your start-up for $100 million. (Probably because of the entrepreneurship course you took !!). Now you are an angel investor. You are pitched with the following:

   You invest $1 million in a start-up and 5 years later your payout is $5M. For this simple case, what is your IRR on this investment?

7. Now, you the angel investor are looking at the following possible investment in a "land-based" salmon farm in Vancouver Island, British Columbia, Canada. After much negotiation, you have negotiated the following terms for a deal:

Pre-money valuation	$	4,000,000
Investment	$	2,000,000
% ownership		33.33%
Preferred stock paying a dividend @		10.0%
Company predicted EBITDA year 5	$	6,000,000
Corporate valuations year 5	$	30,000,000
Ratio to EBITDA		5.00

The final key term you have negotiated is redemptive rights that requires the founders to buy back your stock at the valuation of $30M, in addition to the yearly dividends they are paying you (10% a year on your initial investment of $2M). What is your anticipated IRR if they are able to fulfil the obligations of this stock deal?

8. If the company in question 7, cannot pay you the $10,000,000 redemptive right, what happens next?

9. Your indoor fish company has been considering installing an automatic feeding system to replace the current hand-feeding approach. The automatic feeding system will cost $150,000 and will have a four-year life, zero salvage value, and annual maintenance costs of $20,000. Installing the feeding system will allow you to reduce labor costs by $80,000 per year. Your company's MARR is 20%. Should you buy the automatic feeding system?

10. You've been very successful with your entrepreneurial career, and you do not want to give money directly to your children, but you would like to ensure that your grandchildren and their children have the privilege of a first-class college education. You expect and want to plan for eight grandchildren (that's the number you currently have) to receive $300,000 for their college funds. These first eight grandchildren will start going to college in ten years. How much money should you set aside now to ensure future generations all receive this benefit (i.e., eight scholarship funds), assuming that the generations will be spaced thirty years apart. You expect the investment fund to provide 8% per year annual return.

11 Negotiation

In business, you don't get what you deserve, you get what you negotiate.

Chester L. Karrass

You must never try to make all the money that's in a deal. Let the other fellow make some money too, because if you have a reputation for always making all the money, you won't have many deals.

J. Paul Getty

The fellow who says he'll meet you halfway, usually thinks he's standing on the dividing line.

Orlando A. Battista

11.0 Entrepreneur's Diary

Like it or not, you often will be negotiating something in life, and always will be negotiating something in an entrepreneurial pursuit. The negotiation could be direct or indirect, obvious or subtle, but as an entrepreneur, you always will be negotiating.

As just one example, my colleague Michael (first author of this text) shared with me a very important lesson in this regard from one of the angel investors that was his first major outside investor, Peter. Peter was seventy-one years old when he first became involved in my friend's commercial fish farming company. He was the master at just about everything to do with starting a new business. Peter had left a Big 4 accounting firm (probably was the Big 8 back then) to launch his entrepreneurial career, when he had four children and a wife to support at the time. Among other deal points, Peter negotiated with Michael to purchase a 20% equity stake in his company in exchange for Peter's cash investment. Michael violated a soon-to-be-learned rule, one that Peter was about to teach him the hard way. Peter had originally agreed to invest in the company for 18%, but later simply said that he'd feel a lot better if my friend could "round up" this figure to 20%. It wasn't necessary, but it would be "nice." Wanting to get the deal done, given that the parties were so close, and concerned he might otherwise disappoint or aggravate Peter, Michael said, "OK." Michael didn't ask what he would get for bumping up Peter's equity stake or say something such as, "I'll do that for you if you do *XYZ* in return." Peter later told Michael that this was part of the negotiation, not simply a casual request.

Michael easily could have requested something – even something small – in exchange for the extra 2% equity stake that Peter wanted. After all, was Peter really requesting this just for cleaner math? No, Peter really wanted more ownership. Peter was "nibbling," a common negotiation tactic used often toward the end of the process just to try to gain more value or concessions "for free" before concluding. The rule in this example: In a negotiation, whenever you are willing to, or are about to, give something up, make sure you try to get something in return. My colleague Michael never forgot this lesson.

11.1 Negotiation and Deal Making

The goal of this chapter is to help you understand the negotiation process within an entrepreneurship context.[1] We explore negotiation from the perspective and role of the entrepreneur or his or her team. We do not cover the subject from *A* to *Z*.

Teaching adequate negotiating skills, if not excellent negotiating skills, traditionally has not received the same emphasis as has teaching other business or engineering subjects. This is notwithstanding the fact that businesspeople, engineers, and other professionals negotiate virtually daily. In the entrepreneur's world, virtually everything involves a negotiation. Thus, the subject of negotiations is a large, involved, and essential one. Only recently have entire programs and courses become devoted to this important subject. Unfortunately, exploring it deeply here would be beyond the scope of this book and the course in which the book is used. So, given this reality, the following discussion is purposefully presented in many places with more of a staccato, action-path format, than in a narrative one, to help you gain a quick but accurate feel for the flow of the process. Also, some comments and thoughts are interspersed about various vexing aspects of negotiating. Other resources certainly are available if you would like to explore and practice this subject further. However, you will not learn negotiating only by reading about it. Can you learn to ride a bicycle only by reading a book about it? Similar to bike riding, you actually have to negotiate and experience negotiations firsthand to learn to negotiate. Practice and use negotiating in real settings to get good at it – and to understand your own strengths and limitations as a negotiator.

To start, let's make sure that we all have the same working definition of *negotiation*.

Negotiation is a multilateral, interactive, and often iterative process to accomplish at least two parties' goals. Every word in that definition is important, and

---

[1] © 1992–2013, Rhett L. Weiss, Esq.; all rights reserved.

derives from years of in-the-trenches experience. We'll take a closer look here at some of them.

First, the most critical word in that definition is *process*. Even if sometimes quick, negotiation is not an event. As an entrepreneur, you will see some of your negotiations happen faster than you expected, and others will seem to last for an eternity and to consume every bit of time, energy, thought, money, and patience that you can muster. A negotiation typically comprises a series of events or actions – phone calls, meetings and caucuses (or sidebars), document exchanges, e-mails and/or other correspondence, research and due-diligence reviews, reflections, formal or informal testing and validation, body language, and purposeful silence or other inaction. They all stem from a process that includes each party's mix of one or more strategies and tactics to help the parties achieve their goals. At one end of the spectrum, these strategies can be very carefully designed, articulated, and executed, with elaborate contingency plans (backup strategies or plan Bs). At the other end of the spectrum, these strategies can be simple and obvious, unfolding without much, if any, deliberation. Within each strategy are tactics that can be based on common sense or those that can be counterintuitive in their effectiveness. Because of the range of strategies, tactics, and, frankly, human nature with the vast variations in communication and "chemistry" when people interact, the negotiation process can produce many different and sometimes surprising results.

But the process is largely the same from one transaction to another. It is not only fluid but also has well-defined and recognizable phases or stages, each with important components and traits. More on those stages later. For now, the point is the importance of understanding the process and your place within the process during every negotiation.

Second, the process is *interactive* in two senses of the word. Naturally, one does not negotiate with only oneself. It takes two or more individuals, companies, or other entities to negotiate, just like it does to have a dialogue. Each statement and action by one party during a negotiation may or may not affect the other parties, and may or may not elicit from them a response or other reaction, whether expected or unexpected. Any response or other reaction, in turn, may or may not affect the first party and its "next move." In other words, a true negotiation is a participatory activity, a "contact sport." It does not happen in a vacuum. It cannot be completely scripted. Its result typically cannot be preordained.

Nonetheless, this interactivity produces results that fall within a usually predictable range of general outcomes. It's just that the specific result – and the time and other resources invested to get to that result – may be anything but predictable. An experienced negotiator can ascertain this range of outcomes when possible and can dismiss the "hopes and dreams" that often emerge in the earliest moments of the negotiation process. An experienced negotiator also can estimate well the time that the process should take in a given transaction, whereas a lesser-experienced negotiator often finds that the process takes longer than he or she anticipated. This potential mismatch in the negotiators' time expectations can create an interesting dynamic in the negotiators' relationship with each other.

An interesting exercise is to give different groups of people the same set of facts from which to negotiate a hypothetical transaction. In my experience from running this exercise over many years, the same result has never occurred more than once, and many negotiation teams had to really rush at the end to conclude a deal within the time frame available. By contrast, many teams that finished quickly – or so they thought – often did not reach clear agreement on some important deal points or completely failed to address them at all.

Third, the process is *multilateral*. In other words, the process has at least two channels of activity and communication – because it takes at least two parties to negotiate. The process can be more than two-way communication even though only two parties are directly involved in the negotiation. Why? This is because one or more parties may be subject to the influences, requirements, or control of another entity or individual, who may not be involved directly in the negotiation. For example, a Web-based business start-up may be negotiating to obtain important content for its website from another company. But, the entrepreneur's angel investor is "pulling strings" behind the scenes regarding the entrepreneur's strategic decision making and money that can be spent to obtain that content.

Here is another classic example of the multilateral nature of negotiations, where the number of channels is greater than the number of parties. An entrepreneur's new venture is negotiating to buy a key piece of equipment from a manufacturer (the other party). Simply put, the entrepreneur's goal is to buy this equipment. However, to do so, he or she also needs to raise some money to pay for it. So, the entrepreneur is simultaneously trying to line up financing from a lender. In turn, that lender is asking all sorts of questions, requiring reams of information, and placing numerous requirements on him or her and the venture if the lender is to approve the financing. True, the entrepreneur's venture and the lender likely are having their own negotiation about the financing. But the lender is not negotiating directly with the manufacturer. The lender is not the buyer, is not lending the money to the manufacturer, and does not ever want to operate or otherwise want to be responsible for the equipment. To the contrary, if all goes well, the lender will lend the purchase money to the new venture, the new venture will repay the loan, and the lender will never have to repossess the equipment. Nevertheless, the lender certainly is influencing the buy-sell negotiating process in a very real and practical way. Look at it this way: without the lender's money, the equipment transaction does not happen, at least not unless the manufacturer provides the financing or unless the entrepreneur finds other sources of funds.

The preceding example also highlights the need for some other definitions and clarifications. A *party*, synonymous with *principal,* in negotiation jargon is defined as an individual or entity that is directly involved in *and* is directly affected or benefited by the negotiations, such as a buyer or seller. An *agent* of a party, on the other hand, often is directly involved as a *participant* in negotiations but is not a party itself. An agent is not directly affected or benefited by the negotiation's outcome, except perhaps to the extent that the agent's compensation such as a commission may be contingent on a successful outcome. Agents can include, among others, accountants, professional engineers, attorneys, consultants, other advisors and professionals, brokers (of real estate, equipment, financing, or businesses themselves), and even lenders. Of course, negotiations may include other participants as well, such as the parents or friends of a young entrepreneur. To blur these definitional lines, you will find that the jargon in some regions uses the word *party* or *player* to refer loosely to "everyone around the negotiation table," that is, the principals and all other participants, whether those other participants are agents or have other roles. As mentioned in the following, understanding everyone's roles, and their authority and power derived from those roles, is critical in negotiations.

Fourth, in the entrepreneurial world, the goals can be anything from building a team, raising money, buying or leasing critical equipment (as in an earlier example) or real estate, establishing distribution networks and sales representatives, or getting shelf space at a retail store, to instituting an employee stock ownership plan, creating joint ventures and strategic alliances, and so on.

Negotiation goals practically always are met only with teamwork, effective communication within the team and by the team with the other parties to the negotiation, and adaptability to new or changing information. Further, in sophisticated negotiations, sometimes two or more parties form a team, or alliance, to achieve one or more goals or, at a more granular level, to advance one or more interests, receive one or more deal points, or to solve one or more issues. So, one party may be aligned with a second party on one or more goals and aligned with yet another party on yet others.

Thus, a negotiator needs to think and communicate very clearly about the multilateral aspect of negotiations and the goals that the various parties wish to achieve. Know what questions to ask and to whom, how and when to ask them, how to listen carefully to the answers, what to say and how to say it in response to questions, and what to do with all this information. Successful negotiation involves more listening and thinking than talking.

All that said, negotiation is not the same as persuasion, so please do not confuse the two. In fact, ideally, an artfully designed and implemented negotiation strategy would require no persuasion at all. That's because the negotiated result ideally will be abundantly understandable and acceptable to all sides without the need for any persuasion. Again, that's the ideal. However, we do not live in an ideal world. People, not machines, drive the negotiation process. And, well, people are people.

All sides never are identically situated. They never have the identical information, power, time, motivation, resources, and constraints. Rather, invariably, one party will have more or better data, analysis, experts (or expertise), time, money or other assets, or experience – and newer laptops and cooler mobile phones – than another party. And, to be clear, the negotiators themselves never have identical experience, knowledge, skill, career standing or aspirations, and, very importantly, job consequences for making or not making the deal. Further, as an entrepreneur, you may have a very full plate with a big range of responsibilities just to keep your new company afloat, whereas your counterpart across the table may be tasked with negotiating this deal – and other deals like this one – as his or her full-time job at a long-standing, well-known company.

So, well-timed and well-placed persuasion by the parties sometimes is necessary to get each other past their respective obstacles in the process. Certainly, experience matters in negotiation, and knowing whether or when to persuade comes with experience.

Whether you find yourself needing to persuade your counterpart(s), your standards and your company's standards – or principles or values – such as fairness and proper business ethics should always be maintained in the process. For instance, *do not ever lie* about something in a negotiation, so don't be tempted to misstate or misrepresent a material fact in a negotiation to persuade or otherwise influence another party. Also, maintain your side's values in the other sense of the word, meaning, how your side places a financial or economic valuation on something. For instance, an entrepreneurial initiative based on a new or emerging technology probably values its trade secrets or know-how more than anything else, so do not unnecessarily "sell out," trade, or disclose proprietary information for something else during a negotiation without clear approval and authority from your own side. By the same token, an effective negotiator will strive to learn and understand the other side's standards and valuations, not to mention its goals, information, time, and power to effectuate a successful negotiation, all discussed in more detail below. So, know what your side and the other side values and by how much.

Before we go further, a few terms that repeatedly are used in negotiations should be defined here. An "interest" is what the deal is actually about to a party, that is, what the party really, truly needs or wants from the negotiation. A "position" is what a party *says* or *acts like* it needs or wants from the negotiation. Sometimes a position is coextensive with or matches (that is, accurately represents) an interest; other times, they diverge, intentionally or inadvertently. An "issue" is a party's problem, obstacle, challenge, or constraint that needs to be moved, solved, or eliminated – by someone or something – for that party to reach a successful result in the negotiation.

One final definitional note before exploring the phases or stages of the negotiation process: "YS," used in the following, is the acronym for "your side," meaning the start-up, other company, team, or organization of which you are a member or which you represent. "OS" is the acronym for the "other side," which is a collective, shorthand reference to one or more *parties* (including their entourages) with which YS is negotiating. Notice that "with which" is used in the previous sentence, not "against which." The point is that negotiation should be facilitative and cooperative, not adversarial, whenever possible to increase the probability of success for all involved.

11.2 Four Phases of the Negotiation Process

Phase 1: Preparation

As is the case for most of life's successes, negotiation success starts with proper preparation. Think of preparation as assessing at the outset, and then continually monitoring during the negotiation, three critical components: (1) time *and* timing (T); (2) power, influence, or leverage (P); and (3) information (I). The side with the most T, P and I usually "wins" the most, but one side also usually will not have much more of *all* three components than will the OS. Further, the amount of TPI will vary from side to side and will change within a side during the process. So, think in terms of understanding and adapting to changes in both YS's TPI and the OS's TPI.

TIME AND TIMING

Develop your game plan: determine the steps required, the compromise and concession strategy and tactics, and the planning and goal setting. Avoid time pressure. How much *time* do you have from now till when the deal has to be done? In fact, does the deal need to be done, or can you live without a deal? Are there critical or "drop dead" deadlines along the way? Consider *timing* in the context of the Pareto Principle, or 80/20 rule, as applied to negotiations: 80% of the important points and progress are made during the last 20% of the negotiation and vice versa. So decide when and how often to raise your points or make requests relative to anticipated changes in attitudes, receptivity, information, and time constraints. Here's a useful analogy: time is the period in which a football team has to play during a football game (or during a given increment of it, like a quarter or a two-minute drill), whereas timing is when and how often to make (or attempt to make) certain plays during that period.

Be careful with notions such as "we will work out details later" versus now. Details can bog you down early. They also can blow up a deal when you think that you are near or at the end of the negotiations. Experience will help you sort out when to resolve which details.

POWER, INFLUENCE, OR "LEVERAGE"

Know YS's and the OS's strength and influence. Create leverage with time and knowledge by impressing on the OS that it always has something to gain with a deal and something to lose without a deal, and by increasing the consequences of losing if possible. Create leverage by making the OS believe that you can give it what it wants. Do you or the OS have to make a deal, or can either walk away? Naturally, if you have to make this deal happen, don't let the other side know! In simple terms, you greatly limit your negotiating range, or the range of possible outcomes, by disclosing this or any other constraint (like a time constraint), unless absolutely necessary to disclose it (such as legal requirement to do so).

Understand the different basic types of power: (1) legitimate or title – the power one gets from one's job title and company name on the business card, (2) reward and punishment, (3) referent (or consistency in one's statements or positions), (4) charisma, (5) expertise, (6) situation, and (7) information. Strength in the first four types of power, when combined together, is hard to beat. So, understand the results of combining various types of power. Recognize when power is being exercised on YS and should be exercised by YS. Although the first six factors in the list might be somewhat beyond your control (you either have charisma or you don't), there is no reason you can't excel at the seventh factor, information.

But be careful about thinking you can – or even should – excel in the fifth factor, expertise. Expertise is important but not enough. Frankly, it is usually better to keep your expertise close to the vest, use it sparingly. Plus, the negotiator does not need to be the expert on everything. Bring experts onto the team to address business, legal, or technical (engineering, scientific, or the like) matters as needed. A common temptation of entrepreneurs is to be their side's lead negotiator, figuring that he or she is an expert on the start-up, on the underlying technology that the start-up is commercializing, and so on. But rarely is that same entrepreneur – or anyone else for that matter – an expert on all business, legal, and technical matters that comprise the transaction being negotiated. Effective negotiators are experts in the negotiation process. They do not need to be experts in all aspects of the subject matter at the core of the negotiation. They can and should bring in subject matter experts (or SMEs) as needed. Plus, distributing the team's expertise among various team members enables useful strategies, tactics, and latitude or flexibility, whereas otherwise placing all expertise in just one or two people on a negotiating team would not enable them. Example of tactics that make effective use of distributed expertise, responsibility, and authority include Higher Authority, Good Cop/Bad Cop, and Act Smart/Play Dumb, which are discussed later.

INFORMATION

Always gather and analyze data and other information sufficiently so that they become strengths for YS. Some information is as basic as clearly knowing what you, the entrepreneur, want or need. Do you want or need to raise money? Why? What are your desired results? Do you want to develop a piece of innovative technology and build a business around it or just sell or license the technology once it is developed and proved? What exactly is that technology anyway, what are its differentiators, and what does it do that would have market appeal?

What does the OS want? Know the OS, its proposal, and its goals. An excellent preparation approach is to write the OS's proposal. What would you do, say, or request if you were the OS?

You should try to pin down the OS, by getting as much information out of the OS as possible, while minimizing your information flow to the OS.

There are four different kinds of information. YS

1. has it, and so does or should the OS (for example, common industry knowledge, information in the public domain in the Internet, or other publicly available software languages, codes, or application programming interfaces (APIs) – that is, it's "out there already");
2. has it and can share it with the OS (for example, a business plan, a summary, or excerpts that the OS agrees to protect under a mutually signed nondisclosure agreement or "NDA");
3. has it and cannot share it with the OS (for example, trade secrets or other proprietary information, which is often a valuable leverage point for an entrepreneur's emerging enterprise that may have little else to leverage in its early stages. Depending on the nature and goals of the negotiation, something in class 3 may become available "for the right deal" in class 2, again, protected under an NDA.); or
4. needs it from the OS or another source (likewise, it could be technical know-how from the OS).

Ask questions to get information, organize it into the preceding four classes in order to help you get a sense of the big picture. As you gather information, try to do the following:

- Ask open-ended questions.
- Repeat statements or questions.
- Elicit and summarize restatements, responses, clarifications, reactions, and feelings.
- Talk to people who have dealt with the OS – its competitors, former employees, previous or present vendors/suppliers or customers, its advisors or other agents (to the extent they are at liberty to talk), and the OS's other business relationships. For instance, if you are trying to raise money for your venture or a particular transaction, consider talking with executives in ventures into which the OS, if an equity source, has invested equity capital or to which the OS, if a commercial lender, has extended a loan or line or credit.
- Get your experts and consultants together with theirs; for example, get the parties' peers, engineers, attorneys, and accountants to mix and mingle – they love to swap stories.

- Perform research online and in trade and professional publications, journals, newspapers, and magazines.
- Review applicable documents.
- Learn and understand both YS's and the OS's "culture" and way of doing business (including negotiating), particularly if dealing with foreign parties. This point about culture is critical. Make no assumptions here. This is a huge subject in and of itself. It is particularly important in our global economy. This is not a cliché; rather, this is a reality. Business cultures, norms, laws, and regulations can vary wildly from country to country (and within a country), sometimes in very explicit ways and other times in very subtle ways.

On the flip side is the issue of whether, when, and to what extent YS should answer questions or volunteer information. Having good business ethics is vital, and they should guide your behavior here. But a full discourse on this subject is beyond the scope of this book. Here is a framework for determining whether you should disclose information in response to questions or scenarios arising during the negotiations. A good negotiator is not ethically required do the work for the OS. Similarly, a good negotiator is not ethically required to make the OS's work easier for the OS to do. However, a good negotiator also is fair and honest. The distinctions should not be fuzzy if you maintain the proper core values and rules of conduct you have learned over the years. If you do, then you will make the right judgments. In addition to disclosing or withholding information when legally or contractually required to do so, the following two rules of thumb should suffice for good guidance in the specific context of negotiations.

First, do not volunteer a constraint (such as a time constraint or financial constraint) or other "bad fact" unless you know that the OS is relying on the absence of that fact to the OS's material detriment. For example, let's say you are recruiting a prospective production engineer for your new venture and are in the throes of negotiating the prospect's compensation package, which will include some equity over time in the venture. You know that this prospect, from his or her statements, not only is highly proficient technically and has relevant experience, but also is assuming from some drawings that the venture plans to obtain some of the world's best equipment for the first production line. You also know from his or her repeated statements that the venture's having or getting top-flight equipment is material to his or her decision whether to join your team. But the prospect has not yet asked you to confirm the equipment details. You just listened patiently to all the good reasons that he or she gave for why the venture should have the best equipment, doing nothing to dispossess his or her of these views. Do you disclose that the venture will be lucky to cobble together a bunch of used equipment, in various states of function and disrepair, to get through the first two to three years of production? In a word, yes.

Second, when asked an honest question, such as "What is your deadline?" give an honest answer, and do not give a dishonest one. It also is acceptable to give a range of answers or an answer with "wiggle room." For example, let's say your venture's lease on some office space has expired, that your venture has continued to pay rent but is holding over, and that the rather prickly landlord promptly will throw the venture's furniture and equipment onto the curb if the venture does not vacate by the end of the current month. Meanwhile, you are negotiating with a different landlord of another building for new space. If that second landlord asks you whether you have a certain date by which your venture would like to be in the new space, then answer with the desired date, or, if reasonable, with the additional comment that you can extend the date. You should avoid the temptation to act as if you are under incredible pressure, however real, to make a decision and to act on it quickly (see the Reluctant Buyer tactic discussed in the following). If you say that ideally you want to move your venture in tomorrow, then, naturally, you have just limited the venture's negotiating range, at least with respect to reducing the rent or getting the landlord to make certain improvements or repairs to the premises. Similarly, you do not need to answer the second landlord's question with a statement such as the venture, down to its last paper clip, is about to be thrown out of its current space. Withholding all that information is not dishonest toward the second landlord. After all, he just cares about when your venture wants to move into the new space and, presumably, whether it will pay the rent.

The examples are endless. So, here's the bottom line on disclosure. Entrepreneurs (except perhaps serial, successful entrepreneurs) start with little or no credibility and need much time and many business dealings to build it. One of the best ways to build credibility, and in turn trust, is to negotiate with others fairly and honestly, even when doing so might be painful to you in the short term. The long-term benefits of doing so vastly outweigh any short-term pain, if any. Conversely, an entrepreneur instantly can destroy credibility – for a long time if not forever – either by giving deceitful, deceptive, or downright dishonest or fraudulent answers or by failing to volunteer material information if reasonable people on the OS would view the entrepreneur's nondisclosure as tantamount to a lie, misrepresentation, fraud, or the like. If you have a close judgment call about an honest versus a cagey answer or about disclosure/nondisclosure, please put yourself in the shoes of the OS and imagine being at the receiving end of the situation. What would you, if you were on the OS, consider fair and honest treatment? Just remember everything you have learned up to this point in your life about the value of being fair and honest. Also remember the adage "What goes around comes around." Even if you choose to lie or deceive and actually get away with it – or you think you do – for the time being, then chances are that, sooner or later, the truth will come out, the wronged party will never do business with you again, likely also may use every opportunity to tell others about your conduct, and quite possibly may sue you and/or your venture if that wronged party was damaged by your lie or deception. So, simply put, when asked an honest question, give an honest answer. The nuances and gray areas relate to volunteering information when you have not been asked for it.

Some impetus – a desired transaction, an unmet need, a suddenly recognized opportunity, a potentially rewarding business relationship, or other reason – gives rise to a negotiation. Consciously or unconsciously, you or your team will start gathering information and otherwise will start preparing, as will the OS, to explore the impetus. As the saying goes, the best place to start is at the beginning. Preparation should start at the beginning of the negotiation process. But keep in mind that preparation really never ends. You and the OS will continue to prepare, plan, and revise plans throughout all phases of the negotiation as the parties use, gain, and/or adapt to time, power, and information. You will get to a point in the initial preparation, however, when you feel ready to "do something," that is, to engage the OS to start making tangible progress toward your goals. This brings us to phase 2.

Phase 2: Setting the Stage

Now, YS is making contact with the OS, whether in person, by phone, e-mail, or otherwise. Phase 1 was performed by YS, without direct contact by YS's principals – but possibly with preliminary indirect contact via advisors, agents, or other intermediaries – with the OS. Although YS continues to prepare in phase 2, the fundamental difference in phase 2 compared to phase 1 is that now YS is taking that preparation to the next level by directly engaging with the OS and its principals. Setting the stage involves determining the context, parameters, and possibilities for the negotiation. Essentially in this phase, you are designing, vetting, and determining strategies and tactics for, logistics surrounding, and possible outcomes (desirable and undesirable) from the negotiation. First, determine whether YS and the OS are negotiating or posturing. Determine whether you really have the OS's attention. How likely is an agreement? What happens if negotiations end? What if there is no deal? Generally, always focus on setting up the OS for the next step so that your actions position the OS for an easy acceptance of your deal.

Determine the upside and downside for each side. Determine YS's bottom line and the OS's bottom line. Determine each side's issues, positions, interests, and objectives. Recognize the difference between issues, interests, and positions (these words are not synonymous). What is the real logic or commitment behind the assumed positions? Get to the real issues. Understand the business issues versus nonbusiness issues (such as legal, accounting, or engineering issues). Determine any time constraints for all parties (not just your own). Don't reveal your own unless impossibility/necessity factors are present. Determine any physical constraints, for example, design, space, and bandwidth, and any legal or regulatory constraints, for example, zoning requirements, operating permits, intellectual property license limitations, or regulatory product approvals.

Before entering a negotiation, make sure to establish the ground rules. This may require an NDA (again, short for nondisclosure agreement), pre-negotiation agreement (which, despite its name, often is itself negotiated), term sheet, commitment letter, and/or letter of intent before ever getting to the main, substantive contract that governs and consummates the intended transaction or relationship.

Here's a tip about the NDA: the more time you spend negotiating the NDA, the less time, energy, and interest one party or another may have to negotiate the intended transaction or relationship. I have seen countless situations in which new entrepreneurs, inexperienced businesspeople, or their attorneys or other advisors pour too much energy and sometimes contention into an NDA, whether as a power trip or as a sincere interest in protecting their side. But this often is to the peril of the nascent opportunity or relationship that the parties otherwise want to explore. Most companies have developed a standard form NDA. Your entrepreneurial venture should develop one, too, both for efficiency and to look and act like a "real player." So, unless there really are showstopper issues in a party's NDA, please don't dwell too much on negotiating its provisions (and don't get sucked into a trap of debating or of letting the lawyers debate, whose choice of law or venue provision is better). After all, parties rarely sue each other even if one or the other actually breaches the NDA. NDAs are definitely important to provide both broad coverage of the protected confidential discussions and sensitive information and also valuable legal remedies, if needed, against improper disclosure. Plus, they symbolize and memorialize the parties' good faith of entering into meaningful and serious explorations and negotiations. The more important NDA provisions deal with scope (what discussions, inventions, deals, plans, and other information the NDA actually covers) and the term (how much time, usually measured in years, the parties are bound to the NDA). So, try to get the NDA and other preliminary ground rules done quickly so that the parties can focus on the bigger picture.

The location where the negotiation takes place is actually a form of power, namely, situational power. Here are some questions to consider: Where will the negotiation take place? The entrepreneur's office (which could be a basement or a garage)? The office of the company with which the entrepreneur wants to do business? The office of the entrepreneur's prospective angel investor or venture capital firm? The office of the law firm that represents one or the other? On whose "territory" will this negotiation be conducted? Or neutral territory, that is, belonging to none of the parties, such as a hotel's meeting room? Will the negotiation move around? Will one party or another need to travel long distances to meet? Will they take turns? When will the negotiations take place? What is the time frame? How will the negotiations be conducted, for example, in-person, by phone, by correspondence, or in a combination of these? Who will be involved?

Then there is the actual decision making. Determine who the decision makers are for both sides. Determine each side's advisors and what advice they are providing. Differentiate between professional advice, business advice, and personal advice. Professional advice includes legal (including often critical advice concerning entity formation, intellectual property protection, and the legal and regulatory environment that governs different kinds of business contracts and arrangements), independent appraisal, design, architectural, engineering, scientific or other technical, market, accounting, tax, cultural (such as foreign business customs and norms), and regulatory advice. Business advice includes management, operational, financial (profit and loss, return on investment, return on assets, terms of an equity investment or commercial loan, valuations, and the like), strategic and tactical, deal structuring (which good law firms and accounting firms also can provide), and risk management advice. Personal advice includes personal investment, career, or risk tolerance advice for an individual entrepreneur, for example, personal goals, finances, and lifestyle issues. Determine the extent to which you, the entrepreneur, want advisors or other team members involved in each aspect of the negotiations. Do you want professional and/or business advice? Determine what channels are available to reach the OS's decision makers and advisors, as well as your own, and when to use them.

There are many questions and details to resolve in phase 2 as described earlier. Each one, if handled seriously, represents an opportunity to increase both your chances of a successful outcome and the amount of that success.

Phase 3: Making the Deal

THE PROPOSAL

Now you should be ready to go to what most people probably view the "real negotiation" and to make the deal: to articulate further and confirm each side's goals and deal points, to bargain, to give and take, to explore the extents of various deal points, to offer and counteroffer possible trades of deal points and their extents, to make concessions, to solve issues, and ultimately to reach a workable, acceptable compromise. Indeed, you have been negotiating all along, under the definition we have been using. But, here in phase 3, you bring everything together to earn, make, or win the desired result or as close to it as possible. And this cannot be done well without having artfully passed through the first two phases.

Negotiation, particularly in this phase, is about positioning the OS to believe that it can receive, or to actually receive, and what it wants or optimally can get, while you are going after what you want or optimally can get. It requires a careful and thoughtful balance.

Important note: phase 3 may, and sometimes does, create results that neither side envisioned going into phase 3. While sometimes worse, those results actually may be better than either side or both sides planned or defined as the optimal results in the earlier phases. So, (1) keep your focus on YS's (and the OS's) stated goals, but also (2) keep an open mind to the possibility of creating and achieving something even better. Here's where an entrepreneurial mindset and innovating thinking can be extremely valuable.

In phase 3, for starters, create a proposal (meant here to include a business proposition, investment pitch, offer to buy or otherwise engage in a transaction), however informal or formal the circumstances require. However, creating one is different from presenting one. You may never need to present your proposal to the OS. But have the proposal and a presentation of it prepared. Have them with you at the "negotiation session," meeting, conference call, or other forum in which the parties get together face-to-face literally or figuratively (electronically). The proposal should

- be kept simple,
- present the strongest and best points first, and
- be accompanied by a visual presentation.

Try to get the other side to present its proposal first. Why? You always want to maximize your negotiation range and to avoid limiting or reducing it whenever possible. For all you know, the OS's proposal offers one or more deal points that have greater value or benefit than those contained in your proposal. But if you have to present or propose first to the OS, then also provide a summary outline of the benefits, substantiate the proposal, and offer some bait (enticement) for closure.

COMMUNICATION

If YS really wants to negotiate toward a successful result, then during the entire negotiation process keep all lines of communication open. Know how and when to communicate with others. During phases 3 and 4 (phase 4 will be discussed more in a moment), face-to-face meetings – either live or via Web cam or video-conference – arguably remain the preferred way to negotiate whenever possible, even in this "connected" business world, in that more gets done in a face-to-face meeting during these high-impact phases. True, phone and e-mail communications are virtually unavoidable and, particularly during phases 1 and 2, and periodically in phase 3, can be more efficient and less costly than face-to-face meetings. Try to use phone and e-mail judiciously during phases 1 and 2. Try to minimize using them in phases 3 and 4 because, via phone or e-mail,

- misunderstandings occur easily,
- you can't observe reactions,
- it's easier to say no or be distracted, and
- you can't use documents, presentations, or other visual aids easily (especially via phone, although videoconferencing helps to minimize this problem).

If you must use the phone,

- listen very carefully,
- confirm discussions in writing,
- silence is golden,
- have face-to-face meeting to finalize points if possible, and
- consider recording phone-based negotiation sessions, or the audio portion of negotiation sessions that take place through videoconferencing by using a Web-based recording and transcription service.

If you must use e-mail, mainly avoid the temptation only to skim an incoming message, to "fire back" a response hurriedly, cryptically, or inattentively, and otherwise to treat e-mail communications casually. And be extra careful with features like To, CC, BCC, Reply, and Reply to All. E-mail communications, if sloppily drafted or sent, can really backfire or otherwise come back to haunt you.

If you are having a live or virtual (Web-enabled or videoconference) negotiation session or similar meeting, in which you can see the other participants, there are certain roles that must be played effectively:

- Try to control the agenda (this applies in any form of meeting or communication).
- Keep meetings/communications on track (ditto).
- Determine who will present the arguments or proposals.
- Determine who will respond to or answer questions from the OS.
- Know when to caucus.

In all the preceding scenarios, take notes. Better yet, when possible have someone else take notes for you if you are in the lead role for YS. Having been the lead negotiator on a variety of deals in a variety of settings, I often take some notes – whether about what is transpiring, what is not transpiring, or ideas that are occurring to me in real time. But I know that my own ability to listen, watch, think, ask questions, and make statements can be affected when I am the only one taking notes.

As mentioned earlier, and particularly so as phase 3 progresses, you need to recognize the difference between issues, interests, and positions. Their meanings in a negotiation context are the same as in a regular English dictionary. Get behind the positions to learn the real interests and to address and resolve the real issues, business and professional. Make sure that the appropriate persons address both the business and professional issues. During this phase, expect a few problems to occur:

- Miscommunication
- "Retrading" (when one party brings up an issue again that the other party thought was resolved or dead, or decides to reopen a deal point to try to do better when another aspect of the negotiation is not going as favorably as that party would like)
- Missed deadlines
- "Dropped balls," such as a meeting that had to be cancelled at the last minute because someone was unaware of or unprepared for the meeting

Just keep your composure, your wits about you, and your focus on the positive (for example, accentuate the progress made to date and contain or isolate the problem if possible for further review). You still must expect problems, but they simply need solving. Very few problems are truly "deal killers." So, try these following problem-solving approaches:

- Be creative.
- Know the pros and cons of each alternative solution.

- Know whether you or the OS wants to solve the problem.
- Know when to try to *persuade* the other side.
- Keep sight of the big picture.
- Understand the economic impact of the problem and the solution(s); that is, understand the business ramifications to all parties.

Your negotiating skills will take on their own style over time. Know your own style, and determine the OS's style. But adjust your style as necessary to communicate. You must be able to communicate – and to trade and make concessions – to reach compromise. Be careful because, in addition to the adage "What goes around, comes around," another adage is certainly true in the entrepreneurship world: "it is a small world." Do not compromise or harm your reputation or credibility in the current negotiation for a perceived short-term gain. That invariably becomes a long-term loss. Again, treat the OS well, fairly, and honestly. And always be courteous, act professionally and maturely, and exhibit proper decorum. If you find yourself getting emotionally involved in the negotiations, or others on your team tell you so, be wise enough to take a step back or, if need be, to have someone replace you in your role on the negotiating team. The current OS may be your next partner, vendor, customer, co-venturer, investor, or lender.

Compromise is your most important tool to make a deal. Know when to offer and accept trade-offs and concessions. Know what to offer and accept as trade-offs and concessions. Live with the compromises made within your own team. Do not let any in-fighting or internal friction among your team members be known to the OS; bury it, deeply. You cannot compromise effectively with the OS without first making and adhering to any necessary compromises within your own team. And YS likely will not get most or all of what it wants without giving the OS most or all of what it wants. So trades, compromises, and concessions are progress in a negotiation. They are a good thing, not a bad thing, if you want to get to what YS deems a successful result.

Phase 4: Closing the Deal

By the end of phase 3, YS has framed what it deems the major elements of a successful result as defined by YS's and OS's goals. If this is not the case, then you are not done with Phase 3. If this is the case, then, by now, you have negotiated the deal to a point where your team feels that it is, or is very close to being, acceptable, viable, and attainable – and is consistent with your team's goals with the negotiation. You probably will have been asked to make concessions and trade-offs along the way. And you probably have asked for some from the OS.

In this final phase, phase 4, you try to close the deal, meaning, to conclude the negotiations and to finalize the deal's terms. But, expect some further and seemingly "last minute" concessions or retrading on earlier settled points to bring the process to a conclusion. Don't lose your cool over this. Know when and how to stop conceding and trading off. Phase 4 is about bringing the negotiation process to a logical end, memorializing the agreement or other results reached in the proper documentation or other means appropriate for the situation, and proceeding to consummate the desired deal or relationship. So, phase 4 often requires work beyond simply "writing up the deal."

Sometimes bringing the process to an end is as simple as doing a reality check on the OS. You can do this by restating where you understood the parties to be on the terms, and saying that enough is enough: the deal is fine as it is for all parties; otherwise, people can spend the rest of their lives negotiating over at best marginal returns on their efforts beyond this point. Believe it or not, the OS just may agree with you. Indeed, the OS may have thought, perhaps incorrectly, that *you* were the one trying to keep the negotiations going. Even if things may have not quite gone your way on all fronts, you may be better off stopping now than jeopardizing some or all of the progress up to this point. An essential part of this reality check, therefore, is to frame and measure the results within the stated goals of all sides.

But what if the process seems to have no end in sight? In other words, what if you are in phase 4 but the OS is still in phase 3? Is the OS just so wrapped up in the negotiations that the OS just doesn't know when to stop? Now what? Do you "go to war?" Perhaps you do, but first you need to know when and how to fight. Start (or, better yet, end) the fight by first carefully acknowledging the possibility either to fight about one or more deal points (basically, by digging in your heals while assertively pushing the OS to cave in), to bring some outsider into the fray (such as the OS's current or prospective money source), or to terminate the negotiation. You do not need to be hostile; be nice, matter-of-fact, but firm and deliberate in showing that you would prefer no fight but certainly can carry out your part in a fight. Then, outline a series of actions which are incrementally more consequential as the series progresses, indicate that you are authorized (if true) to take all such actions, and, if the OS still is not "playing nice" and negotiating responsibly at this point, take the first action as immediately and noticeably as possible. However, *do not threaten what you are unprepared to do*. If you go to war correctly, the war will be short – as in, maybe only one battle, so to speak – but effective. The OS may get practical about the negotiations and about concluding them.

> Do not threaten what you are unprepared to do

Or perhaps phase 4 seems to be stuck in a churn because the parties simply are deadlocked on how to resolve one or more issues? If so, then know how to break deadlocks by using some of the following techniques:

1. Set aside an issue for later (maybe it will go away).
2. Use humor.

3. Take a recess or "time out."
4. Invite a reasonable solution from other side.
5. Withdraw an offer that you previously made and is still on the table.
6. Introduce a change in facts, for example, a third-party offer or newly imposed deadline.
7. Simply walk away.

The most powerful negotiating tool, especially in phase 4, is the ability for you to walk away. Even the most obtuse, self-absorbed OS will notice that you are politely, but literally, terminating the discussion, closing your briefcase or laptop, putting your smart phone away, getting up, and physically leaving the room. Always maintain the willpower and ability to walk away from the negotiations. The party that maintains this walk-away power the longest is the party that will gain the most in the negotiation. Conversely, the party that has less walk-away power, or loses it completely, will make more unnecessary concessions to get the deal done, and experienced negotiators for the other parties will recognize and exploit this. Entrepreneurs especially need to be cognizant of this. All too often, they act as if they really need to do the deal, that they have few or no alternatives, and that the life of their young enterprise depends on this deal. Even if true (and in most situations it really is not true), the entrepreneur does not need to reveal this feeling or act in desperation. A desperate negotiator will get a desperate result. To the contrary, the entrepreneur should keep in mind that something – or more than one something – has brought the OS to the table and kept it there. The OS very well may be under its own pressures to get the deal done.

So, know when to walk away. It is that last point at which you are still better off terminating the negotiations without a deal than staying with the negotiations and ending up with an undesirable deal. Once you are passed the point of being able to walk away, you better like the deal you have struck or are about to strike, because you limit your negotiating range when you cannot walk away. Further, as mentioned, an experienced negotiator knows when you cannot walk away and may take full advantage of this. So, likewise, learn when the OS has lost its ability to walk away.

Either way, walking away is not necessarily the same as terminating the negotiation forever. The parties just may need a rest, an opportunity to clear their heads and rethink the goals in light of everything that has transpired during the process up to this point, after which they will be refreshed and recharged to negotiate productively, hopefully to bring the process to a successful conclusion for all parties. And, although "win-win" negotiating concepts are in vogue, win-win negotiating results are not always practical or possible within the context of the various parties' goals, needs, and constraints. So, practically speaking, if a party is willing to accept less than originally sought, that is, a loss, but is nonetheless pleased with the result, then even a win-lose result is a successful one.

Preparation
- Prepare Then, prepare some more
- Develop YS's Game Plan goal setting, values, interests, issues, and positions
- Strategy and Tactics: link, de-links, and concessions
- Assess 3 Critical Components; time, power leverage & information

Setting the Stage
- Contact w/OS searts (building relationship)
- Likelihood of agreementing agreement identifying each side's issues
- Determine/frame everyone's upsides, downsides, goals, values, interests, issues, positions
- Reassess YS's time, leverage and information; Assess OS's time, leverage, and info
- Constraints/parameters+ opportunities (incl. piece of action, future details)
- Set ground rules; protocols: agendas
- Decision making authority vs. responsibility for doing deal

Making the Deal
- Proposals
- Communication watch for charges, signals
- Rapport or 'Chemistry'
- Roles in Communication and at Meetings; watch for changes
- Separating interests and positions; Solving Issues
- Start Compromising Trading (Make/Trade Interim Concessions and Tentative Agreements)

Closing the Deal
- Tentative to Firm Agreements, Interatively
- Conflicting Pressures to Close
- Common Types of Closes
- Getting Through Slow Times/Breaking Deadlocks/Impasses
- Walkway power (Real Ability vs. Theatrical)
- Going to War; How to Fight in Tough Situations
- Retarding
- Confirming/Memorializing the Deal

11.3 Negotiation Basics: Common Tactics

Now you should have a feel for the overall negotiation process and some of the strategic considerations. So, how do you actually work the process, sometimes progressing, sometimes regressing, and sometimes holding firm? What are some useful and generally acceptable *tactics* – distinct actions, inactions, or behaviors (in negotiation jargon, "plays" or "moves") – to use at appropriate times to achieve progress, send a specific signal, or gain (or rebuff) a specific result? Some negotiating tactics seem counterintuitive regarding their viability or likely success, but they generally do work at least periodically when used properly. Of course, they do not all work as expected all the time; nothing does when human nature is involved. A tactic's success is always a function of your ability to execute it, the skill and experience of the OS, the mix of personalities, mannerisms, or "chemistry" among the parties, and, of course, the complexity of the transaction or other subject of the negotiation. Various useful and frequent tactics, but not an exhaustive list of all tactics, for phases 2 through 4 are outlined in the following.

Before delving into the tactics themselves, first understand that you should guide your decision whether and when to employ a particular tactic by the following objectives:

- Agree instead of argue.
- Maximize areas of agreement and commonality.
- Maximize your negotiating range and avoid reducing it unnecessarily.
- Minimize competition and adversity.

Each tactic has one or more counter tactics. Recognize the OS's tactics and let the OS know that you recognize them. This helps to diffuse game playing, builds your credibility as an effective negotiator, and enhances your confidence and power. For instance, if the OS is great with doing flinches, then tell the OS "great job with that flinch." You can do this humorously or with an admiring smile on your face. Yet, don't act smug and don't irritate the OS needlessly with a know-it-all attitude. Sometimes, the OS may not realize that it is using a particular tactic or that it even has a name. Let the OS know, but be a good sport about doing so.

One of the most fundamental and often counterintuitive tactics throughout a negotiation is this: do not show enthusiasm for getting the deal done. A lack of enthusiasm on your part invites accommodation and concession from the OS. Your enthusiasm or eagerness invites demands from the OS, because it will sense you really want the deal and therefore figure you will agree to less favorable terms for YS. Often, the party that cares least wins the most. Why? Because that party is signalling that it may be willing to terminate the negotiations and walk away. Even if you are enthusiastic, try to keep it inside of yourself as much as possible while outwardly being professional, calm, sincere, and matter-of-fact.

To take this point about enthusiasm one step further: silence is golden. It is amazing how things often can turn in your favor if you just say nothing, regardless of whether you like what you just heard or you are disappointed or even angry with the OS's latest proposal. But if you are face-to-face with the OS, look as if you are thinking; do not look dazed and confused. For emphasis, you even can add an occasional shake of your head back and forth, as if to be saying "no" or "I don't think so" or "don't do that." Nonverbal communication can be louder than words. If you are communicating by phone or e-mail, indicate that you are thinking or possibly "rethinking this whole transaction," but otherwise say nothing substantive. You may need to be theatrical at times to make your point. In the United States and several other business cultures (but not all), many negotiators, particularly the inexperienced ones, cannot stand silence. They get so uncomfortable with it that they start to talk again just to break the silence, often with the result of unilaterally softening their positions or making concessions. Remember that the best negotiators listen more than they talk.

> Remember that the best negotiators listen more than they talk.

Tactics for Phase 2 – Setting the Stage

THE LIST

Use a term sheet or other outline of deal points to help frame or set parameters for the negotiation. Better yet, have the OS prepare and present its term sheet or outline. If you are concerned about limiting your negotiating range by using a list, then be vague and/or use ranges. Additionally, the list can have an "escape clause" that says something to the effect of "such other additional terms that we feel are appropriate based on our discussions or due diligence going forward."

RELUCTANT BUYER OR RELUCTANT SELLER

For example, the reluctant buyer says, "We're not sure that we're interested in buying your [the prospective seller's] widget, but, just to be fair to you, what is your very lowest price?" By contrast, the reluctant seller says, "We have a lot of choices [or prospects] under consideration right now, but, just so we know where you [the buyer] are coming from, what is your highest price [or offer]. We'll take it up the chain of command [i.e., we'll take it to a higher authority; see the following discussion] for consideration as well." Note that whether you are the seller or the buyer, something got your prospective buyer or prospective seller, respectively, to the table in the first place. So, typically, you may want to use this tactic to test the seriousness of the OS or validity of its latest proposal. On the other hand, you do not have to believe the tactic if you are at the receiving end of it. And, you do not have to answer the question substantively. If you feel that you should answer the question at all, say that you will have to think about what your lowest price (if you are the seller) or highest price (if you are the buyer) would be and that, in any event, it would depend on many factors. If either side must leave or spend time to think or respond to an offer, the side using the tactic also can impart some final selling points about the added value of the widget or of doing business with that side, to encourage the exiting party to alter its current position/offer by the time it returns to the table.

FLINCHING

If the proposed terms from the OS are just unrealistic to you, or they actually are acceptable to you but you think that the OS can do better, then flinch. This means to visually react if in person, or audibly if over phone, in a surprised and disappointed way when the proposal is made to you. In addition to flinching or instead of flinching, after hearing the OS's proposal, take a break and withdraw, indicating that you will get back to the OS later. This is especially good in a contentious situation that needs some calming-down time or in a situation that is about to become contentious. Either a flinch or silence can be particularly effective in a heated or increasingly tense situation. It makes the OS rethink what it just said or did, without your needing to give a substantive response.

FIRST OFFER/FIRST PRICE

Try to get the OS to make the first offer or state the first price. In other words, avoid being the party that makes the first offer or states the first price. This is a simple and frequently effective tactic. It starts to limit the OS's negotiating range while keeping yours open. Of course, all parties may try to do this, so it may become the negotiation process's version of the game "chicken. Someone eventually has to make the first move. So, when possible, just ask the OS to state its proposal and, if the OS resists, to ask, "Just so we have something to go by [or just to be fair, or just out of curiosity], what do you think your best price [or other term] is?" If you find yourself needing to make the first move, then try to avoid stating what your actual price point or other proposed term is by instead referring to what YS understands as the "market pricing" or your company's or the OS's "target pricing." But, if that does not apply or work, and you really need to say the terms at which you want to buy or sell, as the case may be, then use as broad and flexible a range as possible, with conditions or contingencies attached.

UPROAR OR "WE WANT IT ALL" AND BEST OFFER

Uproar or "we want it all" is the opposite of first offer. It is commonly employed in the business cultures in some major cities, such as New York City. Basically, the party using this tactic acts all blustery and demanding, laying out "this is exactly what we want, yes we want everything our way, and we won't take anything less [or won't give any more, as the case may be]." A related tactic, the best offer tactic, is a gentler way to make the same point as does the "we want it all" tactic but without the uproarious theatrics. Basically, the party using best offer is saying, "Just to be clear [or just to make sure you understand where my team is coming from], our best offer [selling price and other terms, or purchase price and other terms, as the case may be] is *XYZ*." If you use either tactic, or if you are at the receiving end of either tactic, be prepared to calmly and patiently pick apart what you just heard or else to walk away. The OS, if using either tactic, may settle down if you just wait out the performance (remember that silence is golden). If you use either tactic and the OS walks away, then the OS may reflect on your proposal and may call you in a couple of days either to accept your proposal or to make a counterproposal. If the OS makes a counterproposal, just forget that you tried either tactic (after all, the OS obviously has forgotten it) and move on with the process.

ACT SMART/PLAY DUMB

Remember an earlier point about doing more listening than talking. This tactic typically is very effective at gaining concessions and/or information. Here, you act smart by playing dumb. Ask a bunch of questions about the OS's proposal, counterproposal, or one or more deal points, even if you fully understand everything the OS presented and have deep technical knowledge of the subject matter at hand. One result might be that the OS volunteers some critical but previously unbeknownst-to-you information, either to be helpful or, more likely, out of frustration or carelessness. Another result might be that the OS actually starts retreating from its own proposal or counterproposal if it has trouble explaining or justifying it to you. At worst, you easily may have gained some clarifications and other valuable information. At best, you now have the OS "negotiating against itself" for the time being. This tactic requires you to leave your ego elsewhere. At this moment in time, do you really care whether the OS thinks that you are a smart entrepreneur, a brilliant engineer, or a leading expert in the subject? You are being extremely smart by playing dumb – by eliciting information and possibly concessions. The OS may not realize how smart you are being at that moment, but later the OS will realize this – once you take advantage of the concessions and/or information gained from this tactic. You will win in the end. Similarly, never underestimate the OS. Assume that it is smarter, savvier, and more experienced than it may be acting at first, especially if the OS is asking you a bunch of questions.

Tactics for Phase 3 – Deal Making

Assume now there is a deal or offer on the table. But, assume also the deal is not acceptable as presented, or the OS, in your judgment, can improve its proposal materially. Now what? In addition to the tactics above, here are some additional tactics, best tried in phase 3, to help make the OS's deal acceptable or better for you or, conversely, to help minimize YS's concessions to make the deal acceptable or better for the OS.

THE VISE OR THE SQUEEZE

If you do not like the OS's current proposal, then try saying, "You'll have to do better than that," instead of feeling compelled to make a counterproposal. Sometimes, simply with that amount of pushback and without YS's substantive counterproposal to the OS's current proposal, the OS actually will "sharpen its pencil." But, other times, a savvy OS will say, "Exactly how much better do we need to do?"

HIGHER AUTHORITY

Earlier in this material, the point was made that, from the negotiation's start, you want to identify and negotiate directly with those people who have the responsibility to negotiate *and* the authority to make decisions, approve issuing the check, sign the documents, and the like. Try as you might to get all the parties' decision makers to the negotiating table for efficiency's sake, this may not and often does not happen, especially the greater the deal's sophistication and the OS's size or experience. The OS may say that it needs to present your offer to a boss, its board, or some other higher authority for review and, if acceptable, approval. So, if the OS insists on taking the offer to a higher authority, then try to appeal to the OS negotiator's ego by asking whether the higher authority typically values and accepts the OS negotiator's advice and recommendations. If the OS negotiator answers no, then ask (preferably politely) why he or she is even involved in the negotiations and whom else you can contact from the OS to involve in the negotiation. But, chances are the OS negotiator will answer either yes or something else similarly positive. If so, then you want to ask what this negotiator plans to recommend to the higher authority to give you the opportunity to refine what exactly the negotiator will be taking back to the higher authority and, just as important, to get the negotiator to take more "ownership" for what happens behinds the scenes within the OS and away from you. Keep negotiating with that representative until he or she acknowledges that he or she will recommend YS's proposal. That representative now is trapped into working implicitly for you. If he or she comes back to you with an outright rejection or with material changes to your proposal by the higher authority, then that representative has just lost a considerable amount of "face" and credibility.

Note that the typical entrepreneur, at least in the early stages of a business venture, may have a hard time using this tactic (but you may be subjected to it frequently). This is because the OS may find implausible the claim that there even is a person, a board, or another entity with more authority than the entrepreneur has. Maybe there is a higher authority to which the entrepreneur reports, but this will be seen as doubtful in the early stages. It is much more likely in the later stages of venture development. So, meanwhile, if you are faced with disbelief that you report to a higher authority, simply state who or what that higher authority is. It could be an angel investor, other early stage investor, or cofounder. Or, if there really is none, then consider substituting an investor, a trusted friend (or family member), and/or a valued advisor (mentor, accountant, attorney, or other outside professional) for the higher authority when using this tactic.

GOOD GUY/BAD GUY OR GOOD COP/BAD COP

This in essence is a variation of Higher Authority. One of the OS negotiators, the one with whom you mainly are dealing, acts like he or she understands if not agrees with your proposal and wants to work to make the deal happen for you. But, this same person, the "good guy," also sends two warning signals. First, a higher authority or co-decision maker (if your main contact is in the role of decision maker and not of subordinate) needs to approve the proposal. Second, the higher authority or co–decision maker is the "bad guy," a person who often is difficult about approvals, may not like these terms and, basically, is not as nice and cooperative as the good guy. Thus, the good guy will attribute any rejection or material changes of your terms to the bad guy, perhaps with apologies or words to the effect that "I tried, but I told you this might happen. My partner [or colleague or boss] is difficult when it comes to these things."

Have you ever bought a car at a dealership? Then the preceding description of this tactic should probably sound familiar: the good guy is the salesperson, and the bad guy is the salesperson's manager. If you are at the receiving end of this tactic, then take two steps. First, be sure to qualify and/or quantify exactly what the concerns of the bad guy are so that you can address them. Get a complete explanation from the good guy about the bad guy's response. If you are not satisfied with the explanation, send the good guy back to the bad guy for more details. Second, insist if possible that the bad guy be present in any future rounds of negotiations or else you will leave or, time and patience permitting, wait until the bad guy is available. Basically, try to get the good guy out of the middle. You can try to use this tactic yourself, but, like with higher authority, the OS may not believe from the outset that there even is a second person to play the role of the bad guy.

SPLITTING THE DIFFERENCE

Many negotiations will come close to agreement but a gap between the parties' positions still exists. Splitting the difference is sometimes used to eliminate the gap. For you, there is a right way and a wrong way to close the gap. The right way is subtle and includes the following steps:

1. Stress that the parties are apart by only a small amount, even if you think that the amount is big. You are trying to minimize the impact that this particular deal point has in the overall picture. This is part of emphasizing agreement and minimizing disagreement.

2. (a) Get the OS to offer to split the difference either equally 50/50 or in some other proportion (basically, back to the First Offer tactic), and to obtain any necessary internal approvals to do the split; (b) then, *without* making the corresponding offer or commitment to the OS to split as well, indicate that you will go to your "higher authority" (or co–-decision maker) or will caucus to discuss the OS's proposed split; (c) then come back saying that the parties are really close – even closer than before – yet are still apart, but do *not* say that YS can do your part of the split that OS offered in step 2(a); and (d) try to get OS to offer to split difference again, that is, to offer a second and deeper split before YS responds substantively to either the first or second offered split.

3. Never offer likewise to split the difference in either step 2(a) *or* step 2(d). Believe it or not, the OS sometimes will "split twice" before you have to "split once." In other words, the OS may split the difference in steps 2(a) and 2(d) before you have to split the difference yourself in, basically, a step 2(e) or afterward. The result is that you are giving up much less than if you had agreed to do your part of the split in step 2(a), 2(b), or 2(c). Not surprisingly, a good negotiator on the OS will see what you are trying to do in step 2(a), 2(b), or 2(c) and either may challenge you on this or may still do one or both splits just to move the deal forward. For instance, the OS's negotiator may have the authority to reduce its negotiating range to or beyond the effect of one or both splits in steps 2(a) and 2(d), so the OS may not particularly care if you are doing your share of the splits in each step.

4. If the OS challenges you about why you came back in step 2(c) with a request for the OS to make another split instead of just doing your corresponding part of the split in step 2(a), then simply point out that you never agreed in the first place to do your corresponding part of the split in step 2(a).

5. This is the nuance in splitting the difference correctly: it's all in separating steps 2(a) and 2(b) instead of combining them, how you handle these steps, and what you say or do not say in these steps. You want to avoid the natural temptation to indicate that, if the OS does the split, then you will do the split simultaneously. In step 2(a), the OS offered to split the difference, and you merely encouraged the OS to get any necessary approvals to do so. But, in step 2(b), you never agreed to get the corresponding commitment from your own side. You merely agreed to take the OS's proposed split back to YS for consideration. This tactic, similar to any other tactic, does not work every time. However, it works more often than your intuition may tell you. And it can save YS some very real money.

6. Another key point here is that splitting the difference, whether done in the nuanced way as described earlier or in the common simultaneous way, is much easier when there are at least two deal points still in discussion. When there is only one deal point to resolve, and the parties naturally are trying to resolve it via splitting the difference, both parties will be more reluctant to split (or to concede more of the gap than the other party concedes). Once the splitting does occur, if it occurs at all, the both parties will feel like they lost, or else one party will feel like the winner and the other like the loser depending on who gave up more ground. Having one or more parties feel like they lost in a negotiation never is a good thing if the other parties want the negotiated result to last. By contrast, with at least two deal points still in play, each party can win one of those points. Please see Trade-Off, discussed next, for more on this principle.

TRADE-OFF

This tactic has a bunch of negotiation methodology wrapped up in it. Essentially, you want to avoid making a concession on a deal point that is material to the OS (but may be less material or even immaterial to YS), simply because the OS asks you to do so, without getting something in return. You otherwise might be inclined to make the concession, because you really do not care about what you are giving up (what is valuable to the OS may not be valuable to you) and/or because you perceive that your concession will lead to progress. Do not think that way. Instead, try to (1) always get something in return for giving up something valuable to the OS and (2) keep at least one other issue or deal point open per party; that is, do not get down to just one issue or deal point to negotiate so that you actually have something to trade and each side can feel that it got its way (or close to it) on at least one issue or deal point.

> **Trade-Off Rule:**
> If you give up something, then get something in return.

Just remember who raised the request for the concession in the first place and explore why. The OS may withdraw its request for the concession when you try to match it against something you want. You may gain insight into what is going on with the OS due to its request for this concession. Also, even if you do not push the trade-off and instead just make the concession, you could do so with the proviso to the OS that you are "banking" that concession. In other words, you will make this concession now, but, later in the negotiation, you may need a concession from the OS in exchange for nothing in return from you, and you will expect the OS to make that concession by recalling what you did for the OS in this current situation. And, if the OS doesn't make that future concession for you, then you will withdraw your current concession for the OS.

So, trade-offs can have some very positive results. First, you might actually get something valuable or useful for your concession. Second, a trade-off raises the value of your concession. Third, a trade-off prevents or minimizes the OS's further whittling away or "nibbling" (that tactic is discussed in the following) at the deal points. Follow this simple rule: If you are ever asked to give up something, try to make sure you get something in return. And keep a record of the concessions and trade-offs. Consider bringing a laptop or a note taker with you to the meeting so that people can see what is being compromised, conceded, and traded. When various parts of a deal start flying back and forth, capturing these agreements quickly on paper or electronically will be much more efficient than trying to remember them later will be.

FUNNY MONEY

Funny money is the phrase used to describe a party's attempt to minimize a difference among the parties on a particular deal point measured in dollars, such as price, interest on a loan, or rent on real estate, by dividing that difference into another unit of measurement. For example, let's say that you are negotiating to rent 100,000 square feet of industrial space for a production and testing facility. Let's also say that you are willing to pay $5,000 less per month than the landlord wants, and that difference is very real and significant to you, given that your new venture is on an extremely tight budget. The landlord may come back to you with a comment such as "Oh come on … we are only talking about a difference of $0.05 per square foot per month. Surely you don't want to blow this deal over a nickel, do you?" Your response should be "Yes I do, if need be. This still adds up to $5,000 per month, and we can't afford that in our budget."

DECOYS AND RED HERRINGS

There is a difference between decoys and red herrings. Be alert for both. Decoys are real but minor issues to the OS, but the OS starts creating a lot of commotion out of them to divert your attention from a real issue. Red herrings are essentially created or concocted so they can be used for a trade-off later in the negotiation. Just calmly and quickly explore their merits, if any, determining whether they are truly consequential to the negotiation. If you determine that they are inconsequential, let the OS know this politely, but firmly, indicate that they are "off the table"; that is, you will not spend any more time or energy on them in this negotiation, and move on. This is all part of "keep your eye on the ball."

NIBBLE

A nibble involves a small, seemingly insignificant request by the OS to make a "tweak" to a deal point that previously the parties fully decided or addressed. This is one of those annoying little tactics that we often endure, just because we do not want to argue about the point. An example is the OS, your prospective new raw materials vendor, asking whether it can ship your raw materials by three-day ground transportation instead of by the two-day ground transportation that you and the OS previously agreed on earlier in the negotiation. Either way, you do not need the raw materials for ten days from when you anticipate placing each order. A reasonable negotiator's inclination is to answer, "Sure. That's fine. It doesn't affect us." A reasonable *and* savvy negotiator will say, "Sure, if you discount the price by $XX to reflect the reduced shipping cost imbedded in the negotiated price, or if you throw in an extra one-month's warranty." In other words, just nibble back. This will tend to stop the OS's attempts at nibbling.

Tactics for Phase 4 – Deal Closing

By now, the deal's major points have been discussed, analyzed, reviewed, and tentatively accepted to be finalized and memorialized in the appropriate way, for example, contract, purchase order, or similar proof of agreement. The deal's minor points also have gone through the same ringer, too. Or have they? In a perfect or near-perfect world, all parties truly are done bargaining and are in complete agreement about the deal, and the only thing tentative about it by phase 4 is simply the need to memorialize it. But we do not live in a perfect or near-perfect world. So, oftentimes, major and minor points start to look as if they are becoming unglued because one party or another cannot leave well enough alone and wants to keep negotiating "just because." Or, one party or another may be caught up in the process and cannot let go. That party may have lost sight of or forgotten its goals but actually has achieved them tentatively through phase 3. Or, the draft documentation raises other issues that simply did not get addressed previously or states one or more deal points in a way that one party thinks is correct but another party thinks is incorrect. The reasons go on and on. Whatever the reason, YS has gotten to a point when, frankly, it is done. You like the deal. You want the negotiation to be over. You want to finalize the deal, document it, and consummate it. You have had enough talk, and now you want to proceed to the day when the money and goods actually change hands, the lease or technology-licensing agreement actually gets signed, the angel investor actually invests some money in exchange for equity, or whatever the contemplated deal is supposed to do. So, here are a couple of additional tactics, building on the ones discussed earlier, to bring the negotiating process to a close and to proceed to consummating and executing the negotiated transaction.

WRITE THE CONTRACT: FAIT ACCOMPLI

YS should take the initiative to write the contract or other memorializing documentation. This is very important for maximum control of the transition from negotiating to having a "real deal" or a "done deal." It will position the details to YS's advantage and will no doubt slant the overall agreement to YS. The OS will need effective advice and skill, plus the necessary resources, to completely un-slant or neutralize the advantages attached to your "controlling the documents." The document drafting likely will produce further negotiating, which is not necessarily a bad thing because you may win more concessions – or YS truly may have forgotten something that the document process highlights. Fait accompli here means that you present the documentation, already signed by you, with the check (deposit, full purchase money, or whatever the requirements would be under the contract once signed by all parties). This puts pressure on the OS just to accept the documentation as you prepared and signed it, to accept the check (or product, content, code, etc., as applicable) that YS just tendered, and leave well enough alone, because the OS now has what the negotiations were all about. If the OS rejects the signed agreement, just cancel or stop payment on the check or otherwise take back what YS tendered. Fait accompli works best either in simple transactions for which you feel that multiple rounds of document drafting is not warranted or in complex transaction for which several rounds of document drafting have occurred already and you want to avoid further rounds.

SET ASIDE

Negotiations may not go as smoothly as you hoped, and there remain one or more lingering issues that are preventing a deal to be completely made. Try the "Set Aside" tactic, which means literally what it says: you simply set aside issues that cannot be resolved. This is an excellent way to break a deadlock. Just "agree to disagree" about those lingering issues for now and move on to resolve other loose ends. If you actually get to the document drafting stage with some set-aside issues still open, then simply indicate (usually in boldface) in the contract either where those issues would go or, if the draft contract already addresses them, that the parties have not yet agreed on them. More often than not, the negotiation's mere momentum at that point will make small set-aside issues go away or will put the parties in a positive and productive frame of mind to resolve big set-aside issues as the documentation is being drafted and distributed for review and comment.

HOT POTATO: SOMEONE ELSE'S PROBLEM

Sometimes an open issue at this point in the process is still unresolved because, frankly, the issue is not one that the parties can resolve among themselves. Oddly enough, they did not quite realize this until now. At other times, it is a bluff. The OS may claim that it cannot resolve or negotiate an issue or deal point because it is someone else's problem when it really is the OS's problem, but you cannot tell which is correct without further investigation. Or, the OS may claim that it cannot handle the issue because it is a "hot potato"; that is, the OS claims that its raising this issue internally in essence would be taboo; that is, it would be contrary to established corporate practice or policy ("we always do it this way" or "we never do it that way"), or basically brings up really bad memories of another deal that that the OS has or had with another party. Or, the OS claims that the issue has to be resolved the way OS is proposing because, otherwise, it will trigger regulatory oversight, legal concerns, or adverse tax consequences. The OS basically is saying that resolving the issue the way YS wants would create all sorts of problems and could blow up the deal at this very late stage. So, whatever the hot potato the OS is trying to throw at YS to catch and solve, you quickly need to inquire into who or what allegedly controls or is influencing the resolution of the issue, hunt down that "owner" by yourself or with the OS, and approach that "owner" for alternatives and an informed resolution. You either will get a good or bad response from the owner or will call the OS's bluff when you say that you want to talk or meet with the hot potato's owner. A hardball counter to the OS's use of Hot Potato is to say, in essence, "Sorry about your [OS's] issue [or constraint, challenge or obstacle], but it's not our [YS's] problem to solve. In fact, it would be inappropriate of us [YS] to get involved with your [OS's] internal [or regulatory or legal] affairs. Good luck solving it. We'll just stand by for the time being and give you N days to resolve it. If you don't have it solve by then, we'll probably need to look at our alternatives. In the meantime, please keep us in the loop."

Along the way, if the OS is bluffing and in fact does own the issue (or the issue was not real or was easy to solve internally), then the bluff will become apparent, and the OS magically will come up with a resolution. Of course, you might want to try using Hot Potato on an issue that you really do not want to deal with yourself, but do not be surprised if the OS likewise chases you around on it to test its validity.

TURN DOWN/WALK AWAY

This tactic has been touched on previously in this material. If the OS will not stop its negotiating or posturing, if nothing seems to be working to bring the process to a close, and if you believe that your further negotiating will lead to no or only marginal benefit, then you may have to resort to the Turn Down/Walk Away tactic. Turn down, that is, reject, whatever the OS's last proposal was; close your briefcase, laptop, or whatever else can symbolize that you are "packing it up"; and walk out of the room, hang up the phone, or otherwise terminate the communication. If you do walk away, try to leave OS with a reason why and with a possible opening to resume the communications, assuming, of course, that YS really does not intend to slam shut the door to further negotiations and lock it on the way out. Avoid walking away just to be theatrical for no apparent or explainable reason.

Remember that maintain your ability to walk away for as long as possible. If you give it up early, or give the OS the impression early that you will not walk away from the negotiation, then you will lose some of the potential gains from the negotiation, because you will have trapped yourself into a tight negotiating range. You may use this tactic more than once in a negotiation. Of course, if you do it too often or for no apparent reason, then you will begin to look like "the boy who cried wolf," and this tactic will lose its effectiveness. On the flip side, be alert if the OS is about to use this tactic on you. Watch for changes in behavior: shuffling papers, closing a briefcase or laptop, uninterested looks, frequent side conversations among the OS but within your presence, frequent caucuses by the OS out of the room, changes in the frequency of e-mails or in the persons included in the e-mails, and the like. Then, do what seems natural at that point (this takes practice and experience), such as politely and earnestly asking the OS about its actions: "Are you leaving? Have you all lost interest in doing a deal? Is there something wrong at your end? Why? If we upset you in any way, we certainly did not mean to do that. I thought that we were close to an agreement. Let's just go about this a different way." Try to use some humor to cut any tension. If you want the negotiation to move forward, encourage the OS to stay, summarize areas of consensus among the parties, and build on them and other progress made to date.

WITHDRAWN OFFER

There are two versions of this tactic. In the first version, one party makes an offer that it feels will address all remaining open issues and points to everyone's satisfaction. The other parties monkey around with it to the point of frustrating the first party or causing the first party to lose its patience. As a result, the first party simply withdraws the offer entirely, and returns to that party's previous offer, with very short or no notice to the other parties. This sometimes has the effect of "sobering up" the other parties and getting them to treat the first party's most recent, and hopefully final, offer more seriously. The other parties may just take a fresh look at the withdrawn offer and decide that it is just fine the way it is, assuming the other parties want to do the deal at all. In the second version, the party making the offer includes a provision that the current offer expires and will be withdrawn (or "taken off the table") if not accepted by the other parties by a specific date and time, at which point the previous offer will be reinstated (or "put back on the table"). This has the effect of keeping the process moving forward and hopefully toward a conclusion without further delay.

There is an important difference between Walk Away and Withdrawn Offer. The Walk Away sends the signal that the negotiations are terminated, at least for now if not forever, without any agreement among the parties. The Withdrawn Offer sends the signal that the negotiations are still alive but that they just took a big step backward unless the OS accepts the current proposal.

THROW-AWAY CONCESSION: SO WHAT?

This is similar to a nibble, but the opposite. You give the OS a throw-away concession – a concession on a deal point or issue that you can easily live without – as a small enticement to conclude the negotiations. You do not ask for or expect anything in exchange from the OS. You may view your concession as a small gift. However, the OS may view either the concession, your gesture in making it, or both as much larger. And this may be just enough to get the OS feeling good about the deal and therefore beyond any remaining concerns about it or desires to continue negotiating. This is the negotiation world's version of throwing a dog a bone. You can further say that this is the last bone that you will throw; that is, this is the last concession that YS will make and is doing so in exchange for OS's commitment that it will not bargain, trade, or haggle further and that the parties now will finalize the deal and its documents.

11.4 Best Practices

Being an effective negotiator is not easy. Not all of us will be one, but negotiating skills can be learned, cultivated, practiced, and refined. As you develop yours, here are ten attributes of an effective negotiator, ten dos, and ten don'ts that you should emulate.

Ten Attributes

Negotiators' styles and personalities vary. Negotiators' backgrounds vary – business, engineering, scientific, governmental, military, and so on. Negotiators' experience varies – from "this is my first deal" to "I've been doing this since before you were born." There is an infinite variety of negotiators because there is an infinite variety of people. Regardless of this variety, an effective negotiator develops and uses these ten attributes:

1. Remembers that negotiating is a multilateral, interactive, and often iterative process to accomplish the goals of two or more parties
2. Controls the negotiation by knowing when and how to use the various techniques and by positioning the other side for "easy acceptance." Keeps emphasizing areas of agreement to build momentum
3. Maintains sensitivity and adapts to different personality types and negotiation styles, kinds of and changes in communication, cultural differences and nuances, and differences in standards (what's important as a matter of principle) and valuations (something's worth) among the participants.
   - Recognizes hidden meanings and agendas
   - Learns to understand verbal and nonverbal communications but doesn't overanalyze them. Changes in communications or other behaviors, rather than a particular communication or behavior at a particular time, may signal impending movement (forward or backward), an evolving proposal, a new strategy, or the like from the OS. The OS's steady-state conduct – like having meetings only in the afternoon, flinching at anything you say, or always or never having their attorneys present in a meeting – may be relatively unimportant.

4. Conducts the negotiation fairly, honestly, and reasonably, endeavoring to explain and substantiate his or her side's position. In other words, plays fairly. Plays by the rules.

5. Cares sincerely about the objectives of the OS, and show this care by creating open and direct lines of communication. The OS will tend to give YS what it wants if you give the OS what it wants.

6. Prepares for the negotiation, before and during it. Researches, analyzes, integrates, and updates information to master the facts, issues, and their interrelationships. Avoids the natural temptation to "just walk into the negotiation" because he or she knows the technology, is the inventor, has done this kind of deal before, or negotiated with the OS before.

7. Creates an environment in which all parties feel that they can "win" and lets them feel that they in fact have won. This enables all parties to have a sense of accomplishment and gain.
   - Leaves his or her ego at home; doesn't think about being the winner
   - Is cooperative, facilitative, and agreeable when you can, at the right time, but temper your enthusiasm. This is one of those fine-line balancing areas. You don't want to go so far with cooperation as to show too much enthusiasm or eagerness to make the deal at all costs. This will cause the OS to make otherwise avoidable demands on you. Rather, you want to err on the side of a lack of enthusiasm, to elicit concessions and accommodations from the OS. But you can be unenthusiastic or lukewarm about the OS's proposal(s) while also being cooperative, facilitative, and agreeable with the flow of the process and maybe with selected deal points.
   - All sides should feel they have won. This ideal is achievable.

8. Concludes the negotiations (and close the deal) in such a way that the OS would want to engage in a "rematch" – that is, all sides would want to negotiate again because it was an enjoyable and productive experience.

9. Gives the OS good reason to believe that his or her side will uphold the deal
   - Remember the value of credibility and trustworthiness: "what goes around comes around" and "it's a small world." So, be a straight shooter.

10. Is willing to acquire, continually develop, and practice negotiating skills. Basically, an effective negotiator is a lifelong learner about negotiations.

Some Dos and Don'ts

The following lists of dos and don'ts are provided to help you develop and use the above attributes. Being an effective negotiator takes time and requires a blend of strategic and critical thinking ability, personality traits and adaptability, and communication skills. You easily can trip up along the way, in both your development and your effectiveness as a negotiator. So, as you move forward and continue to learn to be an effective negotiator, practice these basic dos and don'ts. In a nutshell, they boil down to thinking clearly and adaptively, maintaining common sense, and communicating effectively (sometimes very clearly, sometimes deliberatively vaguely) in a fluid situation.

Here are the ten dos:

1. Remember the importance and changing nature of
   - time and timing;
   - power, influence, or leverage; and
   - information throughout the negotiation.
2. Remember that both sides are under pressure to deal or compromise and that generally the side under the most pressure "loses."
3. Remember that the location and method of negotiating (by phone, by e-mail, in-person, at YS's or the OS's place of business) can make a big difference in the negotiation process and outcome.
4. Use tactics appropriately and carefully. Know when and how to use them. There are many scenarios, too many to list. But, for example, if you are on the side of the buyer or equivalent party, play the reluctant buyer and keep your offers low but flexible. In appropriate circumstances on the seller side, consider playing the reluctant seller.
5. Keep YS informed of all negotiation communications and follow all marching orders, boundaries of responsibility and/or authority, and internal agreements.
6. Keep your eye on the ball. Adhere to YS's goals, recognize that different ways to reach them may emerge during the negotiation, and know when YS has reached them. Focus on both the OS's verbal and nonverbal communications, but more importantly also on changes in them.
7. Pay attention to the progress made and not just the differences that remain by focusing on areas of agreement, not disagreement. Movement of concessions equals progress. Maintain some sort of written or electronic record of the concessions made by all sides during the negotiations.
8. Know your stuff, throughout the negotiation. Do your homework and stay on top of the subject matter. Know YS's and the OS's decision makers and influencers, alternatives, facts, issues, constraints, and "baggage." But you cannot be expected to know everything about everything, so make good use of advisors and other experts.
9. Stay mindful of new, changed, or changing information on both YS and the OS, and adapt to it as the negotiation progresses.
10. Congratulate the OS upon the conclusion of the negotiations, even if YS got its way. Avoid the temptation to gloat about YS's results or about what the OS's results could have been.

Here are the ten don'ts:

1. Don't assume all sides want the same thing.
   - Different parties have different goals, interests, and perspectives. Plus, different sides may value the same deal point, asset, or other item differently.
   - What does the OS really want? Help the OS get it.

2. Don't assume money is the all-important deal point. There is probably something else as important as or more important than price.
3. Don't assume that YS has the weaker position.
4. Don't disclose YS's time constraints, deadlines, or other pressures, unless there is a compelling reason to do so.
5. Don't unnecessarily narrow your negotiating range or flexibility. Further, don't narrow the negotiation down to only one issue. If parties get down to one issue, then there will be at least one loser and at least one winner.
6. Don't be the first side to make an offer or name a price, if at all possible, in the context of the particular transaction.
7. Don't become emotionally involved, egotistical, or condescending.
8. Don't get greedy. Don't blow a big deal over a little or nitpicky issue, request, money amount, or deal point.
9. Don't leave the details until later. You may find that you have no deal at all.
10. Don't exceed your authority to negotiate. You may have responsibility to negotiate the deal, but you may not have the authority to approve it or otherwise agree to it.

11.5 Summary

By now, you hopefully are gathering that the negotiating process is a large subject. You also hopefully are starting to see the benefit of having a useful framework for approaching and participating in a negotiation, especially one involving an entrepreneurial business matter. At this point in your experience and studies, the process elaborated earlier may seem too detailed or cumbersome for all matters that may come up in your entrepreneurial pursuits. In fact, the contrary is true.

Remember, negotiation is a multilateral, interactive, and often iterative process to accomplish at least two parties' goals. And all sides are going through the same process. The process components themselves, that is, the phases and their interrelationships, are essential in any negotiation, whether that negotiation is simple or complex. Certainly, the sophistication of the subject transaction, the time available for the negotiation, and the number of participants (and therefore the number of channels, communications, and "moves") can vary greatly from negotiation to negotiation. So, the thought, time, and effort spent on each phase – and the number of times each phase might be revisited or iterated – will vary from negotiation to negotiation. But the fundamentals of a successful negotiation remain the same.

As an entrepreneur, please do not approach this process as if you are outgunned by or are otherwise inferior to the OS. Sure, you may want or need something from the OS, but you also bring something that the OS wants or needs. So, your challenge is to be more "scrappy," more clever, more strategic, more creative, and more efficient with the process than the OS is. Last, each negotiation is a learning opportunity. After each negotiation, you need to debrief YS about what went well, what didn't, and what YS can learn from it to apply in the next negotiation.

11.6 Video Clips

Directions for viewing clips on the Prendismo Collection (formerly from www. eclips.cornell.edu):

- Visit http://prendismo.com/collection/. (You must subscribe to the site for full access to all features.)
- Click on the **Subscribe** at the top of the site. Students receive reduced rates.
- Once you have subscribed, type in the name or title of the clip in the **Search** at the top right corner of the screen.
- Click on the clip you wish to view.
- Click on the play icon to view the clip. If you have not become a subscriber, you will only view the first 20 seconds of the clip before being prompted to log in.

1. Jessica Bibliowicz (National Financial Partners)

   - BS, Cornell University
   - Founder and CEO of National Financial Partners (NFP), an independent financial services distribution system
   - NFP was created with $124 million of capital from Apollo Management, a leveraged buyout firm.
   - Title of video clip: "Jessica Bibliowicz Discusses Importance of Becoming a Good Negotiator"

2. Bill Trenchard (LiveOps and Jump.com)

   - Currently CEO of LiveOps, a teleservices company
   - Founder of Jump.com, which was acquired by Microsoft in 1999
   - Lecturer to undergraduate business class at Cornell University
   - Highlights experiences in successful and unsuccessful start-up ventures
   - Discusses transition from leader of a small company and CEO of a large business
   - Graduate of Cornell University
   - Title of video clip: "Bill Trenchard States Importance of Setting the Stage to Negotiate"

11.7 Problems

Although the following questions will not appear on a FE exam, give them careful consideration and reflection. Your answers need not be lengthy, and try to keep them to fewer than 200 words as a guideline.

1. Describe a successful negotiation you have seen on national or business news. Describe an unsuccessful one. Why was each one successful or unsuccessful, as the case may be?

2. Discuss the likely types and ranges of negotiating points in an angel investor–type negotiation and contrast this with a negotiation with a venture capital firm. Which party do you believe has the leverage over which point or points? Which party has or needs what information?

3. Develop a strategy sheet for negotiating with a named "other side" for a plausible technology licensing transaction (in-bound or out-bound) in a start-up. Describe your side's goals, interests, issues, and positions and then describe the characteristics of the OS, and develop that OS's strategy sheet also.

12 Management

I believe the single most significant decision I can make on a day-to-day basis is my choice of attitude. It is more important than my past, my education, my bankroll, my successes or failures, fame or pain, what other people think of me or say about me, my circumstances, or my position.

Charles Swindoll[1]

12.0 Entrepreneur's Diary

I teach a course on entrepreneurship, and I always invite numerous successful entrepreneurs to give guest lectures on the subject. One of my favorite speakers is Greg (PhD from Cornell). Greg was raising money for his start-up venture about the same time I was raising money to start my fish business. Greg is now worth millions of dollars (that is another story), but he retains a casual attitude toward his success. When Greg comes to lecture, he typically wears jeans and a knit sport shirt with a fleece jacket. Greg describes the early beginnings of his company and how he hated big-company corporate structure. You know ... having to report to so-and-so ... following this and that procedure ... properly documenting this and that. Greg started his own company so he wouldn't have to follow all those rules and do a lockstep with corporate ways of doing things. Well, initially, this approach worked okay for Greg. But, then as his company went from 3 employees to 20, to 100, to 300, he found that he had to follow many of those same corporate rules that he hated before. He was having trouble managing under this new structure. It wasn't as much fun as it used to be. I asked Greg if it had to be this way. He responded with a slouch to his shoulders and a roll of his eyes.

The majority of this chapter was written by my oldest brother James. He was always to me the smartest person I have ever known. It seems appropriate he would author this chapter on management. Enjoy the chapter; I think it has lots of nuggets of wisdom that you will find helpful sometime in your entrepreneurial career. And thanks, Jim, for sharing your management knowledge.

---

[1] Charles R. Swindoll, *Attitudes*. Dr. Swindoll is currently Chancellor of Dallas Theological Seminary and has been the regular speaker since 1979 on a worldwide Christian radio program, Insight for Living.

12.1 Overview

We have come a long way in our journey from Chapter 1. You now should envision your company as an up and running entity that is starting to feel some growing pains. You no longer have just a single person on the payroll; most likely you have two to twenty employees at this point (it all depends upon where you are in your ramp up or business model). Even if you only have one or two (or a hundred), you will need to successfully manage your people or your company will not be nearly as successful as it could be and most likely will fail.

You may not realize this, but almost all of you (students) reading this book will very soon become managers of other people as a part of your next job, even if it is not in entrepreneurship. Managing people is probably the most important component contributing to a successful business. (The other is managing cash flow.) If the business requires more things to be done than can be done by one person, then additional people must be hired to perform specified tasks. The larger the business, the more people that must be brought in to perform the total number of specified tasks required to operate the business. In large businesses, there are layers of "management." Layers of management inherently result in drops in efficiency. Of course, a one-person company can be very efficient. But is it really? There is an old saying that one man can do a one day's worth of work, but that two men working together can accomplish the work of three men working independently. Effective management should synergistically make all the company's parts even more productive than if the divisions worked independently from each other.

This chapter reviews some of the basic tenets of management and human resource skills, conducting effective meetings, and some anecdotal stories. Also, you might want to read the chapter supplement on why strong companies eventually fail or the Peter Principle (see www.CambridgePress.com/Timmons Chapter 12, "Why Strong Companies Fail: The Peter Principle"). Of course, this – the failure of your successful company – is something you should strive to avoid.

Is This Good Counsel?

You might be asking what entitles me to give advice on management when there are countless books on the subject by authors who are a lot more famous on this subject than I. Well, that's a good question! I am simply providing you advice from my perspective of managing divisions within large companies of a thousand people. All learning is an extension of current knowledge based on analogy or association. If you were asked to imagine a giraffe but you had never heard of that particular creature, someone could tell you that it looks something like a big horse but with a super long neck, and hind legs shorter than its front legs. It's not precise, but now you have a basis for continuing to learn about giraffes. Further telling you of other physical characteristics that distinguish a giraffe from a horse could refine your new understanding, and pretty soon, you would know all that your advisor knows about the subject. Similarly and hopefully, you will soon know all the author knows about management, and can progress from there.

An analogy comes to mind: my twelve-year-old son asked to be taught to play chess. I was happy to oblige, as this game is regarded as a good way to develop mental skills. My son learned the specific moves of the various pieces quickly, and soon we began playing actual matches. Of course, I (the teacher) easily won the first few. While teaching my son, it was my practice to critique each of my son's moves so that he could understand what it was that he had actually done in comparison to what he had intended to do. If my son made a particularly bad move, I would let him have a "do-over." My son steadily improved, making fewer mistakes in each succeeding match. I had to critique fewer and fewer of his moves, and then it happened. My twelve-year-old son defeated me. How could this be? Simple. The student was playing the teacher using 100% of the teacher's knowledge of chess, plus the tiny additional bit of knowledge he possessed. Regardless of how little my son knew, he still knew more than I, because I had taught him everything I knew. The same applies here to the advice I give in this chapter. I will tell you all I know about management, and you'll be better at it than I have been if you practice and apply your additional knowledge basis to the situation.

Engineering versus Art

Responsibility for success: where does it lie? It all boils down to management. This is true in any ongoing enterprise. The larger the activity, the more critical and difficult the management process becomes, due to the increasing difficulty of communications and the interpersonal actions among the individuals within the activity.

"Engineering" is not management. "Engineering" is vital to the success of a business, or there will be no product to sell or use. In fact, Albert Einstein said about engineers that "scientists investigate that which already is; engineers create that which has never been." However, "engineering" is design on paper (or screen). Designs are static and timeless. A Roman catapult designed and built 2,000 years ago and one built today to the same design would work exactly the same. Engineering is measurable and testable. "Management," however, is a best guess as to what the right thing to do is. So, how does management relate to design?

> Management Definition:
> The best guess as to what the right thing to do is.

The implementation and employment of some operating plan requires management. All companies large or small have reasonably equal access to the same mix of people, as we are in a free society and can choose to work for whomever we want. Why, then, do some companies operate with greater efficiency than others? It has to be a function of management. Motivating employees to do their best is a management function. Determining an employee's skills and interests will result in better task assignments, resulting in better performance. Management of the financial resources of the business is essential. No, this doesn't mean "don't spend money." I once worked for a company (listed on the New York Stock Exchange) that actually required a manager's approval to get a new light bulb for the desk lamp. How successful could that company be? Was the manager's time better spent in counting light bulbs and toilet paper (yes, that too) or working on supporting the customer's need for a support structure for a complex mechanical system? In this particular case, the answer was "Well, yes," as this particular manager was a total incompetent and perfect example of the Peter Principle. Was it his fault that he spent his time counting light bulbs? No. It was his bosses' fault. A severe shortfall of management effectiveness produced this incompetent. Were there more like him in other areas of the company? Yes. Was the company successful? Answer: not very.

Make a Decision

In any ongoing operation, the manager will be continually faced with the task to evaluate the current and potential future conditions of the operation and make decisions about allocation of effort and resources. It would be nice if the manager could base these decisions on full and unbiased data pertaining to the situation being considered. This never happens, or management would be a science and not an art. A manager never has all the relevant data he or she would like in order to make the best possible decision. Also, the manager may not have complete confidence in the data that are available.

One of the distinguishing characteristics of good managers is that they can determine when enough data of sufficient quality are available for them to make a decision. A less capable manager might wait for additional data or for collaboration of available data before choosing an alternative among the several that may be available. Meanwhile, the situation is evolving, and the facts, circumstances, and outcome probabilities continue to evolve as well. It is at this point that an effective manager makes a decision. One possible decision could be to make no decision. The risk is low enough to maintain the status quo or to let the situation continue to develop while additional data are gathered or certain data are validated to permit a no decision. However, in many cases, a no decision is the default condition regardless of the quality and validity of existing data. In these cases, many of the factors that influence the progress and success of the operation are the results of making no decision at all.

One of the most important tenets of management is hereby set forth: a "no decision" is in fact a decision. Something is going to happen, even if the manager can't or won't decide what to do. The principles, cautions, and warnings set forth in this chapter will enable the manager to gather better data and make better decisions, thereby experiencing better results than would be experienced without application of these observations and techniques.

> A "no decision" is in fact a decision!

What Does a Manager Manage?

Managers manage more than employees. In fact, the manager of a one-person operation doesn't manage employees at all. All managers manage two basic elements: money and activities. Everything else that can be managed is a component of one of these two elements. Let's break it down. "Money," or profit, is why a business exists. Entrepreneurs see a path to obtaining money by starting some particular business. That requires that the entrepreneur offers a product or products or a service or services, or both – and that customers will pay more for this product/service than the provider spends to provide the product and/or service. It is as simple as that. The efficiency at which the company performs these activities will determine the amount of money that the business will make. Inefficient management of money or activities will result in a failing business. For a profit-seeking enterprise, the goal is to receive the maximum amount of revenue for the minimum expenditure necessary to provide the product and/or service that generates the revenue

12.2 The Art of Effective Management

What Is Management?

Management is what keeps the activities and functions of your job going in the right direction. It is something we all do, whether we have the official title of manager or are considered to be one of the peer groups. Although management is an art, there are specific tools and techniques that are extremely helpful in quantifying a given situation so that the better course of action becomes more apparent. However, no tool or aid, however good, can overcome the negative effects of bad management practices. It is easy to spot the results of bad management. It is less easy to define exactly what went wrong, and it is never possible to know what would have been the result if a different set of decisions had been implemented rather than those that were.

There have been thousands of books written on the many aspects of management. All of them are just somebody's opinion. This doesn't mean that the "management" concepts and principles set forth in them are bad or invalid, but just realize why the book was written: to present an opinion and to sell books. On the other hand, engineering doesn't have "opinion" books written about it. This subject has manuals and handbooks. No opinions, just facts based on physical principles. Engineering is science. It is quantifiable, measurable, and repeatable, but to do anything well is difficult.

Management is art. It is not directly quantifiable or measurable, nor can "good" results be necessarily made repeatable. These attributes of management can be assessed only by the results, when observed after some period of time. For example, was Enron a well-managed company? The answer depends on when the question was asked. In 2000, the answer would be an enthusiastic "Yes!" In 2002, when the company collapsed, the answer would be a resounding "No!" If anyone says that "management" is quantifiable, then be assured that the statement is an opinion, not a fact.

What's the point? Management, similar to medicine, must be practiced. This topic of management should be an essential part of any engineering curriculum. Remember that after studying this topic, you will have to practice at it to become a good manager.

12.3 The Five Aspects of Management

There are five aspects of successful management. Whatever you are managing has to

1. be planned,
2. have plans that need to be coordinated,
3. have plans that need to be implemented,
4. have its implementation monitored, and
5. has results to be evaluated.

The primary standard of measurement against which these aspects of management will be judged is the budget. Get used to it. The reason to be in business is to make money. Money must be managed by a comprehensive budget. The success or failure of your business will be judged in terms of financial performance. All business decisions are, in some form or another, a budgetary decision. What will it cost to do that? What is the risk? What will it cost not to do that? What is the risk?

> Your management effectiveness will be mostly evaluated as to how you performed against budget.

A word of explanation: the following text and Figure 12.1 show the management process neatly allocated into five parts. In practice, you will not find it to be so neatly divided. The term *multitasking* describes this situation. Typically, a manager will be involved with many activities simultaneously. One project will be in the planning phase. Another might be in the monitoring phase, and so on. Additionally, within one specific project, there are seldom any clear

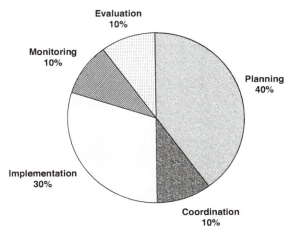

Figure 12.1 Five aspects of the management wheel chart with time requirements as percentages (these reflect author's bias).

distinctions as to when one phase stops and the next begins. The edges of the activities blur and blend, and you will find that you are simultaneously performing activities described in one section (e.g., "planning") and another section (e.g., "organizing"). This is not a problem per se; just be aware of the conditions and communicate accordingly. You would not want someone with whom you were "brainstorming" an idea during the planning process to get the impression you were attempting to actually implement what is obviously an incomplete and perhaps flawed process.

Planning Aspect

The five aspects of good management apply to both a start-up as you write the operations plan of your business plan or how you manage an existing business. In either case, you need to first make a management plan to implement the idea. Your first attempt at writing this plan will probably be very short; perhaps it will be only a page or two in length. The actual contents of your plan will be greatly influenced by the specific means you have chosen to make money. Your idea is going to provide either a product, a service, or both.

We'll skip all the preliminary qualitative assessments that you personally went through to arrive at the decision that your start-up business or this particular idea for an existing company was worth investing time, effort, and money (but you should have done this first as your own personal homework!). Most managers will need to convince someone else (your investor?) that the idea is good and is worth the investment. Therefore, the first thing to do is describe your idea and approach to implementation in a general concept plan. For a business start-up, this is your business plan. In an existing company, you may be describing some expansion effort related to a product or a service. Summarize the idea and describe what is involved, who will participate, how the idea will work, and what the expected result is. Estimate how much the idea will cost to implement, and how much you expect to profit from it, over what period.

The scheduling portion of the ***planning process*** is difficult to do accurately, but essential to success. To establish a realistic budget, one must establish a realistic schedule for a given project. It is absolutely necessary to identify all of the actions that must take place in order for the project to be successful. Each of these actions takes time and money, in some form or another. If you overlook some of these actions and don't schedule them for performance and completion, your project will be delayed and inevitably be over budget.

A very good method for developing this schedule is the "Program Evaluation and Review Technique" (PERT). This technique, or variations of it, was developed by the Department of Defense for managing the acquisition of large, complex weapons systems. This general approach to the scheduling of activities and events is widely known and utilized in many forms and under a few different names. Most often, if not referred to as PERT, this process is identified as the "Critical Path" method (CPM) of program scheduling. When you learn the process for this technique, you will easily recognize and be comfortable with variations of it. Please refer to more specific information and instruction on utilizing this powerful tool presented and discussed at length in Chapter 13.

ORGANIZE

This phase of the planning activity is when you arrange all of your material and prepare whatever presentations or similar products you need to explain your program and plan to your bosses. This is basically an administrative process, but it is important to success. It is an opportunity to review your work, to see if you objectively agree with yourself, and to catch your breath for the next step.

Coordination Aspect

During the planning process, you probably identified several individuals with whom you must communicate to assure your plan has the best chance for success. Let the others know what's going on. In a large organization, these individuals may include managers of other departments, such as shipping and transportation or manufacturing. They will also include individuals in staff support functions, such as the company attorney and accountant. If you are the boss, seek input from specialists; if you are the peer, let your counterparts in other areas of the business review your plans and develop their own plans in the areas where they are affected by your plan.

It is vital to the success of your plan that the key people in all areas of your business that are affected by, or that are affecting, your plan understand what it is that you are planning to do, what their responsibilities are in that implementation, and that they agree to support the implementation. You should also coordinate your plan with individuals outside of your company, insofar as they are affected by or can influence the results of your implementation. This could include suppliers of components and materials, construction businesses, and potential customers for the product you are contemplating. Of course, if you are a one-person activity, all of your coordination activities will be with outsiders. From these individuals, you need feedback that tells you that they understand the plan and can, and are willing, to support the implementation. In the case of the professional advisors, you need to know that you are on solid legal and financial ground and that the necessary permits, studies, and other regulatory authorizations are identified, addressed, and satisfied.

In this phase of management, pay particular attention to the feedback you are getting to assure that there is no misunderstanding about who is responsible for what, particularly in the area of budget allocations. Who pays for what must be defined and most importantly agreed to. Does everyone that is expected to pay for some aspect of the project actually agree to this? If not, you should not proceed until agreement is reached.

Implementation Aspect

Now you implement the plan. This is the shortest and usually the most visible of all aspects of the activity. Everything is set, all is coordinated, and so we begin. This is the time to – c*ut the ribbon – put on the hard hat while wearing your best suit and shiniest shoes and pick up the chrome bladed shovel and smile for the camera – bounce the bottle of champagne off the bow of the ship* – you get the idea. Now, it begins.

Monitoring Aspect

You should monitor activities by collecting indicators of performance. You will certainly need to keep a close eye on what is happening with your new activity. What is really happening, in comparison to what the plan and budget expected to happen? To monitor objectively, you must have hard bits of data to evaluate, to gauge progress, and to measure results.

The means of defining what these bits of data, often referred to as **metrics**, ought to be and how you collect them must be given careful thought. What bits of data would portray something that you would wish to know about the actual status of the activity? Here are some easy examples:

- A fish production company would want to know its variable costs of production (feed, oxygen, fingerling, heating and cooling costs) and its fixed costs of production (labor, rent, office, and management) and at what stage these costs are being incurred to decide when the optimal selling point is to determine the size of the fish to be harvested.
- A taxi company might want to know things like average time between fares, gas mileage of each vehicle, average gas mileage for the fleet, what sections of town produced the most fares, what section of town produced the largest revenues, and so on.

- A retail-clothing store might want to know the number of people entering the store on a daily and weekly basis, the average revenue per day of the week, and the revenue per square foot of store area. Other activities are very difficult to quantify.

A management consultant activity would choose different metrics probably from any of the preceding examples. It must be clear, however, that the metrics must be chosen and collected, or results of activity cannot be quantitatively measured; even harder is selecting the metrics because they are relevant to the evaluation process and not selecting them merely because they might be easily collected. Collectively, the metrics must produce data that describe progress of the activity. Are we on schedule? How do you know? Are we within the budget? How do you know?

Evaluation Aspect

Now that the new project is implemented, and progress is being monitored, the results of these activities to provide products and/or services must be evaluated. Are things progressing as you had envisioned? If not, assessments must be made and plans and schedules must be adjusted. When adjustment becomes necessary, this puts you back at the beginning of the management cycle for this particular aspect of what has to be done. You must plan and schedule it, arrange to accommodate any budgetary impacts, coordinate the plan, organize to implement, and so on. This particular specific adjustment to the overall plan must be incorporated and then continued. Don't ignore bad news. Don't ignore good news. Either can cause problems.

The ultimate evaluation of success or failure of your activity will not be made by you, and that evaluation will be cold and possibly brutal. Either the project is generating revenue in excess of costs required to generate the revenue or it is not. However, this ultimate evaluation may not be revealed for several months or years, particularly if this project is one of many in a large organization's business. Some start-up companies will not make profits for several years, yet they continue. Where does the money come from to make up the difference between what it costs to provide or deliver a product and the revenue that selling the service/product generates? Answer: the investors or shareholders. Patience of an investor only lasts for so long. Be careful.

Laws of Engineering: Is the Project Ever Finished?

We have just gone through the five aspects of successful management. The last aspect was the evaluation phase. A common problem in this phase is that you can sometimes revert to the rut of never really finishing any project. This is one of the "laws of engineering." You have probably noticed already in your career that designs are never ever really finished; hence, the reference to this being a law. As choices are presented, one makes a decision. That decision, in turn, generates several new choices and/or problems. New opportunities crop up in the form of better but yet unincorporated technology. New information puts prior decisions (designs) into question. You face this fact of engineering life every day. So does everybody else who does development work.

So, how do you know when you're finished? The answer: when your design meets the original criteria you established for success, you just stop (the last aspect of management: evaluation). This "law" is applicable to art, yard care, personal hygiene, and every other aspect of an ongoing operation. Typically, the entrepreneur will have established the criteria to tell you when you are done. Sometimes previous criteria for partnership relationships cannot be met; that is, the other partner must share equal financial risk, or equal work effort, so you must modify your "management design or program plan" – to come closer to your original goal. Now the ever present conundrum: *the closer it gets, the clearer the details become*, which often puts you back to the point of "never-finished engineering" when you see a detail that heretofore was too far away to be noticed. Now you need a new design, and you start the process over using the five aspects of management.

System Specification

You will eventually have to manage entire projects and realize a project is made up individual systems that together make up the project. The "system" specification, in all cases, is uniquely developed for a particular purpose and application. It is the functional requirements statement of the entire project. This entire project is composed of various pieces that work together to provide the total function. This "working together" level of indenture is called a **segment**. Each segment has a stand-alone function, separately identifiable and measurable, and without which the system won't work. Each segment is comprised of items and materials. The definition of **material** is easy: it is something from which something else is made. It has no independent function within the system other than its material properties. The definition of **item** is more difficult, and begins to become intuitive, based on contract structures, and so on. Generally, however, items also have a specific performance function that can be specified and measured. Items can also include other items. A jet engine can be a system, a segment, and an item, depending on who uses it for what. The item "jet engine" can be assembled from numerous other items, such as "fuel control." The fuel control can be assembled from other items, such as pressure valves and electric switches, and so on and so on, until finally we get to the very bottom of the pyramid and we are dealing with materials and processes (put a block of such-and-such aluminum on a lathe and carve out something that looks like this) and screw it to the other fabricated component with such and such a screw (catalog item #123–987 from the Acme Screw Company). You will need to accurately define what a segment is made up of and how to describe whether a segment has met its requirements. Additional discussion and further examples of this topic are given in the Chapter 12 supplementary materials (see Chapter 12, Building A Project: Specification Trees).

> A system specification is uniquely developed for a particular purpose or application.

12.4 The Five Management Competencies

As a manager, here's something you need to watch out for. It is a natural tendency for groups within an organization to visualize their work environment as an "us" versus "them" within the context of the organization. Unless the manager is very careful, this tendency will create an adversarial relationship between subordinate internal groups that the manager views as "us." For example, a manager might manage both the production facility and the shipping/receiving operation, so the presumption is that all of "us" are happy and work together well. Really? On the other hand, the sales and marketing operations are the responsibility of another manager, and they are "them," and they sure mess us up by taking rush orders and giving outrageous discounts. We need to straighten those guys out. How do you think the big boss perceives this relationship? And so starts the conflict between groups within an organization, which all should be working towards the common good of the company. A good manager will figure this problem out.

You have to be an very good manager to fix problems such as the one just discussed. Management is a complex set of activities. To become a good manager, you will need to have a solid understanding of how motivate both the individual and to also channel them into team contributions. One of my favorite writers on the subject is Dr. Richard Boyatzis, a faculty member at Case Western Reserve University, who recently published a paper on the subject.[2] Borrowing from that paper and his much earlier text,[3] I developed the following summary to help you recognize the competencies required of a good manager. Try to recognize what you are already good at and where you might need to improve your current capabilities.

Cluster 1: Specialized Knowledge

1. Thorough Technical Knowledge

- Is recognized as an expert and continues to maintain expertise in the field of knowledge

---

[2] See Richard Boyatzis, "Competencies as a Behavioral Approach to Emotional Intelligence," *Journal of Management Development* 28, no. 9 (2009): 749–770.

[3] Richard E. Boyatzis, *The Competent Manager: A Model for Effective Performance* (New York: John Wiley & Sons, 1982). See Table 12–1; this material is reproduced with permission of John Wiley & Sons, Inc.

Cluster 2: Goal and Action Management

2. Efficiency Orientation

- Sets goals or deadlines for accomplishing tasks which are challenging but realistic and has a realistic budget attached
- Plans the action steps, resources needed, and means for overcoming obstacles involved in achieving a goal
- Matches specific people to jobs with the explicit intent of increasing efficiency or productivity
- Expresses a clear standard of excellence for specific task performance

3. Proactivity

- Initiates action rather than waiting to react to the situation as it develops
- Seeks information on own initiative from a wide variety of sources concerning an issue or problem
- Takes calculated risks and admits and accepts personal responsibility for success or failure

4. Concern with Impact

- Explicitly expresses a need or desire to persuade others
- Explicitly expresses a concern with his or her image and reputation or the image and reputation of his or her organization or its products and services

5. Diagnostic Use of Concepts

- Uses an explicit framework or theory to distinguish relevant from irrelevant information in the situation and/or
- Explains or interprets an event, person or case in terms of how the specific aspects of the present instance differ from, or are similar to, the way another event, person or case has transpired in the past or the way it would ideally occur

6. Communication

- Communicates to both upper management (boards, investors) and to project personnel
- Keep your key people informed of where you are in the project's completion

Cluster 3: Directing Subordinates

7. Developing Others

- Gives subordinates or others performance feedback to be used in improving or maintaining effective performance

- Invites subordinates to discuss problems with the explicit purpose of improving their performance
- Helps a subordinate to accomplish a task while allowing them to take responsibility for completing the task

Cluster 4: Focus on Others

8. Self-Control

- Explicitly denies a personal impulse, need, or desire (i.e., makes a personal sacrifice) for the good of an overriding organizational need
- Acts not to show anger or other kinds of emotional upset when being verbally attacked
- Acts patiently and calmly in situations of continuing high pressure
- Changes a course of action, plan, or activity to one that is more appropriate, according to major stressful changes in the situation

9. Perceptual Objectivity

- Describes another person's point of view on an issue when it differs from his or her point of view
- Accurately states the differing perspectives that each of the parties in a conflict brings to the situation
- Recognizes or explicitly regrets another person's loss of status or injured feelings resulting from actions taken for the good of the organization

10. Managing Group Process

- Stresses team goals when holding staff meetings or assigning responsibilities
- Relates the importance of one person's position to another person's position and to the success of the group (people respond if they think their efforts play a key role)
- Acts to involve all parties concerned in openly resolving conflicts within the work group
- Explicitly allows the work group to take responsibility for certain task accomplishments and does not assume personal responsibility for them
- Actively promotes cooperation with or provides help for another work group

11. Use of Socialized Power

- Acts to build a political coalition or potential influence network in order to accomplish a task
- Acts to influence others involved in a conflict or dispute through building coalitions, which will affect the resolution of the conflict or dispute

Cluster 5: Leadership

12. Self-Confidence

- Presents themself, verbally and nonverbally, in an assured, forceful, impressive and unhesitating manner
- Expresses little ambivalence about decisions that he or she has made
- Expresses the belief that they will succeed at a task
- Willing to try new things
- Willing to change when necessary

12.5 Roles of Management

All of your management activities will be devoted to trying to motivate someone else to do what you want them to do (otherwise you are not managing). You will be managing people whether they are employees, peers, organizational superiors, or people outside the organization. How can you best get them to do what you want? This is an extremely well-studied area of the management discipline, and expert opinions and guidance are easily obtained. We do not attempt to supply a specialized technique for use in all management situations. We do, however, provide some basic guidance that will be applicable in all instances. The success you will have will depend on the skill that you use and the discipline that you impose on yourself.

Correcting Performance Errors

There are several statements that are true but not necessarily obvious and a successful manager will remember them and incorporate them into his or her management activity. The first of these is that nobody makes a mistake on purpose. The most incompetent person that you can think of does not make errors on purpose. People do what they believe is the correct thing to do, all the time, from their perspective.

> Mistakes:
> nobody makes a mistake on purpose

It may be a fine point, but if someone does make an "error" on purpose, it isn't an error, is it? The person's action or inaction may result in adverse impact to the project or budget, but if it was done purposefully, it is not an error. It may be sabotage, but it is not an error. Therefore, if a manager is encountering behavior that exhibits what appears to be "mistakes," then the person's view of what is correct and proper must be modified. This would be done through education, instruction, direction, positive and/or negative motivation, or a change of assignment.

Positive Motivation

A not particularly obvious observation about management is that disciplinary actions are effective only on a positively motivated person. If a manager assigns someone the task of trimming all the hedges around the corporate headquarters building, and the manager didn't like the results, it would do no good at all to tell the worker,

> "You did a terrible job trimming the hedges and you will never be given that task again!"

Unless the worker has a particular fondness for hedge trimming, this managerial statement would probably result in an unspoken

> Thank you. I hate trimming hedges.

This person is negatively motivated to trim hedges, so the following disciplinary action will not result in this person trimming the hedges better the next time:

> You'll never trim another hedge around here, by golly.

However, the worker may be more positively motivated to keep his or her job. So, if the manager were to say,

> You did such a terrible job trimming the hedges that I can't rely on you to do anything else right, so I'm going to have to let you go.

The worker may respond now with a plea for another opportunity. However, if the person had no particular regard for continued employment with that organization, that statement would produce no better hedge trimming from that person either. To enforce effective discipline, you must understand the motivation of the individual.

Treating Employees with Respect

Closely related to the creating superior performance through positive motivation is obtaining the same result by showing respect to the employee. I have observed how some managers seem to get the most out of their people. Is it team atmosphere, chemistry, company picnics, and so on? It might just be that old simple ingredient of personal respect for the individual. Let me relate a true story about a supervisor I once had (this is a thank-you letter I wrote to my supervisor Roy):

Hi, Roy [the supervisor] –

I'm getting forgetful in my older age, so if I mentioned this before, please excuse me. Are you familiar, perhaps, with the famous football player Jim Brown? He was the fullback for the Cleveland Browns of the late fifties and early sixties, a very good and powerful team in those days. After that, he went on to a career in the movies. In his playing days, Jim Brown was really great. The opposition held him in great respect, and his own teammates supported him steadfastly.

He had a particular manner and style about his ball carrying technique. It was not that he did anything unusual while actually carrying the ball. He was a big, fast, powerful man and could be relied on to gain at least three or four yards every time he carried the ball, sometimes, way more. His distinguishing characteristic, though, was that every time he was tackled, he was extremely slow to get up. He would just lie there absolutely still, in the position he landed, for five or ten seconds, then he would slowly roll over to his knees and wait another few seconds, then he would slowly rise to his feet, then he would slowly walk to the huddle. He did this every time he carried the ball, for years.

There was much speculation among the fans and sportswriters and sports broadcasters about why he did this. Some said he was resting because he was not in very good condition. This was obviously not the case, or he couldn't have done what he did. He played nine seasons and never missed a game. Some said that he was seeking attention. This was not the case, either, as he couldn't possibly have gotten more attention than he was already accorded as the premier running back in the entire league. When asked about it, he would never say. Then, finally, after he retired, he grew weary of the same question time and time again, so he answered. The reason he was always so slow to rise and return to the huddle was that sometimes he was hurt, and that was as fast as he could go. However, he didn't want the opposition to know that he was hurt, or they would try to take advantage of that knowledge. So, if he couldn't move faster when he was hurt, he could certainly move slower when he wasn't, and so by adopting this characteristic, the opposition (and his own team) never knew when he was hurt. Therefore, he was always taken at face value, and accorded the proper respect he had earned by virtue of his skills and ability.

What has this to do with you? You (my supervisor) have developed a similar technique, so to speak. You are unceasingly thoughtful and respectful of everyone. I am sure that you know and work with people that you regard as too dumb to blow their nose, but I don't know who they are. They don't know who they are, either (maybe I'm one?). But because you treat everyone with respect, you also treat these people with respect. Therefore, everyone works with you to the best of their ability (whatever that is) and as a result, you tend to get the best possible results from the team of people you are given. Everyone feels that you value them and their work, so they reciprocate by doing their best. As you progress through your career, I urge you to retain this approach. I enjoyed working with you, and hope to do so again.

There is no excuse for not treating all the people with the same respect that you expect to receive yourself from those that you interact. This is especially true from managers who manage people. You might say that this is simply practicing the golden rule (do unto others as you ...) or the old saying that what goes around comes around.

Choosing Managers

Recruiting effective management personnel is not easy, and recognizing talent is a talent in itself. But there are managers, and there are followers. And, some of the followers will have an attitude. Watch out for indications of something along these lines.

I once had an employee, John, whom I was intending to make my general manager for a new start-up company. Some of John's efforts at work were not going so well, and I made some reference to military structure and order and timing and such. In response, John said something to the effect of

> *I've dealt with colonels before and they did not scare me; they can read me the riot act all they want and it won't impact my behavior or performance one bit!*

John's reference to the "colonels" and the riot act is very, very insightful. If you are faced with a similar-type individual or attitude, be very careful. This type of individual will never be adequate to your task in a management role as you envision them to be. You probably will have had some other warning signs in the past if you've had direct experience with this type of individual. This person has an authority problem. You and your company will be headed to disaster if you keep this type of a person in a management position. Be wary of the type of individual who cannot conceive of being in a leadership role but resents all those who are. Although these individuals may have many positive attributes, this type of person still cannot and will not handle a leadership role. It is just not in their dimension.

Before hiring a management person, review their previous performance and check with other people who have worked with/for this individual. Does this person meet your standards for the position for which you are considering them? If the answer is "no, but," that is not a yes, so therefore it is a no. If the answer is no, then you need someone else in that position. That is the bottom line.

Your entrepreneurial venture is your professional life and heartbeat. You would not install a marginal mechanical component or other questionable piece of essential equipment or support system into your venture. You should certainly not install the most vital of all assets unless you are confident in the results. This advice also is relevant to colleagues with whom you intend to go into business. Look out for these warning signs of an aversion to leadership.

Take Responsibility

As leader of your company, you will be faced with difficult decisions. The toughest are often related to personnel because personal relationships may be affected. Personnel decisions as described previously are never an easy thing. You need to take full control of your company and responsibility for decisions. You will not gain any respect from the affected party by inferring that you are doing "something" (such as making a hard choice decision such as fire, demote, or not hire a person) because someone else told you to, for example, the board of directors. Even if you are, it is no business of the person to whom you are speaking how your decision was developed. It gives the other party (them) a false sense of control and it gives the person you are talking to the feeling that closure (about whatever you are discussing) with you is not possible, as agreements could be overruled by "them." If absolutely necessary, you could indicate that your partners agree with whatever it is you are saying, but don't shift the focus to them. People need to know with whom they are dealing. If they think they are actually dealing with a shadowy "them," they won't respond as you think they should. Your decisions and actions should be based on "I want/need" rather than "they." Make business actions on a personal basis, for example, "I agree to do …" or "I am going to do such-and-such."

Managing Compromise: Solving Arguments

If you are one who is adamant about maintaining professional autonomy, you are very likely to find yourself in a very difficult place, ethically speaking, when a significant disagreement arises. You may have to compromise. It may help if you are able to compartmentalize such dilemmas in a way that helps you avoid the pitfalls of allowing every conflict to degenerate into a moral undertaking of good versus bad. After all, just as not all compromise is good, not all compromise is bad either.

Respect of the other side, the adversary, is always a requirement. There may be times that you just cannot come to a compromise over what seems like an irreconcilable disagreement with the other party. Before walking away and perhaps burning the proverbial bridge, consider these five reasons why you may want to arrive at a compromise as an integrity preserving decision even when you think it may violate some of your own moral ethics.[4]

1. There is some significant degree of factual and/or conceptual uncertainty. We may not, for instance, really know the empirical probability of the failure of some device, nor how safe is safe enough.
2. There is genuine moral complexity surrounding an issue. There are reasonable differences of opinion, and perhaps even individual questioning, about what factors are morally relevant, and equally intelligent and conscientious individuals on due reflection continue to disagree. Technical risk assessments are often of this character.
3. Cooperative relationships exist that are not desirable to rupture. Such may be the case, for instance, in a corporate team, and among friends and family and other significant personal and institutional collaborations.

---

[4] This is taken from Carl Mitcham and R. Shannon Duvall, *Engineer's Toolkit. Moral Compromise: Bad and Good* (City: Publisher, 2000), 111–112.

4. There is an impending and nondeferrable decision. For example, some engineering decision regarding design, maintenance, or operation simply has to be made now; it cannot be put off any longer.

5. Limited resources require that something less than the perfect or optimum decision be made. The perfect option cannot be afforded, or not all those options that are desirable can be adopted. A choice must be made among less than ideal alternatives.

So review these five reasons carefully the next time you think that a compromise is impossible when you and the other party must collectively decide on some course of action. This is particularly true when avoidance (no decision) of the issue or a resulting rupture of the current cooperative bond between the parties is really not viable options. Compromise may be the moral choice that results in a preservation of integrity for both sides.

12.6 Types of Leadership

You've started your new company and you are in phase 2; you've made it through the tricky start-up period, you have a prototype and sales, and you now have a staff of eight people. You received a phase 2 cash injection, and you hired a CEO to replace you in your current role. The CEO has now been on the job for a couple of months. You have tried to keep a "hands-off" approach with the new CEO, so as not to micromanage. Your CEO is fifty years old and took early retirement from another very successful career. You are dealing with a strong individual whom you hired because of their demonstrated successes, background, and decision-making capability and because he or she is accustomed to running the show. Now, however, you are starting to have some problems with the CEO because it seems to you that you are being shut out of the decision-making process in the company. How do you approach this problem?

> Speaking of possible choices and determining which to choose, Woody Hayes, the famous football coach of Ohio State University for thirty years, was once asked why his teams didn't choose to pass the ball more often. He said, "There are only three things that can happen when you pass the ball, and two of them are bad."

There is an easy answer and a hard answer. The easy answer is seldom the best solution. The hard answer is often correct but unsatisfactory. The easy answer is "do it my way, because I control the majority of interest in the company." The hard answer is outlined in the following.

Assume that you have brought your concerns to the CEO. Now there is a specific project for your company, one that may play a critical role in its financial success. Several approaches or solutions could be pursued. And let's assume that the thing you want to do is *not* something the CEO wants to do. Let's explore the possibilities of doing what you think should be done:

- You are correct, and this would be a good thing to do.
- You are incorrect, and this would not be a good thing to do.
- If you do it, the outcome could be good.
- If you do it, the outcome could be bad.
- If you do it, the outcome could be neutral.

When working with individuals, there are four leadership situations who they are willing to accept direction, but these situations have different degrees of experience and capability.

1. **Directing.** You must tell the individual exactly what to do and when, and what not to do. Full disclosure of the total vision of the program is not necessary. The military analogy would be the drill sergeant instructing the recruit.
2. **Controlling.** You must make the long-term and midterm plans for the operation and fully disclose them to the individual, but leave the details of implementation to him. The military analogy would be the colonel providing instructions to the drill sergeant.
3. **Coaching.** The individual has the basic skills to do what is to be done, but does not have the specific background and experience to be optimally effective at the task. So, you provide the game plan, the strategies for implementation, and the details of information that the individual doesn't have, and identify techniques that are more effective than others might be, and so on. The analogy would be a coach leading a football team.
4. **Mutually Participating.** You turn the individual loose and have him or her provide you with his or her own proposals for activity, long-term and midterm plans, implementation techniques, and so on, which you review and comment on. The analogy would be a two-person LLC in which you are the majority shareholder and you provide feedback to your other member's plan and iterate until you have a mutually acceptable plan.

The relationship between you and the individual is constantly shifting throughout these four situations. In most cases, you will tend to be in two of the four more than the other two: depending on the specific situation, directing and controlling, or controlling and coaching, or coaching and participating.

As you can probably see, it is easy to be in modes that adjoin each other, depending on the situation at hand, but it is difficult to operate in situations that do not adjoin. For example, you wouldn't tell the CEO sweep the floor with the big push broom and scoop up the trash and put it in the covered barrel and set the barrel outside, then be sure to shut the door and ask for his or her plans and schedule for the correct approach to self-generated electrical power. However, you might suggest to them a course of action and points of contact for the design and installation of an automated meat-processing system and ask for his proposal and ideas for organic waste disposal.

Readdressing the problem at hand (you hired a new CEO to replace you, and your input on a particular activity or project is now being ignored), carefully examine why you want to do this thing. If you are convinced that something good will occur if it is done (or the opposite, that if you do this something bad will be prevented which would otherwise occur), then you can proceed. If you are certain that this is what you want to do, you should proceed as follows: draft a memo *detailing* your proposal, and give it to the CEO.

The word *detailing* is emphasized because it is important to get beyond the basics, which the CEO probably already understands, has considered, and has deemed to be such that this is something they don't want to do. Your task is to show the CEO the reasons for doing this thing, and, ipso facto, it is a given you think that the CEO doesn't understand the details, or they would not currently have the position, so you must show this person the details. You are now operating in a coaching role and must proceed accordingly. If you do a good job, the CEO will see the logic of your proposal and will begin working on strategy for implementation.

If, however, all this does not work, you will have a more difficult decision to make, having more to do with the relationship of the CEO to the company than with the specific situation about which you are concerned. If you are absolutely sure that this is something to do and the CEO doesn't want to, then it will become the issue wherein you give a person the authority to make independent decisions affecting the welfare and financial health of the company, and then what do you do if you don't agree with some of their actions. Be prepared; when you delegate responsibility to others, this is bound to happen.

Having said all this, you must remember one thing: you are the only founder of your company, and nobody else has your technical background and expertise as related to your company. You are the one who launched the company, and you are the reason everyone else is there. You have the right to expect that your opinions and desires will be recognized and respected. You likely will have more clout in the business than anyone else, and the CEO knows that. The CEO probably knows that more than you know that. So, if you and the CEO continue to disagree, it is perfectly reasonable for you to ask them for the detailed reasons, rationale, and logic to defend his or her position, and the CEO should comply. Then, you should find the solution to the situation at hand. The CEO may have errors or faulty logic in one or more areas in which you are an expert, or perhaps you will find that the CEO is correct in an area in which he or she may be more of an expert than you are, and it is sufficient to cause you to change your thinking.

In any case, you will be evaluating the other person for attitude (the single most important character aspect in all situations), depth of thought, and practicality of approach. Such situations as these will be a good test of the love in this marriage between you and your company/board/CEO.

12.7 Attitude

Let me spend just a bit more time on attitude. This is the most important characteristic of any individual. The attitude you bring to the job will determine success for both you and your employer, or your start-up. As you continue to evaluate your future and how you will attack or retreat or how to face your next big challenge or disappointment, look internally. The following quotation by Charles Swindoll should be framed and placed on your office wall where you will see it every day; you might even put it in your employee handbooks.

ATTITUDES

Words can never adequately convey the incredible impact of our attitude toward life. The longer I live the more convinced I become that life is 10 percent what happens to us and 90 percent of how we respond to it.

I believe the single most significant decision I can make on a day-to-day basis is my choice of **attitude**. It is more important than my past, my education, my bankroll, my successes or failures, fame or pain, what other people think of me or say about me, my circumstances, or my position.

Attitude keeps me going or cripples my progress. It alone fuels my fire or assaults my hope. When my attitudes are right, there's no barrier too high, no valley too deep, no dream too extreme, no challenge too great for me.[5]

12.8 Holding a Successful Meeting

Meetings can be good, particularly if the meeting is to fix some problem or to define a strategy or procedure. For an important meeting, it is similar to rabbit hunting with a single shot shotgun. If you don't do it with this first meeting, that is, clear the air and set the tone of cooperation, mutual respect, and top-level understanding, then forty-two more meetings won't fix it. To run a successful meeting, you must control four aspects:

1. Agenda
2. Conduction (no deviation)
3. Leadership
4. Minutes

FIRST: AGENDA
Establish and write down an agenda; that is, here are the points of discussion:

- Individual responsibilities clarification/limits of authority
- Problems encountered (brief list, i.e., schedules/activity coordination, etc.)

---

[5] Swindoll, *Attitudes*, Ibid.

- Problem solution
- Agreements

This is just a quick example, and the example topics are not important. What you need to understand is that you must have an agenda. Ideally, you can get it out for everyone to review before the meeting. If you need to have the first part of the meeting for the purpose of defining the agenda for the rest of the meeting, do so, but go in with *your* agenda ready.

You must have an agenda and then stick to it.

SECOND: CONDUCTION (ADHERE TO THE AGENDA)

Once your agenda is agreed to, you *must not* deviate. You must conduct the meeting according to the agenda. This is especially critical with folks who like to rationalize and excuse themselves. These people will knock you off your agenda almost immediately, and you'll never get back because you'll be answering their complaints, among other things, and your one and only meeting will be wasted. If necessary, and someone insists on deviating from the agenda, schedule another meeting to talk about his or her sore thumb. *Do not* deviate from the agenda. Basically, only those meetings conducted under these rules will have any hope of accomplishing something.

THIRD: LEADERSHIP (WHO IS IN CHARGE)

Important, but not quite as much (assuming you stick to the agenda), is who is in charge and leads the meeting. Who is the chairperson of the meeting? It may sound egotistical, but you must develop the attitude that you are in charge of every meeting you go to, especially when you call the meeting, unless you choose not to be. If it is your meeting, then you must be in charge. You run the meeting. You say who speaks and when. This does not mean that you do all or even most of the talking. It does say that you control the meeting to keep it focused on the reasons for having a meeting, and that no one introduces any red herrings that will cause the group to lose focus.

In a start-up, remember that you are the entrepreneur. The others would not be there if not for you, and none of the others knows what you know, nor can they do what you do. Everyone in the meeting is there because you brought them into the project to assist you to accomplish your goals. Each has already signed up to that condition, consciously or subconsciously, or he or she wouldn't be there at all. Therefore, you are the natural and rightful chairperson of the meeting. Grasp it and go.

FOURTH: MINUTES

Last, and extremely important, someone needs to keep written notes that become minutes. Task someone not normally expected to attend to be the recorder and take notes. This means that they will not become a part of the meeting, and thereby subject to missing taking some important notes while involved in speaking or listening to someone else. However, this person should be familiar enough with the project and the language of the project so that correct and accurate notes are kept. Empower him or her to stop the discussion to ensure that he or she understands the point made. Then, you take the notes, edit as necessary, and publish the minutes as quickly as possible. The person who publishes the minutes controls the outcome of the meeting.

> Meeting Minutes:
> The publisher of the minutes controls the outcome of the meeting.

Keep the meeting part of the meeting as short as is reasonable. Fifteen to twenty minutes per agenda item is plenty. If discussion goes beyond that time, ask yourself if the discussion is relevant to the agenda item or if the discussion is digressing from the topic. The tendency will be that the meeting wants to turn into a complaints session and/or a blame session, which accomplishes nothing. If extended discussion is relevant, OK. If not, get back to the agenda.

12.9 Basic Human Resource Skills

Job Descriptions

A job description is the word picture of what the job is, what it requires of the incumbent, and the type of skills necessary to satisfactorily accomplish the job. Job descriptions have two distinct components:

1. Responsibilities: a complete listing of the duties and responsibilities of the position that combine to identify all aspects of the job
2. SKAPs: the requirements for doing the job: Skills, Knowledge, Attitude, and Profile

In organizations large enough to have several individuals performing the same basic function, you will find that SKAPs are useful in providing job growth and a career path for the employees. Using SKAPs, a job may then be divided into two or more "levels," wherein the higher the level, the greater the responsibilities and rewards. This is the career ladder, or an equivalent term, at your company. Following this logic, the job description can be generic. Consider this very simple job description for an aircraft pilot:

- Fly plane
- Understand navigation
- Responsible for moving aircraft from A to B

The SKAPs will depend on level of responsibility and complexity, for example,

- Level 1: Cessna pilot
- Level 2: Twin engine
- Level 3: Small jets
- Level 4: 747 pilot

Of course, pay would go up from the lower level to the higher level. More details on SKAPs are given in the chapter supplementary material (see Chapter 12, Job Skills and Advancement).

To be a good manager, you must understand that the most fundamental block of a successful employee and management relationship is to have a good job description for each position in the company that is filled by an employee. Many conflicts between a superior and subordinates come from each having a different expectation of what responsibilities and functions a person is to perform. Simply because you hire someone to be a warehouse foreperson does not mean that you both agree on the scope of responsibility, the tasks involved, the degree of authority delegated, the assignment of task priority, and so forth that are all a part of running a warehouse. This set of mismatched expectations will inevitably result in you having a perception of poor performance on the part of the foreperson and the foreperson having a sense of frustration and lack of support from you (the manager or the management).

One method of assuring that both you and the employee agree on the scope of the employee's duties and responsibilities is to have the *employee write his or her own job description*. Using this technique, you provide a brief, broadly stated description of the duties of the position such as follows:

> The warehouse foreperson runs the warehouse. Our company receives goods, stores them, and ships them. Now, please provide me a detailed description of the functions and responsibilities of the foreperson in successfully accomplishing all aspects of this job as you the new employee see them.

You are asking the prospective or newly hired foreperson to document their duties and responsibilities as they envision them. You (or whoever the foreperson reports to) already know what you think the duties are, so when reviewing the job description written by the foreperson, you can see if they share your concepts about how to do the job. For example, if you are particularly concerned about ensuring that all goods leaving the warehouse are properly packed and are packaged to prevent damage in transit, but the foreperson doesn't mention that aspect of the job in the job description they write, then you can discuss it and reach a common understanding of what you expect in this area of activity. When this process is completed, both parties will agree on what the job entails.

Job Descriptions:
Have the employee write their own, and then review together

Job Performance and Review

Implement a clear and coherent process by which every employee's job performance is reviewed on a regular basis. This should be done formally and without any confusion as to its regularity and purpose. Formal reviews should be done on either a semi yearly or a yearly basis. Informal reviews should be done much more often, to reinforce what is being done right and to discuss what is being done wrong. Correct unfavorable trends right away. If someone who is normally a dependable person is procrastinating on doing something, it may be that they don't know what to do or how to do it and are too proud to ask for help. Be alert for this, particularly in a start-up situation. An informal review will bring this situation to a point that whatever problem or problems that exist can be known and addressed.

For formal reviews, establish a regular cycle for considering cost-of-living-type raises. This could be annually, or semiannually; it would be easier on you and the accountant if you did it annually, say each July or, whatever the fiscal year is for the company. This, then, would become the annual raise that everyone of us looks forward to, truly believes they deserve and have earned, and will be upset if not tendered. It can be discussed and established that this raise consideration is for everyone and will occur at the same time each year and that the raises are generally based on the rate of inflation. *But do not publish or discuss* a company average percentage (this reflects the extra compensation that certain employees received for doing something beyond normal expectations) or anything else that anyone can compare specifically to what he or she got (or didn't).

> Do not *publish* or discuss a company average percentage raise.

Each person must be made to realize that their compensation must be covered by the money they make for the company. You have to create the general understanding that compensation is connected to company profits and that profits are directly connected to production and cost avoidance. Don't buy it if you don't need it. Do buy it if you must have it. If you buy it, use it. Waste little. Stuff like that.

Management studies generally recommend that discussions on job performance should be kept distinctly separate from discussions on money. I don't completely agree, because if you want to get someone's attention, money is a good way to do that, in the negative context.

We recommend establishing job performance reviews on a quarterly get-together to accomplish the following.

Quarter 1: Mutually establish and agree on the coming year's work plan for that individual; establish realistic, doable, but growth oriented, valuable goals for the company that will translate to increased company profit when accomplished. Mutually agree on how to measure progress to plan and agree on success criteria that are objective and measurable. Try to keep your evaluation yardstick to a one-year period, although you may extend occasionally for two years. Anything beyond that will probably be beyond the ability of the individual to implement or control its outcome.

Quarters 2 and 3: Review and discuss work plan and how it is coming along.

Quarter 4: Measure and position for closure. This should be done in private and should be recorded and filed.

Wages and Motivation

Do wages serve as motivators? There are numerous theories about this general topic. All of them have merit and all of them are flawed. The first and most important thing to know and realize about wages is that "good" wages do not motivate people to do good work. An individual worker might use the absence of "good" wages as an excuse to justify bad work, but that is bogus.

People are not motivated to do good work by the prospect of higher wages. A reasonable wage must be paid. Reasonable is defined however you choose, for example, area prevailing, typical for the industry, and so on. This is not to say that management – you, in this case – can't tie an essential person to the company through financial arrangements; indeed, you can and probably should, for your key people. People are motivated to do good work by other things:

- the challenge of a difficult task,
- the satisfaction of performing well,
- power, and
- recognition and fame.

Getting more specific, what you want to make happen with your workforce won't happen because of wages. Therefore, you must devise a wage policy that will prevent what you don't want to happen – that is, dissatisfaction, no work, bad work, and sabotage, among others – from happening. This is possible, but difficult. What you don't want to happen is to have someone to perceive injustice in the compensation system and react unfavorably.

> Wages:
> What you don't want to happen is someone to perceive injustice in the compensation system and react unfavorably.

Obviously, the overall wage structure must function within the cash flow capabilities of the business. Assuming that is not a problem, then, the next thing is that the wage structure should consider the individual performance scenario as much as possible.

Let me digress a bit. I was in the Air Force, as an aircraft maintenance officer, and worked in harsh, uncomfortable conditions, with a workweek that averaged seventy hours. I was responsible for lots of things. I made the same pay as a contemporary officer working in another field. Both of us had the same amount of time in service. He worked forty hours (or less) a week in an office. He was responsible for virtually nothing. Was it fair? No. Was I de-motivated to do less than good work as a result? No. Did either of us have any idea how to make the system better? No. The system was too big. However, I did occasionally think that the difficulty of a job ought to be reflected, somehow, in the wages paid. I rationalized that if I were to start my own air force, I'd list all the jobs that paid this, all the jobs that paid that, and so on, then let the folks choose. The easy jobs at a certain wage would fill up first, and so on. Then, for the unfilled – that is, hard – jobs, raise the wage, until some balance was achieved. So, generally, the difficulty of the job and the prevailing wage for that job should somehow be connected. So what's the answer to how to structure your company's wage program and should you consider bonus programs as a key part of your salary compensation packages?

Do not make an awards or bonus program an expectation on anyone's part. Awarding a bonus to an individual should be kept absolutely secret because others will wonder why they didn't receive a bonus (this is in addition to the company wide salary improvement awards, which are usually percentage raises). You might consider some form of bonus or reward other than cash. But, if the company acknowledges that there is a monetary bonus program, someone will certainly ask about it, and by the way, where's mine and who got one already and how much was it? Pretty soon, your bonus program will become just part of your wage program, and you will lose whatever effectiveness you had at one time in using it.

Be as open and liberal as you want on perks that everyone can get. Offer rewards and praise for things that are beneficial to you that everyone can control and contribute to with reasonable effort (such as the perfect attendance certificate, 100 accident free days, etc.). Make it as clear as possible to everyone that the profit sharing program is the main way the team members can increase their collective cut of the money.

Secrecy of Wage Rate

It is generally agreed that a person's wage should be strictly confidential between the person and the employer. No one should know exactly what the other guy makes. This information is none of anyone's business and can lead only to dissension somewhere on somebody's part. Keep your company's wages and bonuses secret. However, somehow, the information will leak out. It always does. When it does, it is important not to confirm. Just don't discuss it. In lieu of specific information, it should be made known that equivalent jobs pay about the same. When a defense contractor hired me, my boss, who was also the man that hired me, said, "We managers and section leaders all make about the same thing. But, your specific salary is confidential between you and me, and if someone else finds out, it will be from you, and that is a firing offence." Such a policy should be clearly noted in the employee handbook.

12.10 Business Etiquette

Proper etiquette is critically important in all of life's venues, but especially so in the business world. During your meetings with potential investors (or interviewing for a new job), you will undoubtedly be invited to meet over lunch or dinner. There is more going on at a business lunch then you may realize. The investor will also be evaluating you on how you present yourself in terms of dress and manners.

Although a business plan and angel-investor letter may help you get your foot in the door, how you present yourself and act in person is what will either seal the deal or send you packing. Investors (or companies recruiting new employees) will be interested in the presence and interpersonal skills that you bring to the table. While your interview may occur in the typical office set up, more and more such interviews are being held at what is described by the now ever so famous phrase "the power business lunch." The business lunch is a great opportunity for you to win the approval the investor or your future employer. The two keys to a successful lunch are your attire and manners.

More than half of all business deals are finalized over a meal, and a higher percentage of business discussion occurs while out to lunch (officially) or at dinner. Business meals are often used to conduct interviews, to get to know a customer or consultant, to network with a colleague, or to sign an agreement. Still not convinced? Have you ever been on a date and witnessed your companion's manners to be objectionable? Your impression of that person was tainted. This same thing can occur in a business setting whether or not you choose to acknowledge it.

So, in the website that supports this text (www.CambridgePress.com/ Timmons), we give you a quick review of some basic business meeting etiquette rules (just in case you've forgotten some of the manners that your mother taught you). For this chapter supplement, go to Chapter 12, Etiquette.

12.11 Additional Resources

Craig, Betty. *Don't Slurp Your Soup: A Basic Guide to Business Etiquette*. New Brighton, MN: Brighton Publication, 1996.

Dupont, M. Kay. *Business Etiquette and Professionalism*. Rochester, NY: Crisp Publications, Axo Press, 2009.

Fox, Sue. *Business Etiquette for Dummies*. 2nd ed. Hoboken, NJ: Wiley Publishing, 2008.

Morrison, Terri. *Kiss, Bow, or Shake Hands: How to Do Business in 60 Countries*. Holbrook, MA: Adams Media Corp., 1995.

Post, Emily, and Peter Post. *Emily Post's* The Etiquette Advantage in Business: Personal Skills for Professional Success. 2nd ed. New York: HarperCollins, 2005.

Sabath, Ann Marie. *Business Etiquette, Third Edition: 101 Ways to Conduct Business with Charm and Savvy*. Franklin Lakes, NJ: Career Press, 2010.

Stewart, Marjabelle Young, and Marian Faux. *Executive Etiquette in the New Workplace*. New York: St. Martin's Press, 1999.

Tucherman, Nancy, and Dunnan, Nancy. *The Amy Vanderbilt Complete Book of Etiquette: A Guide to Contemporary Living*. Garden City, NY: Doubleday & Company, Inc., 1995.

12.12 Video Clips

Directions for viewing clips on the Prendismo Collection (formerly from www.eclips.cornell.edu):

- Visit http://prendismo.com/collection/. (You must subscribe with the site for full access to all features.)
- Click on the **Subscribe** at the top of the site. Students receive reduced rates.
- Once you have subscribed, type in the name or title of the clip in the **Search** at the top right corner of the screen.
- Click on the clip you wish to view.
- Click on the play icon to view the clip. If you have not become a subscriber, you will only view the first 20 seconds of the clip before being prompted to log in.

1. Mike Pratico (Founder of Davanita Design)

   - Graduated from Cornell in 1998 with honors
   - During school, he formed Davanita Design, an e-strategy and implementation consulting firm
   - Responsible for most of the business operational needs and, later, the sale of Davanita to Avatar Technology
   - Video clip: "Mike Pratico Discusses Challenges of Growth and Handling Personnel Issues"

2. Rosa Sugranes (Founder of Iberia Tiles)

   - Iberia Tiles was founded in 1979 by Ramon Sugranes as a tile manufacturer
   - Received Small Business Person of the Year from the Small Business Council of America
   - Video clip: "Rosa Sugranes Discusses Challenges of Managing People"

3. Bill Trenchard (Founder of Jump.com)

   - Currently CEO of LiveOps, a teleservices company
   - Founder of Jump.com, which was acquired by Microsoft in 1999

- Graduate of Cornell University
- Video clip: "Bill Trenchard Shares Thoughts on Leadership"

12.13 References and Bibliography

Boyatzis, R. E. *The Competent Manager: A Model for Effective Performance*. New York: John Wiley & Sons, 1982.

Mitcham, Carl, and R. Shannon Duvall, R. *Engineer's Toolkit. Moral Compromise: Bad and Good*. City: Publisher, 2000.

Robert, Jackall. *Moral Mazes. The World of Corporate Mangers*. New York, NY: Oxford University Press, Inc, 1988.

12.14 Problems

While the following questions will not appear on a FE exam, give them careful consideration and reflection. Your answers need not be lengthy, and try to keep under 200 words as a guideline.

1. What was your favorite work experience and why?
2. a. Contrast the characteristics of your favorite supervisor and your worst?
   b. Did you learn anything from your bad supervisor for your own future management roles?
3. Can a person in his or her twenties effectively manage people in their forties and older? What will be your major challenges?
4. How do you best motivate a person to do quality work?
5. What ways can you reward a person other than by salary?
6. Develop a quantitative plan of how to reward key employees for contributions that went beyond the expectations for their job. You may have to define some set of characteristics for your hypothetical company, for example, the market focus for the company or the position of the employee.
7. Why is dinner etiquette important in a business setting?

13 Project Scheduling: Critical Path Methods, Program Evaluation, and Review Techniques

Before everything else, getting ready is the secret of success.

Henry Ford

13.0 Entrepreneur's Diary

Planning for my new fish farm start-up was going pretty well. After all, I had twenty years of experience at doing this. I thought I had everything under control. One of the major rules with starting a fish farm was that the backup generator be in place before you stock fish on the farm. This is because in the event of a complete power loss, you have about fifteen minutes to get the pumps working again or you'll lose all your fish. I knew this. We were having one of our preconstruction meetings the week before we were breaking ground. Someone asked me about the backup generator. I said, "Don't worry, these are pretty much stock items and we can get them on short notice." *Wrong!* When I called our intended supplier, he informed me that these units were all custom built and the lead time was ten weeks. Fortunately for me, we wouldn't be stocking fish for at least twelve weeks. I immediately ordered the generator. What's the point? I should have constructed a network showing the interdependencies of all activities and events that would lead up to a successful conclusion of the project. If I had done this, I wouldn't have had the shock I encountered. In my case, I was fortunately able to recover. Don't rely on blind luck. Learn the basics of the two types of networks: the Critical Path Method (CPM) and PERT (Performance Evaluation Review Technique) charts. You won't be sorry or be left up the proverbial creek without a paddle.

13.1 Project Scheduling

In project implementation the major concerns are: time, cost, and scope. In the project implementation phase, time is money. Longer completion times generally lead to higher costs. Mistakes and oversights are unnecessary costs; plus, it takes more time to correct them, compounding the situation. Not properly identifying the scope, or end objectives of the project will also impact cost and your ability to schedule events accurately in a sequential manner. Proper contingency planning can help avoid these unnecessary cost and schedule impacts. Of course, the costs avoided by careful and detailed planning cannot be quantified. Cost avoidance is a nebulous concept. Did you really save $5,000 in medical expenses by getting a flu shot? You'll never know. You do know that the flu shot cost $20, however.

We accept some concepts on obvious merit, and we all agree that good planning and effective scheduling are fundamental parts of project management. We can easily see the after-the-fact results of poor planning and scheduling. At the outset of a project, however, the vision is not so clear. Humankind has been conducting large-scale projects since the Stone Age and has developed some techniques that are very helpful in planning and managing a project. This chapter presents some of these techniques and offers some guidelines for their use.

As you progress in your entrepreneurial efforts and initiate a series of new projects, you will increasingly realize the importance of carefully planning and defining each new project before starting. Defining a project is another way of saying

- how long it will take to implement it,
- what must be done to implement it, and
- how much it will cost?

Without proper planning and definition, nobody knows what the probable costs of a project will be because everybody will define the project in a different way, some with relevant assumptions and some excluding them. Then, when discussing the project, each participant will relate to the discussions based on his or her own set of assumptions about the activities, cost, and schedule associated with the project and make recommendations and decisions accordingly.

One of the major problems of unsystematic project planning is that everyone tends to consider the project from their particular set of experiences and interests and will discount or disregard the relevant aspects that fall outside of their sphere of interests. For example, the construction manager would tend to place more emphasis on the structural layouts, probable construction methods, and sequences than on the timelines required to define and obtain the equipment that is to be installed or the development of operational protocols necessary to operate the facility once it is constructed and outfitted. Therefore, a common definition of activities, events, and timelines that everyone associated with a project can use is essential.

So, before underestimating the importance of this chapter, remember this management trap. It is in our optimistic and egotistic nature to say, "I'm so good at what I do, that (1) I don't need plans and schedules, it just comes naturally, and/or (2) I am so wise and powerful that I can make everything happen in the best case time."

You may be your own boss, or you may be a project leader reporting to another person. In each case, what you do will have to be explained and justified to others. For this illustration, the others are identified as "the boss." For a start-up venture, this might be your primary investor or your board of directors.

Undoubtedly your boss will tell you that the project you are responsible for must be kept under some specific budget number, completed within a certain timeframe, and have the end-product fulfill the given requirements. Because time is money, your boss tells you that as project manager, you have the responsibility of ensuring that these three constraint parameters are met. As an able project manager, you will ensure detailed planning and extremely effective scheduling, resulting in a well-defined, precisely scheduled, and accurately budgeted project. You present your work to the boss. The boss is not pleased. "This costs too much. It takes too long. Go fix it." Can you? You can easily fix the presentation. All it takes is to change the numbers and list of tasks on the charts, and it will look like the boss wants it to look. Have you fixed the project? Of course not. Have you planned, scheduled, and budgeted your project correctly and efficiently and presented it in a way that is clear to others? Perhaps not. Be aware of the very powerful temptation to "manage the charts" as a substitute for managing the project. They are *not* the same thing.

> Be aware of the very powerful temptation to "manage the charts" as a substitute for managing the project. They are *not* the same thing.

Every project carries a certain level of uncertainty, which inevitably results in inefficiency. Inefficiency results in schedule slips and cost overruns. Therefore, it is the job of a good manager to do everything possible to minimize the uncertainty of their project. This is done by creating comprehensive plans and schedules with the recognition that the activities and events that must be accomplished to implement the project. There are useful tools such as CPM (Critical Path Management) available to the project manager that will enable these plans and schedules to be formulated, documented, and understood by others. These same tools and outputs can be updated and used throughout the life cycle of the project to assure that the management decisions remain objective and thorough. Through use of these tools, when uncertainties arise they can more easily be dealt with.

Unrecognized activities and events eat up valuable time and resources. Surprises are most easily tolerated at the outset of a program when projects still have some flexibility. It is very frustrating when a project assumed to be 70% complete all of a sudden turns out to be actually only 50% complete because a major activity was not foreseen and therefore was not planned, budgeted, and scheduled. Examples of these various types of uncertainties can be seen in the following charts:

Characterizing Uncertainty in Projects

TYPE OF UNCERTAINTY	PROJECT MANAGER'S ROLE	MANAGING TASKS	MANAGING RELATIONSHIP

VARIATION

Flow Chart
A linear flow of coordinated tasks (circles) represents the critical path towards project completion. Variation in task times will cause the path to shift unpredictably, but anticipating that and building in buffers (triangles) helps the team to complete project within a predictable range.a

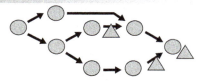

Troubleshooter and expeditor
Managers must plan with buffers and use disciplined execution.

Planning
• Simulate scenarios.
• Insert buffers at strategic points in critical path.
• Set control limits at which to take corrective action.

Execcution
• Monitor deviation from intermediate targets.

Planning
• Identify and communicate expected performance criteria.

Execcution
• Monitor performance against criteria.
• Establish some flexibility with key stakeholders.

FORESEEN UNCERTAINTY

Decision Tree
Major project risks, or "chance nodes" (circles), can be identified, and contingent actions can be planned (squares), depending upon actual events and desired outcomes (Xs).

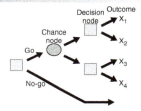

Consolidator of project achievements
Managers must identify risks, prevent threats and develop contingency plans.

Planning
• Anticipate alternative paths to project goal by using decision-tree techniques.
• Use risk lists, contingency planning and decision analysis.

Execution
• Identify occurrences of foreseen risks and trigger contingencies.

Planning
• Increase awareness for changes in environment relative to known criteria or dimensions.
• Share risk lists with stakeholders

Execution
• Inform and motivate stakeholders to cope with switches in project execution.

UNFORESEEN UNCERTAINTY

Evolving Decision Tree
The Project team can still formulate a decision tree that appropriately represents the major risks and contingent actions, but it must recognize an unforeseen chance node when it occurs and develop new contingency plans midway through the project.

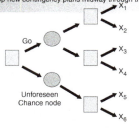

Flexible orchestrator and network as well as ambassador
Managers must solve new problems and modify both targets and execution method.

Planning
• Build in the ability to add a set of new tasks to the decision tree.
• Plan iteratively.

Execcution
• Scan the horizon for early signs of unanticipated influences.

Planning
• Mobilize new partners in the network who can help solve new challenges.

Execcution
• Maintain flexible relationships and strong communication channels with all stakeholders.
• Develop mutually beneficial dependencies.

CHAOS

Iterative Decision Tree
The project tem must continually create new decision trees based on incremental learning Medium-and long-term contingencies are not plannable.

Entrepreneur and knowledge manager
Managers must repeatedly and completely redefine the project.

Planning
• Iterate continually, and gradually select final approach.
• Use parallel development.

Execution
• Repeatedly verify goals on the basis of learning: detail plan only to next verification.
• Prototype rapidly.
• Make go/on-go decisions ruthlessly.

Planning
• Build long-term relationships with aligned interests.
• Replace codified contracts with partnerships.

Execcution
• Link closely with users and leaders in the field.
• Solicit direct and constant feedback from markets and technology providers.

Everyone seeking to be an effective project manager will agree that sound planning, scheduling, and budgeting is essential. Cost overruns and project failures can be minimized by using a structured mode of evaluating, planning, and reviewing projects. Many techniques have been developed over the years to assist with this aspect of management. Each of them requires the application of certain techniques to record and measure the completion of project milestones. In this chapter, two of these techniques are discussed:

- **Critical Path Method** (CPM), which employs a deterministic approach
- **Program Evaluation and Review Technique** (PERT), which uses a probabilistic approach (involving probability of outcomes)

However, it cannot be overemphasized that either technique will not compensate for incomplete or inaccurate fundamentals.

The development of a CPM chart, a network of project activities or tasks showing their sequencing and their estimated start and finish times, is extremely useful at the very beginning phases of a project. It can help with the development of a realistic project schedule, which in turn is necessary for the development of a realistic project budget.

Consider a project involving numerous tasks or activities. Each activity requires resources: people, materials, and equipment to complete. The more resources allocated to any activity, the shorter the time that may be needed to complete it. Implementing a CPM can address questions such as

- How long will the entire project take to complete?
- Which activities making up the project determine total project time?
- Which activity time durations should be shortened based on cost considerations?

So, is a CPM chart necessary from the initial conceptualization of a project? Yes. Absolutely. Without it, the project manager has no means of assessing the relationship of project activity, project schedule, and project budget. The objective of this chapter is to understand how and why these tools will assist in formulating and documenting a presentable and justifiable project plan, schedule, and budget.

Background

CPM and PERT scheduling were created and refined by the US Department of Defense for use in the acquisition of complex weapon systems. These two approaches are both simple to use and effective when applied to planning and managing projects.

There is a general technique associated with developing these tools. The CPM tool uses a ***node-link network diagram*** to show each activity or task within the project and the required sequence of each of these tasks in relation to other tasks. The planner estimates the duration of each task and then the optimum sequencing and the earliest and latest start and finish times for the individual tasks that minimize the total project duration. This results in a probable and realistic schedule for the entire project.

These calculations also identify the ***critical path*** of the project. This is the sequence of tasks where delays in the completion of one task will affect the completion time of the entire project and where addition of resources can speed up the project. This reinforces the fact that CPM and PERT are only appropriate when you know the tasks of the project or process, their sequence, and how long each activity or task takes. It is critical to understand these facets of your project before you start.

CPM and PERT charts are useful for

- planning a complex project or process with interdependent or interrelated tasks and resources;
- analyzing the timing of project tasks;
- allocating resources such as people, money, and time to a project; and
- monitoring project progress against schedule and budget constraints.

The techniques of creating these tools call for

- identifying project **milestones** (defined as major events within a project);
- defining all the actions necessary to accomplish these milestones;
- estimating the time the actions are likely to take; and
- estimating the cost associated with performing these actions.

Overview

CPM is a technique that recognizes two aspects of a project: ***events and activities***. Events are a point in time, representing a completion of an activity or group of activities. Activities or tasks require periods of time when actions are taken and that, when completed, result in an event. For example, "load the boat" is an activity whereas "cast off" is an event after the completion of the activity. An event is either complete or incomplete, whereas activities can be measured in increments of completion. In the preceding boat example, the boat either has cast off or it is has not. However, the activity of loading the boat can be measured by degree of completion and the remaining portion of activity can be estimated. Through this method, one can assess the degree of project completion and compare that assessment to other criteria.

Usually project managers are interested in the degree of completion in comparison to three criteria:

- requirements,
- schedule, and
- budget.

If, through use of the CPM technique, the project manager discovers that the project has met 30% of the requirements, 42% of the project has been completed, and the budget is 68% spent, the success of the project could be in serious trouble. Should such a discouraging but all too frequent discovery be made, the project manager could then evaluate the CPM network to decide where best to concentrate recovery efforts (we address this later in the chapter).

The CPM process works well for any type of project because the same basic methods are suitable for projects of all sizes. Small projects can be outlined quickly, and at the same time, very large and complex projects can be divided into smaller increments. CPM techniques may be applied to those smaller and smaller increments until the project is subdivided into manageable segments. However, CPM is not the only management discipline one needs to assure a viable project. This is a tool and must be used in concert with other prudent management processes and techniques during the life cycle of a project. A bad plan is a bad plan, regardless of how well it may be documented, but CPM techniques will enable a quicker recognition and alteration of problems as they arise.

> A bad plan is a bad plan, regardless of how well it may be documented.

Simple Example for Developing a CPM Chart

Let's revisit the example of building a boat introduced earlier in the chapter. We have been given an order: cast off the boat. That is definitely an event. Did you assume we had a boat to cast off? Well, we don't. We must buy, steal, borrow, rent, or build a boat. Let's say that after completing the appropriate "make/buy/ source of supply" analyses, we decided to build the boat, and we can assume that we know how big the boat must be (our boss told us).

Starting today, what are the events that must take place in order to achieve a successful result? In our example, let's assume we want to build a boat for transporting heavy cargo across Lake Erie. We must (a) buy the materials, (b) get a set of blueprints, (c) hire a crew of workers, (d) build the boat, (e) load the boat, and (f) launch the boat. Is that it? Well, no. Each of these events requires many subordinate decisions and associated activities and events.

   a. Buy the materials: What kind? How many? Where? Delivery or pick-up? How can we do this before we know what the blueprints call for?
   b. Get a set of blueprints: Where? What kind of stuff are we going to haul? Passengers, crude petroleum, cars, or general cargo? Is this an oceangoing boat, or is it to be used on lakes and rivers?
   c. Hire a crew of workers: What skills? Welders or carpenters? Electricians or sailmakers? Where shall I tell them to show up? When is payday?

All of these subordinate events and associated activities will benefit from the application of CPM. Having evaluated the requirements, we have decided to build a steel-hull, diesel-powered boat capable of hauling general cargo on the Great Lakes. So, let's plan this project.

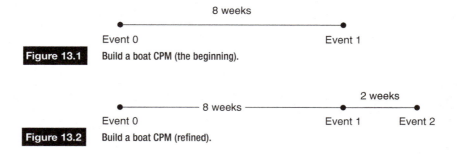

Figure 13.1   Build a boat CPM (the beginning).

Figure 13.2   Build a boat CPM (refined).

We must decide which activities can take place simultaneously and which must be done in a sequential order. In our boat project, we can't buy materials until we know what materials are to be used in the boat's construction, nor should we hire workers. And we certainly can't build the boat until the materials are available and a workforce is in place.

It seems that our first event should be to get a set of blueprints. Now we must discover how long this task is going to take. We should be able to procure a set of prints in eight weeks (per our initial estimate). Therefore, starting today, on our CPM chart, we can line out eight weeks as the activity of "get a set of prints" and, at the end, draw the symbol for the event (Blueprints; refer to Figure 13.1.)

Having a viable set of plans, can we then hire the workers? Yes, after an analysis of the plans and materials required, we can identify the different skills required and the number of people of each skill. Can we also order the materials? Yes, after the analysis tells us how much of what kinds of materials we need. Can these activities take place simultaneously? Yes, they can, after the analysis.

Wait a minute! We've just identified another activity and event that must be concluded before our next activity can begin. We need to add "analyze plans for labor and materials requirements" to the CPM chart and to determine the completion event time.

We should be able to analyze the plans to determine these requirements in two weeks time. Go to the chart and add a two-week activity time after the conclusion of the previous event (event 1, buy plans), with a completion symbol (event 2, analyze plans; refer to Figure 13.2).

Now, we can proceed with two simultaneous activities:

a. Hiring workers: How long will it take to hire our workforce? We'll assume that labor is plentiful and that all skills are readily available. We need only advertise in the local newspaper and/or contact the local union hall. We estimate that our workforce can be hired and in place in one month's time.

b. Obtaining construction materials: We know that we need X sheets of 3/8-inch sheet steel, Y feet of steel beam, Z feet of electrical wire, and so on. We know that the time between ordering and delivery of the steel components is two months. The wire is readily available for immediate delivery.

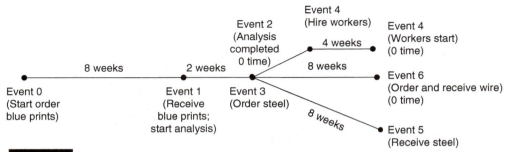

Figure 13.3   Build a boat CPM (refined a third time).

Go to the chart and add a line for "hire workers" and assign a four-week time frame to that activity. Go to the chart and add a line for "order steel." Because the steel won't be available until two months after it is ordered, assign an eight-week time frame for that activity. (Refer to Figure 13.3.)

Now, we have valuable information. We know that to maintain the most efficient schedule, we must order the materials as soon as we know what they are. We also know that we should not begin the activity of hiring a workforce until four weeks after the materials have been ordered, or we will have workers on a payroll with no work to do. Now, we can assign a start date to our "hire workers" activity and still maintain the completion event simultaneous with delivery of materials.

Our next activity is "Build the Boat." This is a big task and should be broken into subordinate activities. Plan for a twenty-six-week construction period. Of our identified activities, no other can take place until the boat is built, so we have only one activity on the CPM chart until the completion event, noted as twenty-six weeks after the arrival of the materials. (Construct your own chart to show the completion of activities.)

Now that we have a boat, we can load it. What kinds of skills are necessary to load the boat? What kind of stuff are we going to load? Do we need cranes or other items of support equipment? Special containers? Special environmental conditioning capability?

All these questions must be answered before we can load the boat. CPM is applicable as a sub-tier activity in each of these. Having given due consideration to these sorts of questions, we determine that we can load the boat in one week's time. Go to the CPM chart and add an activity line annotated as one week, and a "Boat Loaded" completion annotation.

Now, we can cast off, and our project, as we have defined it, is complete. Of course, it is possible that our project is a sub-tier of a larger project, such as "Evacuate Ohio," and the manager of that project can now mark off one of his or her CPM chart events "Cast off the Boat" as complete. Referring to our CPM chart for our particular project, we can see that our schedule should call for a forty-five-week time line from start to completion (Figure 13.4).

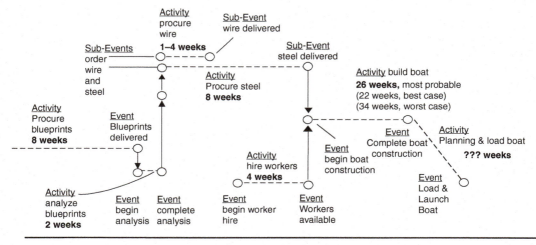

Project Duration Weeks

Figure 13.4 Completed simple example of implementing a CPM chart for activities and events that result in completing a boat and final event launch.

13.2 The Techniques of Implementing the Critical Path Method

CPM Networks

The first step in employing CPM is to list all activities necessary to complete the project. Next, create an *activity network*, a diagram of the sequence of activities, with time flowing from left to right. The example below illustrates how to draw an activity network.

Assume that you have identified five activities that make up your projects: A, B, C, D, and E. In this example, C cannot begin until A and B are completed, D must begin after B is completed, and E must follow C. You can choose between two kinds of generally accepted networks to create the CPM network for your given project:

- activity on node (AON)
- activity on link (AOL)

If *nodes* represent activities, that is, work being performed, then the **AON network** looks like Figure 13.5. The links symbolize events, that is, identifiable completion of a task, within the project such as the completion of a single activity or set of activities.

If *links* represent activities, then the **AOL network** looks like Figure 13.6. Notice that the nodes on the diagram now symbolize the events of the project.

If an activity cannot be completed before two or more activities reach completion, a "dummy link" is added. The "dummy link" is a dashed line, which is required to show proper sequencing and does not indicate any action. In this example, activity C may not begin before both activities A and B are completed.

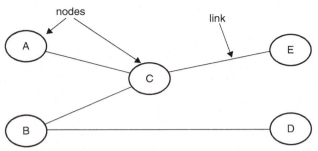

Figure 13.5   AON network (activities on nodes).

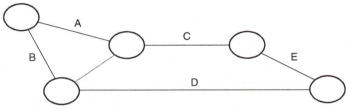

Figure 13.6   AOL network (activities on links).

Without the dashed line, the diagram does not illustrate the need for activity C to wait for the completion of activity B.

Project Duration and Activity Start and Finish Times

Each activity must be assigned a duration. The ***activity time*** is the best estimate of the time that each task should require. Be sure to use one consistent measuring unit, for example, weeks or days. For this example, the duration times for each activity are listed in Table 13.1

To compute the project duration (time) and the sequence of activities that determine the total project time, the first step is to compute ***earliest start time*** (***EST***) and ***earliest finish time*** (***EFT***) for each activity. Calculate the earliest times each task can start and finish, based on how long preceding tasks require. Start with the first task, where the earliest start is zero, and work forward.

Earliest start time (EST) of an event = largest earliest finish of the tasks leading into this event

Earliest finish time (EFT) of an event = earliest start + task duration time

In order to assure that no activity begins until those it must follow are completed, each activity's *EST* and *EFT* values should be written on the network next to the activity. The numbers in the nodes or on the links are the activity durations. In our example, we can calculate these values in a table, shown in Table 13.2.

Representing this as an AON and AOL charts we can see visual representations as shown in Figures 13.7 and 13.8.

Next, compute *latest start and finish times* (*LST* and *LFT*, respectively) without extending the total project duration (twenty-eight weeks in this example).

Table 13.1 Activity Durations of Project

Activity	Duration Time
A	10 time units (e.g., weeks)
B	14
C	6
D	11
E	8

Table 13.2 Activity Durations of Project with *EST* and *EFT*

Activity	Must Follow	Duration Time	Earliest Start Time *EST*	Earliest Finish Time *EFT*
A	–	10 weeks	0	10 = EST + Duration
B	–	14	0	14
C	A & B	6	14 = max(*EFT* A, *EFT* B)	20
D	B	11	14	25
E	C	8	20	28

AON:

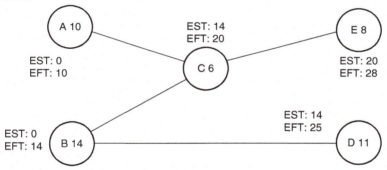

Figure 13.7 AON network with *EST* and *EFT*.

AOL:

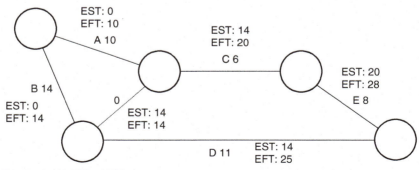

Figure 13.8 AOL network with *EST* and *EFT*.

Table 13.3 Activity Durations of Project with EST and EFT

Activity	Must Follow	Duration Time	Latest Start Time LST	Latest Finish Time LFT
E	C	10	20 = LFT − 10	30
D	B	14	17	28
C	A & B	6	14	20
B	–	11	0	14
A	–	8	4	14

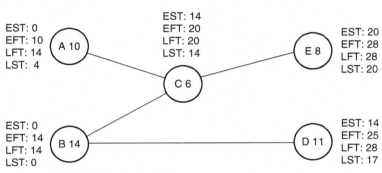

AON:

Figure 13.9   AON network with *EST, EFT, LST,* and *LFT.*

Project Duration = max*(EFT* A, *EFT* B, *EFT* C, *EFT* D, *EFT* E)

Once we have identified the duration, we can begin calculating the *LST* and *LFT* (see Table 13.3). Start from the last task, where the latest finish is the project deadline, and work backward. Do this by working from right to left, that is, from the end to the beginning. At each activity, first determine *LFT* and then *LST.*

Latest finish time (LFT) of an event = the smallest latest start of all the tasks leading out of this one

Latest start time (LST) of an event = latest finish – task time

These caculations with constraints as defined in Table 13.3 are visually represented in an AON or AOL chart as shown in Figures 13.9 and 13.10.

Instead of writing *EST, EFT, LFT,* and *LST* next to its corresponding value for each activity, some people prefer to use the box method as shown in the following table to organize each value. Write all four times – *EST, EFT, LST,* and *LFT* – beside the task arranged as follows:

Earliest Start (*EST*)	Earliest finish (*EFT*)
Latest Start (*LST*)	Latest finish (*LFT*)

AOL:

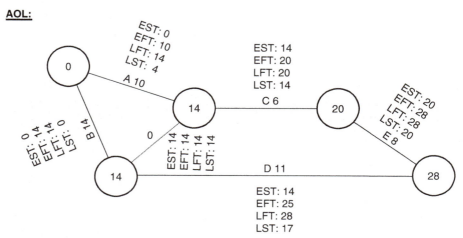

Figure 13.10 AOL network with *EST, EFT, LST,* and *LFT.*

Critical Path

The **critical path** is the "must know" piece of management information associated with any project. It shows how long a total project will take. The critical path is a key factor in establishing any necessary contingency plans. Now, having identified the earliest start, the earliest finish, the latest start, and the latest finish for each event in the project, you can identify the project's critical path.

The critical path is defined as the sequence of activities that collectively determine the total project time. The critical path is that sequence of activities that has identical early and late start times and consequently identical early and late finish times. In this example, it is B → C → E. If the duration of any activity on the critical path is increased, the time to complete the total project will also increase by the same amount(s).

Activities not on the critical path – in this example, activity A and activity D – can start more flexibly. Activity A can start anytime between 0 and 4, and activity D can start anytime between 14 and 17 without delaying the total project completion time.

Total, Interfering, and Free Slack

Some activities can be completed in a shorter time than is allowable based on critical path evaluation. This is because of something called **slack** (sometimes identified as **float**). In general, slack increases a project manager's ability to successfully manage the project.

Total slack (TS) is the time a task start could be postponed without delaying the project schedule.

$$\textit{Total slack (TS)} = \textit{LST} - \textit{EST} \text{ or } \textit{LFT} - \textit{EFT} \tag{13.1}$$

The *free slack (FS)* of a task is the time that task can be postponed without affecting the total slack of the network or change the earliest start time of any future task.

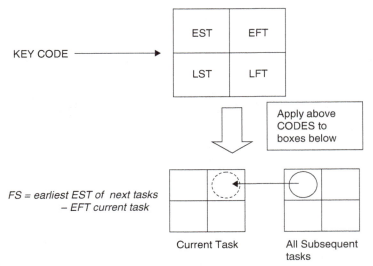

Figure 13.11 Arrow diagram: how to calculate slack times (arrow points to the time you are subtracting).

Free slack (FS) = the earliest EST of all the tasks leading out of (i.e., following) this one – EFT of current task

Interfering slack *(IS)* is any slack that will force a delay in the earliest start time of one or more subsequent activities. The total slack of an activity is the sum of its free and interfering slacks:

$$\textit{Interfering Slack (IS)} = \textit{Total Slack (TS)} - \textit{Free Slack (FS)} \text{ or}$$
$$TS = IS + FS. \tag{13.2}$$

Figure 13.11 shows a schematic way to remember how to calculate slack. First, complete the quadrant methodologies outlined in the previous section to organize the earliest start and finishes and latest start and finishes for each task on the network diagram. After completing this exercise, calculating the different types of slack requires just a few additional steps.

Total slack for each task is calculated by subtracting the earliest start time, *EST* (upper-left quadrant), for a task from the latest start time, *LST*, for a task (lower-left quadrant). Alternatively, you may choose to calculate total slack, *TS*, by subtracting the earliest finish time, *EFT* (upper-right quadrant), from its latest finish time, *LFT* (lower-right quadrant). Completing both calculations is an easy way to check your work. *Both calculations should deliver the same answer.*

Calculating free slack using the quadrant methodology is a little more complicated than total slack. Two numbers are critical in calculating the free slack of a task:

- the *EFT* of the task being calculated (upper right quadrant) and
- the smallest *EST* of the tasks directly following the calculated task (upper left quadrant).

When a task has only one task following it, simply subtract the *EFT* of the task being calculated from the *EST* of the task directly following it. Be sure to note that if multiple tasks branch from the task being calculated, choose the smallest *EST* from the set. Although this may initially sound complicated, after working through a few practice problems, you will find this method makes intuitive sense.

After you have calculated both total slack and free slack for all tasks on the network diagram it is very easy to calculate the interfering slack of each task. Simply subtract the free slack for each task from its corresponding total slack. The quadrant methodology can be very useful in organizing the CPM calculations for complicated network diagrams into an easy-to-read format. In the end, formatting the start and finish times into quadrants will save you time and cut down on mistakes in calculations.

There are numerous texts available that will present a more in depth treatment of different approaches to applying the CPM tool.[1]

Resource Allocations

Once the CPM network has been constructed with the critical path and slack times identified, the project manager can evaluate alternatives for performing the activities of the project. The goal is to determine if more effective resource allocations are possible. For example, if a special machine is needed for activities A and C, and the machine costs $2,000 to bring to the site and $500 per week to operate, then clearly activity A should start at week 4. If the machine were needed for activities C and D, then one could compare and select the minimum cost of

- renting two machines instead of one;
- beginning D at twenty, causing a delay in project completion by week 3; and
- shortening activities C and/or D by a total of three weeks so that they do not overlap.

In this case, activity D will be on the critical path. The activities not on the critical path will not have an impact on the total project completion time. There are no cost savings for shortening the duration of activities not on the critical path. As critical-path activity durations are shortened, the critical path itself may change and multiple critical paths may result.

Examples

EXAMPLE 13.1: INTERIOR BUILDING CONSTRUCTION

QUESTION

Suppose that you are a project manager working on the completion of a building. The exterior structure has already been finished, leaving only the interior to be constructed and interior partition wall frames have been built. Your supervisor asks you to give your best estimate of how much time it will take to complete the job. Using CPM techniques, you break the project into smaller components

---

[1] For example, see S. Keoki Sears, Glenn A. Sears, and Richard H. Clough. Construction Project management: A practical guide to field construction management, 5th edition, (Hoboken, NJ, : John Wiley & Sons, 2008).

Table 13.4 Activities, Immediate Predecessors, and Durations

Activity	Immediate Predecessor	Duration (days)
A – put in windows	–	10
B – do electrical wiring	A	18
C – do Internet wiring	A	16
D – install dry wall coverings on walls	B,C	24
E – paint walls	D	15
F – lay flooring	E	12
G – put in lighting fixtures	E	8
H – place wall hangings	G	6

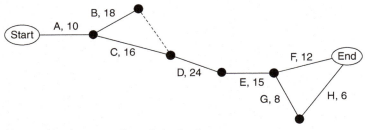

Figure 13.12 Diagram of interior construction project network.

and assign time estimates for each activity. You also label which activities must be completed before a specific activity can begin. These values are found in Table 13.4. (Note: The list of activities is an oversimplified model.)

Based on the estimates given above, determine the following:

a. Draw a project network. Determine the project duration and critical path of the project.

b. Find the total, free, and interfering slack for each activity. Explain the significance of these values.

c. The hiring of additional workers can shorten the duration of some activities. The cost of reducing the activities per day is given in Table 13.5. You are told the job must be completed in 68 days. Which activities should you speed up and by how much?

Solution:

• *Part (a)*

Figure 13.12 illustrates the correct project network using the guidelines from Table 13.5. Notice the diagram is an AOL (activity-on-link) network. The dashed line between activities B and D shows that both activities B and C need to be completed before activity D may begin.

The EST and EFT for each activity may be completed for the project by beginning with activity A and working from left to right. Notice there are two activities at the end of the network that could possibly be the project completing activity. The activity with the largest *EFT* is considered to be the final activity and indicates the project duration. Based upon the largest value of *EFT* in Figure 13.13, the project duration is estimated to be eighty-one days.

Table 13.5 Cost of Shortening Each Activity to Maximum Number of Days

Activity	Total Cost to reduce Activity duration	Reduction Days
A (10 days)	$2,000	1
	$4,500	2 (Hence cost of second day reduction is $2,500)
	$7,500	3
B (18 days)	$1,250	1
	$3,000	2
	$5,000	3
	$7,500	4
C (16 days)	$1,000	1
	$2,500	2
	$4,500	3
D (24 days)	$15,000	6 (Constant cost per day reduced of $2,500)
E (15 days)	$1,500	1
	$3,500	2
F (12 days)	$2,250	1
	$5,000	2
G (8 days)	$2,750	1
	$6,000	2
H (6 days)	$1,000	1

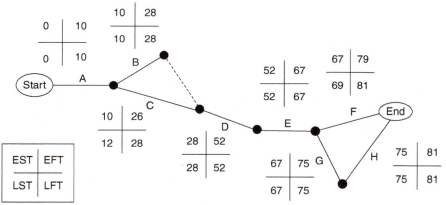

Figure 13.13 Values used to calculate duration and critical path (each node shows the calculated *EST, LST, EFT*, and *LFT* times).

To find the critical path, *LST* and *LFT* values for each activity must be calculated. *The critical path lies along the sequence of activities where EST equals LST and EFT equals LFT.* Based on this defining criterion, you can see that the critical path is defined to be A, B, D, E, G, H (see Figure 13.13).

• *Part (b)*

Using Figure 13.13, you can calculate slack values for each activity. A good rule of thumb to remember is that *any activity* that lies along the critical path *has no slack.* Thus, the total slack, free slack, and interfering slack for A, B, D, E, G, and H are zero.

Calculate total slacks for the remaining activities by subtracting their *EFT*s from their *LFT*s. Activity C has two days of slack (twenty-eight to twenty-six) as well as activity F (eighty-one to seventy-nine). The free float for activity C is

Table 13.6 Summary of Calculated Values for Interior Construction (critical path is along A, B, D, E, G, and H)

ACTIVITY	EST	EFT	LST	LFT	TS	FS	IS
A	0	10	0	10	0	0	0
B	10	28	10	28	0	0	0
C	10	26	12	28	2	2	0
D	28	52	28	52	0	0	0
E	52	67	52	67	0	0	0
F	67	79	69	81	2	2	0
G	67	75	67	75	0	0	0
H	75	81	75	81	0	0	0

calculated by subtracting its *EFT* of twenty-six from the smallest *EST* leading out of that activity. In this case, only one activity leads out of activity C. Activity D's *EST* is twenty-eight. Thus, the free float for activity C is two. Activity F must be evaluated in the same way. However, no activities lead out of activity F. This is where understanding the definition of free slack becomes essential. Free slack is the time allowed to postpone an activity without affecting the slack of any job following it. Because activity F could be postponed for two days without affecting any other activities, it contains a free slack of two days. The problem contains no interfering slack for the reason stated above. Table 13.6 summarizes these conclusions.

• *Part (c)*

To complete the job in sixty-eight days, the project must be shortened by thirteen days. The only way to reduce the duration of a project is to reduce the time taken to complete one or more activities on the critical path, because the sum of activity durations on the critical path determines the total project duration. Shortening durations of activities not on the critical path has no impact on the total project duration, except perhaps by adding to its cost. However, as the durations of activities on the critical path are reduced, it is possible that multiple critical paths will develop. Unless an activity is on each critical path, shortening its duration may not reduce total project duration.

To illustrate this consider reducing the example project by thirteen days. We can do this a day at a time, picking the least cost option. (We can do this to reach a minimum total cost of reducing the project by thirteen days because the marginal cost of reducing each activity by an additional day is nondecreasing.) After each day reduction, we can check to see if a new critical path develops.

The least expensive way to reduce the duration of this project by one day is to reduce activity H by one day at a cost of $1,000. Thus, the sequence of reductions of one day each will start with activity H for one-day reduction. This is followed by activity B for the next day of reduction, then activity E for the third day reduction, and then B again for the fourth day of reduction. At each step, the critical path stayed the same. To reduce another day either, A or E can be picked, for a

Table 13.7 Activities and Number of Days Shortened to Shorten CP One Day at a Time (The activity chosen each day is the least expensive option for reducing the total project time by one additional day.)

Activity	Days Reduced	Total Project Days Reduced	Total Extra Cost
H	1	1	$1,000
B	1	2	$1,250
E	1	3	$1,500
B	1	4	$1,750
E	1	5	$2,000
A	1	6	$2,000
A	1	7	$2,500
D	6	1	$2,500 × 6 = $15,000
Total	13		$27,000

Summary Activity	Days Reduced	Total Cost
A	2	$4,500
B	2	$3,000
C	0	$0
D	6	$15,000
E	2	$3,500
F	0	$0
G	0	$0
H	1	$1,000
Total	13	$27,000

total of five days' reduction. However, at this point of having reduced five days, there are now two critical paths: A, B, D, E, G, and H and A, C, D, E, G, and H.

For the sixth day reduction, either A or E can be reduced by a day, which ever was not picked for the fifth day reduction. Both of these activities are on both critical paths. Once six days are reduced, every activity is on a critical path except activity F. Activity F would be critical if activity G were reduced by a day. If we reduce activity B, we must also reduce activity C by the same number of days. If we reduce activity G by two days, we would have to reduce activity F by one day to reduce the project time by two days.

Reducing activity A by another day and reducing activity D by six days meets the total time reduction constraint of thirteen days at a total minimum cost. Both activities A and D are on each critical path, and therefore, their reduction is not dependent on the duration of any other activity. Table 13.7 shows the sequence of reductions and costs.

Notice that the activities shortened are always on the critical path(s). When these networks contain several hundred tasks, these early and late start and finish time calculations together with least cost time reductions motivate the use of CPM and PER software analysis packages. Figure 13.14 is the AOL schematic for the new project duration of only 68 days. Now, the only activity that is not part of a critical path is activity F.

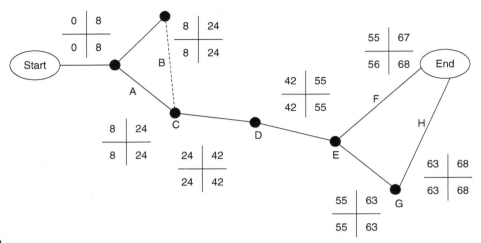

The AOL schematic for the new project duration of only 68 days (now, the only activity that is not part of a critical path is activity F).

13.3 Program Evaluation and Review Technique

The PERT method extends CPM by considering the uncertainty in estimating activity durations in order to estimate the probability of finishing a project in a given time. Three time estimates are required for each activity's duration:

- most likely (m)
- optimistic (a)
- pessimistic (b)

The expected value of activity duration, e_i, is a weighted combination of the most likely (mode of the distribution) and the midpoint of the distribution $\dfrac{a_i + b_i}{2}$ for each activity i. Then we can calculate expected value as

$$e_i = \frac{2m + (a + b)/2}{3}. \tag{13.3}$$

Notice that a weight of two-thirds is placed on the mode of the distribution and a weight of one-third on the midpoint of the distribution. equation (13.3) may be stated more simply by multiplying the numerator and denominator by two.

$$e_i = \frac{4m + a + b}{6} \tag{13.4}$$

It is assumed that the duration distribution is a spread of about six standard deviations, or $\pm 3\sigma$, which is 99.7% of the area of the normal distribution curve. It follows that standard deviation and variance may be described by the following equations:

$$\sigma = \frac{b-a}{6} \qquad \sigma^2 = \left(\frac{b-a}{6}\right)^2. \tag{13.5}$$

Note that the variance, σ^2, is based on our estimates of the worst and best duration times to complete some specific task. It should also make sense that the more you break a large task into smaller tasks, the better your estimates would become.

The process of solving for the critical path is identical to the CPM process given in the previous section. However, the activities are now based on these new expected durations from the uncertainty analysis. Expected project duration, PD, can be calculated by summing all of the expected activity durations along the critical path of the project network.

$$D = \sum_i e_i \tag{13.6}$$

Expected project standard deviation, σ_D, is the square root of the sum of all project variances, σ_i^2, along the critical path.

$$\sigma_D = \sqrt{\sum_i \sigma_i^2} \tag{13.7}$$

Figure 13.15 illustrates the uncertainty of a project's completion time. The expected project duration time, calculated by summing the activity durations along the critical path, will always lie in the center of the normal distribution chart. Most projects have a specified length of time in which a problem must be complete, T. This project completion time can also be expressed in terms of the calculated project duration time and standard deviation.

$$T = PD + z \cdot \sigma_D \quad \text{or} \quad z = \frac{t - PD}{\sigma_D} \tag{13.8}$$

In equation (13.8), notice that another variable, z, is used in the expression known as the z-value. It is a scaling factor for the project's standard deviation. The z-value scales the standard deviation such that the calculated duration (D) and the scaled standard deviation ($z\sigma_D$) equal the time the project must be completed.

To calculate the probability that the project will be completed by the deadline T, look up the value for z on a z-chart, which is found at the end of the chapter (see Figure 13.15). By reading the chart, you can find the probability that the project will be completed before the project due date. Table 13.8 gives select values of z and their corresponding probabilities.

Using Table 13.8 you can see the construct of the z-chart. If the calculated duration time is exactly the same length as the deadline ($z = 0$), there is a 50% chance the project will be completed on time. If the z-value is greater than zero, your chances increase that the project will be done on time. Conversely, if the

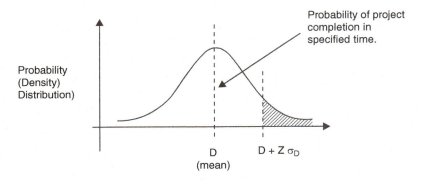

Figure 13.15 Normal distribution of project completion times.

Table 13.8 Select Values from z-Chart

z-stat	−1	0	1	2
Probability	15.9%	50%	84.1%	97.7%

Table 13.9 Activities, Immediate Predecessors, and Durations

Activity	Immediate Predecessor	m	a	b
A – put in windows	–	10	9	12
B – do electrical wiring	A	18	16	19
C – do Internet wiring	A	16	14	19
D – put in walls	B, C	24	22	29
E – paint walls	D	15	14	17
F – lay flooring	E	12	10	15
G – put in lighting fixtures	E	8	7	11
H – place wall hangings	G	6	5	7

z-stat is less than zero, your chances of completing the project by the deadline are less than 50%.

Example 13.2: Interior Building Construction Continued

QUESTION:

Now that you have finished completing the CPM calculations for the interior building construction project, your boss wants you to evaluate the project again using the PERT Chart method. The duration values used earlier are now the most likely (m). You estimate the pessimistic (a) and optimistic (b) values and create the values shown in Table 13.9.

Based upon the estimates given above, determine the following:

a. Draw a project network. Determine the project's expected duration and critical path of the project.

b. What's the probability the project will be completed in 84 days?

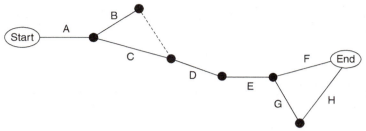

Figure 13.16 Diagram of interior construction project network.

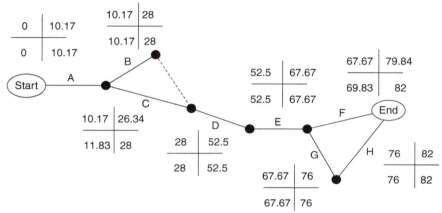

Figure 13.17 Values used to calculate duration and critical path.

Solution:

Part (a)

The project network is the same as in Example 13.1, interior building construction, see Figure 13.16.

However, this time we have to calculate the activity durations using the expected value formula. The expected value for activity A is calculated in the following manner.

$$e_a = \frac{4m + a + b}{6} \tag{13.9}$$

where *m = 10, a = 9, b = 12*.

Thus, the expected activity duration of activity A is 10.17 days. Table 13.10 contains the expected activity durations.

Table 13.10 Expected Activity Durations

Activity	Immediate Predecessor	Expected Duration
A – put in windows	–	10.17
B – do electrical wiring	A	17.83
C – do Internet wiring	A	16.17
D – put in walls	B, C	24.5
E – paint walls	D	15.17
F – lay flooring	E	12.17
G – put in lighting fixtures	E	8.33
H – place wall hangings	G	6.00

The *EST, EFT, LST*, and *LFT* for each activity are labeled in Figure 13.17.

Remember, the critical path is the sequence of activities that contains no slack. Figure 13.17 illustrates that the sequence A, B, D, E, G, and H is once again the critical path. Based on the new data, project expected completion will occur in eighty-two days.

Part (b)

The variance of each activity along the critical path must be calculated to estimate the probability that the project will be completed in eighty-four days. Remember that the probability it would be completed in eighty-two days (the expected duration) is 50%, $z = 0.00$. The variance for activity A can be found be applying the following formula:

$$\sigma_A^2 = \left(\frac{b-a}{6}\right)^2 \tag{13.10}$$

where $a = 9$, $b = 12$.

$$\sigma_A^2 = \left(\frac{12-9}{6}\right)^2 = \left(\frac{1}{2}\right)^2 = .25$$

The rest of the critical path variance calculations are located in Table 13.11.

Table 13.11 Critical Path Activity Variances

Critical Path Activities	Expected Variance
A – put in windows	0.25
B – do electrical wiring	0.25
D – put in walls	1.36
E – paint walls	0.25
G – put in lighting fixtures	0.44
H – place wall hangings	0.11

The variance of the project is the sum of the activity variances along the critical path. The standard deviation of the project is the square root of the project variance.

$$\sigma_D = \sqrt{\sum_i \sigma_i^2} = \sqrt{(.25 + .25 + 1.36 + .25 + .44 + .11)} = 1.63$$

Now you are ready to solve for the z-value.

$$z = \frac{t-e}{\sigma_D} \text{ where } t = 84 \ days, \ e = 82 \ days, \ and \ \sigma_D = 1.63$$

$$z = \frac{84-82}{1.63} = \frac{2}{1.63} = 1.23$$

By using the z-chart found at the end of the chapter (Figure 13.14), the probability that the project will be completed in 84 days is approximately 89% (0.8907 to be exact for a Z = 1.23).

Remember that this type of computation is not robust. A critical path with a short duration and large standard deviation may be very risky. A path along the project network that has a little longer project duration but smaller standard deviation may prove to be the more prudent choice (avoids risk). Z-test evaluations may be used for investigating a series of activities such as example 2, or it may be used for a one-activity project. Example 3 illustrates how the z-chart can apply to a single project or activity.

Example 13.3: Construction Project (PERT)

Consider a construction project. We start the planning of this project in a relatively simple way and add detail as we learn more about what may be involved. To begin, consider the construction of concrete footings that requires earth excavation, reinforcement, formwork, and concreting. The tasks include the following:

A Layout and survey of site for foundations
B Dig holes for foundations
C Construct and place formwork
D Obtain steel reinforcement
E Cut and bend steel reinforcement
F Place steel reinforcement
G Obtain concrete mixing materials
H Mix (add water to proper slump) and Pour concrete

These separate tasks are more or less in the order they need to be accomplished. Yet some of these tasks can be done simultaneously. The network diagram in Figure 13.18, identifies which tasks (nodes, so we are using the AON method here) must precede other tasks and which can proceed towards completion simultaneously.

This network shows that tasks A, D, and G can begin at the same time. The foundation chain of tasks A, B, and C, on the other hand, must be completed in succession; the steel chain of tasks D and E also must be completed in succession. Task F, involving both steel and foundation, must follow the completion of all of these previous tasks. Finally, task H must follow the completion of all other tasks. Usually, the arrows in each of the links of Figure 13.18 are omitted. and it is assumed tasks must proceed as shown from left to right.

The next step in this planning exercise could be the estimation of each task duration time under what are considered "normal" situations. The actual duration will depend to some extent on the number of workers and attention allocated to the task, and of course on unforeseen factors such as weather, unexpected geologic conditions, delays caused by legal problems or in obtaining the necessary

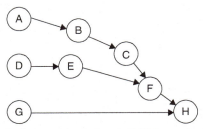

Preliminary network diagram (AON) for constructing foundation, shown with and without arrows signifying which tasks must precede, or follow, other tasks.

Table 13.12 Foundation Construction Tasks, Sequencing, and Expected Durations and the Float for Each Task

Task	Duration	EST	EFT	Float LST	LFT	Total	Free	Interfering
A	14	0	14	0	14	0		
B	7	14	21	14	21	0		
C	5	21	26	21	26	0		
D	2	0	2	14	16	14	0	14
E	10	2	12	16	26	14	14	14
F	8	26	34	26	34	0		
G	4	0	4	30	34	30	30	0
H	6	34	40	34	40		0	

Task	Must Follow	Duration
A	–	14
B	A	7
C	B	5
D	–	2
E	D	10
F	C, E	8
G	–	4
H	F, G	6

steel and concrete, and so on. One obviously must make a best guess and then update as more-current information becomes known.

Assume for this example the estimated durations are as shown in Table 13.12.

Table 13.12 is the result of a CPM analysis, first computing the early start and finish times, moving from left to right along the network, and then, holding the total project time constant, computing the latest start and finish times moving from right to left along the network. Start and finish times for each task shown in the network of Figure 13.18 are based on the durations listed in Table 13.12. Also shown are the free and interfering floats of those tasks not on the critical path. Here it is obvious the critical path, containing no floats, is A, B, C, F, and H, and the total project time is forty.

To obtain a more precise estimate of the total project time, this preliminary analysis can be expanded to examine in more detail subtasks that make up each of the major tasks considered so far. For example, the first task is to prepare for the overall project. This task can be followed by moving onto the site; getting the

Table 13.13 Task Duration Data for PERT Durations

Task	Optimistic	Most Likely	Pessimistic	Expected	Variance
A	10	14	18	14.00	1.78
B	5	7	10	7.17	0.69
C	3	5	9	5.33	1.00
D	1	2	4	2.17	0.25
E	7	10	15	10.33	1.78
F	5	8	12	8.17	1.36
G	3	4	6	4.17	0.25
H	4	6	10	6.33	1.00

Table 13.14 Task Start and Finish Times Based on Expected Task Durations (time units are in days)

Task	E (Duration)	EST	EFT	Float LST	LFT	Total	Free	Interfering
A	14.00	0	14.00	0	14.00	0		
B	7.17	14.00	21.17	14.00	21.17	0		
C	5.33	21.17	26.50	21.17	26.50	0		
D	2.17	0	2.17	14.00	16.17	14.0	0	14
E	10.33	2.17	12.50	16.17	26.50	14.0	14.0	0
F	8.17	26.50	34.67	26.50	34.67	0		
G	4.17	0	4.17	30.50	34.67	30.5	30.5	0
H	6.33	34.67	41.00	34.67	41.00	0		

excavator, the steel reinforcement, and concrete materials; and arranging for the inspection of steel reinforcement. Once on the site, the foundations can be designed and the formwork can be obtained, and so on, until the end of the project.

Even if a more detailed list of tasks is included in this analysis, any estimate of total project time will be uncertain. It is based on uncertain estimates of the duration of each task. One approach to quantifying that uncertainty is to apply the Project Evaluation and Review Technique, PERT. By using PERT, one can obtain an approximate estimate of expected completion times and the probabilities of completing the overall project at any specified time. Now we apply estimates of pessimistic and optimistic start and finish times to the list of tasks and calculate most likely finish times, variances, and probabilities (see equations 13.8–13.10).

The critical path of tasks that define the expected total project duration is based on the expected task durations. Carrying out the same procedure used in the CPM method for Table 13.13, the expected earliest and latest start and finish times can be computed for each task.

As one can see from Table 13.14, the expected critical path is the same as found using CPM in this example, namely, A, B, C, F, and H, and the expected project duration is 41.00 days. This assumes, as the procedure in PERT assumes, that the critical path based on expected task durations will indeed be the critical path. It is possible that in reality, another path could prove to be the critical one. For example, if each task on the critical path is completed in its most optimistic duration the tasks on the expected critical path would require only twenty-seven days. If all other tasks require their most pessimistic durations, the critical path becomes D, E, F, and H and will require a total of forty-one days. Yet, the project is expected to require only twenty-seven days, which is considerably less than the expected time to complete the A, B, C, F, and H path.

Following the procedure used in PERT, the project variance is the sum of the task variances along the expected critical path. This variance is 6.83, and hence its standard deviation is 2.61. Recalling that the sum of expected values is normally distributed, and that the probability of exceeding the mean is 50%, and that the area under the distribution between the mean plus and minus one standard deviation is 68%, and ±2 standard deviations is 95%, this allows one to make some estimates of the probability of completing the project by (at or before) various times. For example,

- the probability of completing the project by 41 – 2(2.61) or about 36 time units is 2.5%.
- the probability of completing the project by 41 – 1(2.61) or about 38 time units is 16%.
- the probability of completing the project by about 41 time units is 50%.
- the probability of completing the project by 41 + 1(2.61) or about 44 time units is 84%.
- the probability of completing the project by 41 + 2(2.61) or about 46 time units is 97.5%.

One can refer to statistics books to obtain other estimates, but keep in mind that this procedure is based on guesses, albeit educated ones, concerning future events, and inevitably, these guesses will prove inaccurate. However, this does not negate the value of planning, for if one does not, surely the projects will take more time and cost more money.

Example 13.4: Project Bidding

QUESTION:

The z-chart approach might be used in a case where you are trying to bid on a project (see Table 13.15 for a z-chart). For example, your boss tells you that an outside client wants a quote on a new project to build a database system. Your job as project manager is to estimate the time to complete the job with an 80% probability of finishing the project for the client within the time frame given. You go back to your team, and after considerable discussion, the team estimates it could take as long as 100 days or as little as 20 days. It will most likely take 60 days. So, what do you tell your boss to quote?

Solution:

Using the z-chart (see Table 13.15), the z-value for 80% is 0.845 (interpolate).

Use the following numbers for the formula:

$b = 100$

$a = 20$

$m = 60$

Table 13.15 z-Chart

z	0.00	0.01	0.02	0.03	0.04	0.05	0.06	0.07	0.08	0.09
0	.5000	.5040	.5080	.5120	.5160	.5199	.5239	.5279	.5319	.5359
0.1	.5398	.5438	.5437	.5517	.5557	.5596	.5636	.5675	.5714	.5753
0.2	.5793	.5832	.5871	.5910	.5948	.5987	.6026	.6064	.6103	.6141
0.3	.6179	.6217	.6255	.6293	.6331	.6368	.6406	.6443	.6480	.6517
0.4	.6554	.6591	.6628	.6664	.6700	.6736	.6772	.6808	.6844	.6879
0.5	.6915	.6950	.6985	.7019	.7054	.7088	.7123	.7157	.7190	.7224
0.6	.7257	.7291	.7324	.7357	.7389	.7422	.7454	.7486	.7517	.7549
0.7	.7580	.7611	.7642	.7673	.7704	.7734	.7764	.7794	.7823	.7852
0.8	.7881	.7910	.7939	.7967	.7995	.8023	.8051	.8079	.8106	.8133
0.9	.8159	.8186	.8212	.8238	.8264	.8289	.8315	.8340	.8365	.8389
1.0	.8413	.8438	.8461	.8485	.8508	.8531	.8554	.8577	.8599	.8621
1.1	.8643	.8665	.8686	.8708	.8729	.8749	.8770	.8790	.8810	.8830
1.2	.8849	.8869	.8888	.8907	.8925	.8944	.8962	.8980	.8997	.9015
1.3	.9032	.9049	.9066	.9082	.9099	.9115	.9131	.9147	.9162	.9177
1.4	.9192	.9207	.9222	.9236	.9251	.9265	.9279	.9292	.9306	.9319
1.5	.9332	.9345	.9357	.9370	.9382	.9394	.9406	.9418	.9429	.9441
1.6	.9452	.9463	.9474	.9484	.9495	.9505	.9515	.9525	.9535	.9545
1.7	.9554	.9564	.9573	.9582	.9591	.9599	.9608	.9616	.9625	.9633
1.8	.9641	.9649	.9658	.9664	.9671	.9678	.9686	.9693	.9699	.9706
1.9	.9713	.9719	.9726	.9732	.9738	.9744	.9750	.9756	.9761	.9767
2.0	.9773	.9778	.9783	.9788	.9793	.9798	.9803	.9808	.9812	.9817
2.1	.9821	.9826	.9830	.9834	.9838	.9842	.9846	.9850	.9854	.9857
2.2	.9861	.9864	.9868	.9871	.9875	.9878	.9881	.9884	.9887	.9890
2.3	.9893	.9896	.9898	.9901	.9904	.9906	.9909	.9911	.9913	.9916
2.4	.9918	.9920	.9922	.9925	.9927	.9929	.9931	.9932	.9934	.9936
2.5	.9938	.9940	.9941	.9943	.9945	.9946	.9948	.9949	.9951	.9952
2.6	.9953	.9955	.9956	.9957	.9959	.9960	.9961	.9962	.9963	.9964
2.7	.9965	.9966	.9967	.9968	.9969	.9970	.0971	.9972	.9973	.9974
2.8	.9974	.9975	.9976	.9977	.9977	.9978	.9979	.9979	.9980	.9981
2.9	.9981	.9982	.9983	.9983	.9983	.9984	.9985	.9985	.9986	.9986
3.0	.99865	.99869	.99874	.99878	.99882	.99886	.99889	.99893	.99896	.99900
3.1	.99903	.99906	.99910	.99913	.99915	.99918	.99921	.99924	.99926	.99929
3.2	.99931	.99934	.99936	.99938	.99940	.99942	.99944	.99946	.99948	.99950
3.3	.99952	.99953	.99955	.99957	.99958	.99960	.99961	.99962	.99964	.99965
3.4	.99966	.99967	.99969	.99970	.99971	.99972	.99973	.99974	.99975	.99976
3.5	.99977	.99978	.99978	.99979	.99980	.99981	.99981	.99982	.99983	.99983
−3.5	.00023	.00022	.00022	.00021	.00020	.00019	.00019	.00018	.00017	.00017
−3.4	.00034	.00033	.00031	.00030	.00029	.00028	.00027	.00026	.00025	.00024
−3.3	.00048	.00047	.00045	.00043	.00042	.00040	.00039	.00038	.00036	.00035
−3.2	.00069	.00066	.00064	.00062	.00060	.00058	.00056	.00054	.00052	.00050
−3.1	.00097	.00094	.00090	.00087	.00085	.00082	.00079	.00076	.00074	.00071
−3.0	.00135	.00131	.00126	.00122	.00118	.00114	.00111	.00107	.00104	.00010
−2.9	.0019	.0018	.0017	.0017	.0016	.0016	.0015	.0015	.0014	.0014
−2.8	.0026	.0025	.0024	.0023	.0023	.0022	.0021	.0021	.0020	.0019
−2.7	.0035	.0034	.0033	.0032	.0031	.0030	.0029	.0028	.0027	.0026
−2.6	.0047	.0045	.0044	.0043	.0041	.0040	.0039	.0038	.0037	.0036
−2.5	.0062	.0060	.0059	.0057	.0055	.0054	.0052	.0051	.0049	.0048
−2.4	.0082	.0080	.0078	.0075	.0073	.0071	.0069	.0068	.0066	.0064
−2.3	.0107	.0104	.0102	.0099	.0096	.0094	.0091	.0089	.0087	.0084
−2.2	.0139	.0136	.0132	.0129	.0125	.0122	.0119	.0116	.0113	.0110
−2.1	.0179	.0174	.0170	.0166	.0162	.0158	.0154	.0150	.0146	.0143

Table 13.5 (*cont.*)

z	0.00	0.01	0.02	0.03	0.04	0.05	0.06	0.07	0.08	0.09
−2.0	.0228	.0222	.0217	.0212	.0207	.0202	.0197	.0192	.0188	.0183
−1.9	.0287	.0281	.0274	.0268	.0262	.0256	.0250	.0244	0.239	.0233
−1.8	.0359	.0351	.0344	.0336	.0329	.0322	.0314	.0307	.0301	.0294
−1.7	.0446	.0436	.0427	.0418	.0409	.0401	.0392	.0384	.0375	.0367
−1.6	.0548	.0537	.0526	.0516	.0505	.0495	.0485	.0475	.0465	.0455
−1.5	.0668	.0655	.0643	.0630	.0618	.0606	.0594	.0582	.0571	.0559
−1.4	.0808	.0793	.0778	.0764	.0749	.0735	.0721	.0708	.0694	.0681
−1.3	.0968	.0951	.0934	.0918	.0901	.0885	.0869	.0853	.0838	.0823
−1.2	.1151	.1131	.1112	.1093	.1075	.1057	.1038	.1020	.1003	.0985
−1.1	.1357	.1335	.1314	.1292	.1271	.1251	.1230	.1210	.1190	.1170
−1.0	.1587	.1562	.1539	.1515	.1492	.1469	.1446	.1423	.1401	.1379
−0.9	.1841	.1814	.1788	.1762	.1736	.1711	.1685	.1660	.1635	.1611
−0.8	.2119	.2090	.2061	.2033	.2005	.1977	.1949	.1922	.1894	.1867
−0.7	.2420	.2389	.2358	.2327	.2297	.2266	.2236	.2207	.2177	.2148
−0.6	.2743	.2709	.2676	.2643	.2611	.2578	.2546	.2514	.2483	.2451
−0.5	.3085	.3050	.3015	.2981	.2946	.2912	.2877	.2843	.2810	.2776
−0.4	.3446	.3409	.3372	.3336	.3300	.3264	.3228	.3192	.3156	.3121
−0.3	.3821	.3783	.3745	.3707	.3669	.3632	.3594	.3557	.3520	.3483
−0.2	.4207	.4168	.4129	.4090	.4052	.4013	.3974	.3936	.3897	.3859
−0.1	.4602	.4562	.4522	.4483	.4443	.4404	.4364	.4325	.4286	.4247
−0	.5000	.4960	.4920	.4880	.4840	.4801	.4761	.4721	.4681	.4641

$$Expected\ Value = \frac{20 + 100 + 4.60}{6} = 60$$

$$\sigma^2 = \left[\frac{100 - 20}{6}\right]^2 = 13.33^2 = 177.8$$

$$Z = \frac{Selected\ Value - Expected\ Value}{\sqrt{\sigma^2\ of\ Expected\ Value}}$$

$$0.845 = \frac{X - 60}{\sqrt{177.8}}$$

Solve for X = 72 days

Answer is to quote at seventy-two days for completion time. In all these calculations, you are assuming a normal and symmetric distribution, see Figure 13.19.

Application to Entrepreneurship: Risk of Losing Company

We have stressed the importance of not running out of cash in your entrepreneurial activities. The obvious way to avoid this is to always raise enough money. However, typically the more money you raise, the more your ownership position will be diluted. In the end, you will have to balance the risk of running out of money (meaning you will essentially lose control of the company) versus the desirability of staying in a majority ownership position.

Generally, if you keep at least a two-thirds ownership position and certainly 70% (called a supermajority and basically eliminates a minor shareholders legal

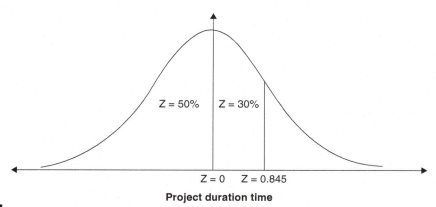

Figure 13.19 Normal distribution curve.

right to raise claims against the company relative to decisions made), you will maintain enough voting shares to control company decisions from a legal perspective. A simple majority is the next best position, but for major decisions such as selling the company, the 51% majority may not be sufficient. Therefore, you should assess your acceptable risks associated with these two ownership positions, for example, 51% and 66.7%, and then define your own allowable risk associated with running out of money for these two ownership positions.

Remember that choosing a 99% level of certainty will require you to raise a lot more money than being at the 50.1% level of certainty. In other words, you might want to take a position that you will be 90% sure of not running out of enough money to maintain your 51% position, but only 60% sure of not running out of money to maintain the 70% position. If you cannot raise enough money to maintain your 51% majority (even after months of seeking), you should probably reevaluate your business plan and scale back so that less investment capital is needed. Also, remember as you approach this type of problem that the expected result of a project is the sum of the products of a particular subevent and its own probability of outcome, or

$$Expected\ Result = \sum_{i=1}^{n} \Pr obability_i \times Outcome_i \qquad (13.11)$$

13.4 References and Bibliography

Tague, N. R.. *The Quality Toolbox*. Milwaukee, WI: ASQ Quality Press, 1995.

13.5 Video Clips

Directions for viewing clips on the Prendismo Collection (formerly from www. eclips.cornell.edu):

- Visit http://prendismo.com/collection/. (You must subscribe with the site for full access to all features.)

- Click on the **Subscribe** at the top of the site. Students receive reduced rates.
- Once you have subscribed, type in the name or title of the clip in the **Search** at the top right corner of the screen.
- Click on the clip you wish to view.
- Click on the play icon to view the clip. If you have not become a subscriber, you will only view the first 20 seconds of the clip before being prompted to log in.

1. Bill Shores (Founder of Share Our Strength)

- Share Our Strength – a national not-for-profit which organizes individuals and businesses to embrace innovative ways to end hunger
- Title of Video Clip: "Bill Shore Shares Thoughts on Strategic Planning"

NOTE to Reader: The following video clip of Frank Farwell is included to illustrate a case in which using CPM/PERT would have been useful (and maybe he did).

2. Frank Farwell (Founder of WinterSilks Inc., Middleton, WI)

- Sold the company in 1990, pocketing several million dollars
- Stresses the importance of customer service and cash flow
- Title of Video Clip: "Frank Farwell Discusses Starting the Business and Adjusting to Setbacks"

13.6 Problems

1. Construct a CPM chart for raising your necessary capital for your start-up company; show key milestones also.
2. Given the following activities and normal durations:

Activity	Normal Duration	Optimistic	Pessimistic
A	10	9	12
B	14	12	15
C	6	4	8
D	11	7	12
E	8	7	10

determine the following:

- *EST, EFT, LST*, and *LFT* for each activity
- The critical path
- Free float, interfering float, and total float for each activity
- The standard deviation of the project completion time
- The probability of completing the project in twenty-nine days

Appendix

Contents

The following tables are provided on the book's companion website (see www.cambridge.org/Timmons). Go to Appendix Supplement, MACR Charts and Class Lives.

- ○ Appendix A. MACRS Percentage Table Guide
- ○ Chart 1. Use this chart to find the correct percentage table to use for any property other than residential rental and nonresidential real property.
- ○ Chart 2. Use this chart to find the correct percentage table to use for residential rental and nonresidential real property.

Summary of Interest Formulae and Compound Interest Tables

There are tabulated values in the Appendix for F/P, P/F, A/F, A/P, F/A, P/A, A/G, and P/G for given values of interest rate (i) and analysis period (n). These tables simply list the values that you would calculate from the equations anyway, but it works well as a fast reference. Note that the values given for A/G and P/G are for *arithmetic gradient*; *geometric gradient* problems require you to use their interest formula.

Single Payment

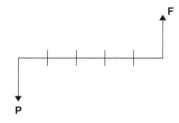

COMPOUND AMOUNT:

To find F, given P:

$$(\text{F/P, i, n}): F = P(1 + i)^n \tag{A.1}$$

PRESENT WORTH:

To find P, given F:

$$(\text{P/F, i, n}): P = F(1 + i)^{-n} \tag{A.2}$$

Uniform Series

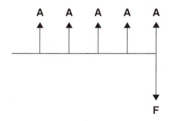

SERIES COMPOUND AMOUNT:

To find F, given A:

$$(\text{F/A,i,n}): \quad F = A\left[\frac{(1+i)^n - 1}{i}\right] \tag{A.3}$$

SINKING FUND

To find A, given F:

$$(A/F,i,n): \quad A = F\left[\frac{i}{(1+i)^n - 1}\right] \tag{A.4}$$

CAPITAL RECOVERY

To find A, given P:

$$(A/P,i,n): \quad A = P\left[\frac{i(1+i)^n}{(1+i)^n - 1}\right] \tag{A.5}$$

$$\textit{or this is same as} = P\frac{i}{1-(1+i^{-n})}$$

SERIES PRESENT WORTH

To find P, given A:

$$(P/A,i,n): \quad P = A\left[\frac{(1+i)^n - 1}{i(1+i)^n}\right] \tag{A.6}$$

$$\textit{or same as } P = A\left[(1-(1+i)^{-n})/i\right]$$

Arithmetic Gradient

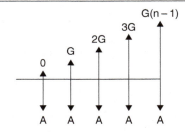

ARITHMETRIC GRADIENT UNIFORM SERIES

To find A, given G:

$$(A/G,i,n): \quad A = G\left[\frac{1}{i} - \frac{n}{(1+i)^n - 1}\right] \tag{A.7}$$

$$\textit{as } n \to \infty \textit{ then } A = G\!\big/\!i$$

ARITHMETIC GRADIENT PRESENT WORTH

To find P, given G:

$$(\text{P/G,i,n}): \quad P = G\left[\frac{(1+i)^n - in - 1}{i^2(1+i)^n}\right] \tag{A.8}$$

ARITHMETIC GRADIENT FUTURE WORTH

To find F, given G:

$$(\text{F/G,i\%,n}) \quad F = \frac{G}{i}\left[\frac{(1+i)^n - 1}{i} - n\right] \tag{A.9}$$

Geometric Gradient

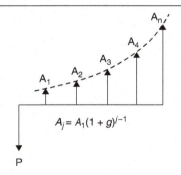

$$A_j = A_1(1 + g)^{j-1}$$

GEOMETRIC SERIES PRESENT WORTH

To find P, given A_1,g:

$$(\text{P/A,g,i,n}) \quad P = A_1[n(1+i)^{-1}] \quad for \quad i = g \tag{A.10}$$

$$(\text{P/A,g,i,n}) \quad P = \frac{A_1}{(i-g)}\left[1 - \frac{(1+g)^n}{(1+i)^n}\right] \quad for \quad i \neq g \tag{A.11}$$

Continuous Compounding at Nominal Rate r

SINGLE PAYMENT (A ONETIME DEPOSIT AT BEGINNING OF FIRST PERIOD):

To find F, given P : $F = Pe^{rn}$

To find P, given F : $P = \dfrac{F}{e^{rn}}$

UNIFORM SERIES

$$A = F\left[\frac{e^r - 1}{e^{rn} - 1}\right] \qquad F = A\left[\frac{e^{rn} - 1}{e^r - 1}\right]$$

$$A = P\left[\frac{e^{rn}(e^r - 1)}{e^{rn} - 1}\right] \qquad \frac{e^{rn} - 1}{[e^{rn}(e^r - 1)]}$$

Compound Interest Variable Definitions

n = number of interest periods (when interest is applied)

m = number of compounding periods per year

i = interest rate per interest period, i = r/m

r = nominal **annual** interest rate

P = present worth of investment or a sum of money at time zero

F = future worth of investment at the end of "n" interest periods

A = uniform payment at end of each interest period

G = linear increase in payment at the end of each interest period, the *arithmetic gradient*

g = geometric (percentage) increase in payment at the end of each interest period, the *geometric gradient*

Remember that i equals the nominal annual rate for annual compounding only. For a nonannual compounding rate, i = the nominal annual rate divided by number of compounding periods per year.

Table C-1 ¼% CFIs ¼%

n	F/P	P/F	A/F	A/P	F/A	P/A	A/G	P/G	n
1	1.003	0.9975	1.000	1.003	1.000	0.9980	0.0000	0.0000	1
2	1.005	0.9950	0.4994	0.5019	2.002	1.993	0.4990	0.9950	2
3	1.008	0.9925	0.3325	0.3350	3.008	2.985	0.9980	2.980	3
4	1.010	0.9901	0.2491	0.2516	4.015	3.975	1.497	5.950	4
5	1.013	0.9876	0.1990	0.2015	5.025	4.963	1.995	9.901	5
6	1.015	0.9851	0.1656	0.1681	6.038	5.948	2.493	14.83	6
7	1.018	0.9827	0.1418	0.1443	7.053	6.931	2.990	20.72	7
8	1.020	0.9802	0.1239	0.1264	8.070	7.911	3.487	27.58	8
9	1.023	0.9778	0.1100	0.1125	9.091	8.889	3.983	35.41	9
10	1.025	0.9753	0.0989	0.1014	10.11	9.864	4.479	44.18	10
11	1.028	0.9729	0.0898	0.0923	11.14	10.84	4.975	53.91	11
12	1.030	0.9705	0.0822	0.0847	12.17	11.81	5.470	64.59	12
13	1.033	0.9681	0.0758	0.0783	13.20	12.78	5.965	76.21	13
14	1.036	0.9656	0.0703	0.0728	14.23	13.74	6.459	88.76	14
15	1.038	0.9632	0.0655	0.0680	15.27	14.70	6.953	102.2	15
16	1.041	0.9608	0.0613	0.0638	16.30	15.67	7.447	116.7	16
17	1.043	0.9584	0.0577	0.0602	17.34	16.62	7.940	132.0	17
18	1.046	0.9561	0.0544	0.0569	18.39	17.58	8.433	148.2	18
19	1.049	0.9537	0.0515	0.0540	19.43	18.53	8.925	165.4	19
20	1.051	0.9513	0.0488	0.0513	20.48	19.48	9.417	183.5	20

n	F/P	P/F	A/F	A/P	F/A	P/A	A/G	P/G	n
21	1.054	0.9489	0.0464	0.0489	21.53	20.43	9.908	202.5	21
22	1.056	0.9466	0.0443	0.0468	22.59	21.38	10.40	222.3	22
23	1.059	0.9442	0.0423	0.0448	23.64	22.32	10.89	243.1	23
24	1.062	0.9418	0.0405	0.0430	24.70	23.27	11.38	264.8	24
25	1.064	0.9395	0.0388	0.0413	25.77	24.21	11.87	287.3	25
26	1.067	0.9371	0.0373	0.0398	26.83	25.14	12.36	310.8	26
27	1.070	0.9348	0.0359	0.0383	27.90	26.08	12.85	335.1	27
28	1.072	0.9325	0.0345	0.0370	28.97	27.01	13.34	360.2	28
29	1.075	0.9302	0.0333	0.0358	30.04	27.94	13.83	386.3	29
30	1.078	0.9278	0.0321	0.0346	31.11	28.87	14.31	413.2	30
31	1.080	0.9255	0.0311	0.0336	32.19	29.79	14.80	441.0	31
32	1.083	0.9232	0.0301	0.0326	33.27	30.72	15.29	469.6	32
33	1.086	0.9209	0.0291	0.0316	34.36	31.64	15.77	499.0	33
34	1.089	0.9186	0.0282	0.0307	35.44	32.56	16.26	529.4	34
35	1.091	0.9163	0.0274	0.0299	36.53	33.47	16.75	560.5	35
36	1.094	0.9140	0.0266	0.0291	37.62	34.39	17.23	592.5	36
37	1.097	0.9118	0.0258	0.0283	38.72	35.30	17.72	625.3	37
38	1.100	0.9095	0.0251	0.0276	39.81	36.21	18.20	659.0	38
39	1.102	0.9072	0.0244	0.0269	40.91	37.12	18.68	693.4	39
40	1.105	0.9050	0.0238	0.0263	42.01	38.02	19.17	728.7	40
50	1.133	0.8826	0.0188	0.0213	53.19	46.95	23.98	1,126	50
55	1.147	0.8717	0.0170	0.0195	58.88	51.33	26.37	1,354	55
60	1.162	0.8609	0.0155	0.0180	64.65	55.65	28.75	1,600	60
65	1.176	0.8502	0.0142	0.0167	70.48	59.93	31.12	1,865	65
70	1.191	0.8396	0.0131	0.0156	76.39	64.14	33.48	2,148	70
75	1.206	0.8292	0.0121	0.0146	82.38	68.31	35.83	2,448	75
80	1.221	0.8189	0.0113	0.0138	88.44	72.43	38.17	2,764	80
85	1.236	0.8088	0.0106	0.0131	94.58	76.49	40.50	3,098	85
90	1.252	0.7987	0.0099	0.0124	100.8	80.50	42.82	3,447	90
95	1.268	0.7888	0.0093	0.0118	107.1	84.47	45.12	3,812	95
100	1.284	0.7790	0.0088	0.0113	113.5	88.38	47.42	4,191	100
200	1.648	0.6069	0.0039	0.0064	259.1	157.2	91.21	14,342	200

Table C-2 ½% CFIs

n	F/P	P/F	A/F	½% A/P	F/A	CFIs P/A	A/G	½% P/G	n
1	1.005	0.9950	1.000	1.005	1.000	0.9950	0.0000	0.0000	1
2	1.010	0.9901	0.4988	0.5038	2.005	1.985	0.4990	0.9900	2
3	1.015	0.9851	0.3317	0.3367	3.015	2.970	0.9970	2.960	3
4	1.020	0.9802	0.2481	0.2531	4.030	3.950	1.494	5.901	4
5	1.025	0.9754	0.1980	0.2030	5.050	4.926	1.990	9.803	5
6	1.030	0.9705	0.1646	0.1696	6.076	5.896	2.485	14.66	6
7	1.036	0.9657	0.1407	0.1457	7.106	6.862	2.980	20.45	7
8	1.041	0.9609	0.1228	0.1278	8.141	7.823	3.474	27.18	8
9	1.046	0.9561	0.1089	0.1139	9.182	8.779	3.967	34.82	9
10	1.051	0.9513	0.0978	0.1028	10.23	9.730	4.459	43.39	10
11	1.056	0.9466	0.0887	0.0937	11.28	10.68	4.950	52.85	11
12	1.062	0.9419	0.0811	0.0861	12.34	11.62	5.441	63.21	12
13	1.067	0.9372	0.0746	0.0796	13.40	12.56	5.930	74.46	13
14	1.072	0.9326	0.0691	0.0741	14.46	13.49	6.419	86.58	14
15	1.078	0.9279	0.0644	0.0694	15.54	14.42	6.907	99.57	15

n	F/P	P/F	A/F	½% A/P	F/A	CFIs P/A	A/G	½% P/G	n
16	1.083	0.9233	0.0602	0.0652	16.61	15.34	7.394	113.4	16
17	1.088	0.9187	0.0565	0.0615	17.70	16.26	7.880	128.1	17
18	1.094	0.9141	0.0532	0.0582	18.79	17.17	8.366	143.7	18
19	1.099	0.9096	0.0503	0.0553	19.88	18.08	8.850	160.0	19
20	1.105	0.9051	0.0477	0.0527	20.98	18.99	9.334	177.2	20
21	1.110	0.9006	0.0453	0.0503	22.08	19.89	9.817	195.2	21
22	1.116	0.8961	0.0431	0.0481	23.19	20.78	10.30	214.1	22
23	1.122	0.8916	0.0411	0.0461	24.31	21.68	10.78	233.7	23
24	1.127	0.8872	0.0393	0.0443	25.43	22.56	11.26	254.1	24
25	1.133	0.8828	0.0377	0.0427	26.56	23.45	11.74	275.3	25
26	1.138	0.8784	0.0361	0.0411	27.69	24.32	12.22	297.2	26
27	1.144	0.8740	0.0347	0.0397	28.83	25.20	12.70	320.0	27
28	1.150	0.8697	0.0334	0.0384	29.98	26.07	13.18	343.4	28
29	1.156	0.8653	0.0321	0.0371	31.12	26.93	13.65	367.7	29
30	1.161	0.8610	0.0310	0.0360	32.28	27.79	14.13	392.6	30
31	1.167	0.8568	0.0299	0.0349	33.44	28.65	14.60	418.3	31
32	1.173	0.8525	0.0289	0.0339	34.61	29.50	15.08	444.8	32
33	1.179	0.8482	0.0280	0.0329	35.78	30.35	15.55	471.9	33
34	1.185	0.8440	0.0271	0.0321	36.96	31.20	16.02	499.8	34
35	1.191	0.8398	0.0262	0.0312	38.15	32.04	16.49	528.3	35
36	1.197	0.8356	0.0254	0.0304	39.34	32.87	16.96	557.6	36
37	1.203	0.8315	0.0247	0.0297	40.53	33.70	17.43	587.5	37
38	1.209	0.8274	0.0240	0.0290	41.74	34.53	17.90	618.1	38
39	1.215	0.8232	0.0233	0.0283	42.94	35.35	18.37	649.4	39
40	1.221	0.8191	0.0227	0.0276	44.16	36.17	18.84	681.3	40
50	1.283	0.7793	0.0177	0.0227	56.65	44.14	23.46	1,036	50
55	1.316	0.7601	0.0158	0.0208	63.13	47.98	25.75	1,235	55
60	1.349	0.7414	0.0143	0.0193	69.77	51.73	28.01	1,449	60
65	1.383	0.7231	0.0131	0.0181	76.58	55.38	30.25	1,675	65
70	1.418	0.7053	0.0120	0.0170	83.57	58.94	32.47	1,914	70
75	1.454	0.6879	0.0110	0.0160	90.73	62.41	34.67	2,164	75
80	1.490	0.6710	0.0102	0.0152	98.07	65.80	36.85	2,425	80
85	1.528	0.6545	0.0095	0.0145	105.6	69.11	39.01	2,696	85
90	1.567	0.6383	0.0088	0.0138	113.3	72.33	41.15	2,976	90
95	1.606	0.6226	0.0083	0.0132	121.2	75.48	43.26	3,265	95
100	1.647	0.6073	0.0077	0.0127	129.3	78.54	45.36	3,563	100
200	2.712	0.3688	0.0029	0.0079	342.3	126.2	83.15	10,496	200

Table C-3 ¾% CFIs

n	f/p	p/f	a/f	¾% a/p	f/a	CFIs p/a	a/g	¾% p/g	n
1	1.008	0.9926	1.000	1.008	1.000	0.9930	0.0000	0.0000	1
2	1.015	0.9852	0.4981	0.5056	2.008	1.978	0.4980	0.9850	2
3	1.023	0.9778	0.3308	0.3383	3.023	2.956	0.9950	2.941	3
4	1.030	0.9706	0.2472	0.2547	4.045	3.926	1.491	5.852	4
5	1.038	0.9633	0.1970	0.2045	5.076	4.889	1.985	9.706	5
6	1.046	0.9562	0.1636	0.1711	6.114	5.846	2.478	14.49	6
7	1.054	0.9490	0.1397	0.1472	7.159	6.795	2.970	20.18	7
8	1.062	0.9420	0.1218	0.1293	8.213	7.737	3.461	26.78	8

n	f/p	p/f	a/f	¾% a/p	f/a	CFIs p/a	a/g	¾% p/g	n
9	1.070	0.9350	0.1078	0.1153	9.275	8.672	3.950	34.25	9
10	1.078	0.9280	0.0967	0.1042	10.34	9.600	4.438	42.61	10
11	1.086	0.9211	0.0876	0.0951	11.42	10.52	4.925	51.82	11
12	1.094	0.9142	0.0800	0.0875	12.51	11.44	5.411	61.87	12
13	1.102	0.9074	0.0735	0.0810	13.60	12.34	5.895	72.76	13
14	1.110	0.9007	0.0680	0.0755	14.70	13.24	6.379	84.47	14
15	1.119	0.8940	0.0632	0.0707	15.81	14.14	6.861	96.99	15
16	1.127	0.8873	0.0591	0.0666	16.93	15.02	7.341	110.3	16
17	1.135	0.8807	0.0554	0.0629	18.06	15.91	7.821	124.4	17
18	1.144	0.8742	0.0521	0.0596	19.20	16.78	8.299	139.2	18
19	1.153	0.8676	0.0492	0.0567	20.34	17.65	8.776	154.9	19
20	1.161	0.8612	0.0465	0.0540	21.49	18.51	9.252	171.2	20
21	1.170	0.8548	0.0441	0.0516	22.65	19.36	9.726	188.3	21
22	1.179	0.8484	0.0420	0.0495	23.82	20.21	10.20	206.1	22
23	1.188	0.8421	0.0400	0.0475	25.00	21.05	10.67	224.7	23
24	1.196	0.8358	0.0382	0.0457	26.19	21.89	11.14	243.9	24
25	1.205	0.8296	0.0365	0.0440	27.39	22.72	11.61	263.8	25
26	1.214	0.8234	0.0350	0.0425	28.59	23.54	12.08	284.4	26
27	1.224	0.8173	0.0336	0.0411	29.81	24.36	12.55	305.6	27
28	1.233	0.8112	0.0322	0.0397	31.03	25.17	13.01	327.5	28
29	1.242	0.8052	0.0310	0.0385	32.26	25.98	13.48	350.1	29
30	1.251	0.7992	0.0299	0.0373	33.50	26.78	13.94	373.3	30
31	1.261	0.7932	0.0288	0.0363	34.75	27.57	14.40	397.1	31
32	1.270	0.7873	0.0278	0.0353	36.02	28.36	14.86	421.5	32
33	1.280	0.7815	0.0268	0.0343	37.29	29.14	15.32	446.5	33
34	1.289	0.7757	0.0259	0.0334	38.57	29.91	15.78	472.1	34
35	1.299	0.7699	0.0251	0.0326	39.85	30.68	16.24	498.2	35
36	1.309	0.7641	0.0243	0.0318	41.15	31.45	16.70	525.0	36
37	1.318	0.7585	0.0236	0.0311	42.46	32.21	17.15	552.3	37
38	1.328	0.7528	0.0228	0.0303	43.78	32.96	17.60	580.2	38
39	1.338	0.7472	0.0222	0.0297	45.11	33.71	18.06	608.5	39
40	1.348	0.7417	0.0215	0.0290	46.45	34.45	18.51	637.5	40
50	1.453	0.6883	0.0166	0.0241	60.39	41.57	22.95	953.8	50
55	1.508	0.6630	0.0148	0.0223	67.77	44.93	25.12	1,129	55
60	1.566	0.6387	0.0133	0.0208	75.42	48.17	27.27	1,314	60
65	1.625	0.6153	0.0120	0.0195	83.37	51.30	29.38	1,507	65
70	1.687	0.5927	0.0109	0.0184	91.62	54.31	31.46	1,709	70
75	1.751	0.5710	0.0100	0.0175	100.2	57.20	33.52	1,917	75
80	1.818	0.5500	0.0092	0.0167	109.1	59.99	35.54	2,132	80
85	1.887	0.5299	0.0085	0.0160	118.3	62.68	37.53	2,353	85
90	1.959	0.5104	0.0078	0.0153	127.9	65.28	39.50	2,578	90
95	2.034	0.4917	0.0073	0.0148	137.8	67.77	41.43	2,808	95
100	2.111	0.4737	0.0068	0.0143	148.1	70.18	43.33	3,041	100
200	4.457	0.2244	0.0022	0.0097	460.9	103.4	75.47	7,805	200

Table C-4 1% CFIs

n	F/P	P/F	A/F	A/P	F/A	P/A	A/G	P/G	n
1	1.010	0.9901	1.000	1.010	1.000	0.9900	0.0000	0.0000	1
2	1.020	0.9803	0.4975	0.5075	2.010	1.970	0.4980	0.9800	2
3	1.030	0.9706	0.3300	0.3400	3.030	2.941	0.9930	2.921	3
4	1.041	0.9610	0.2463	0.2563	4.060	3.902	1.488	5.804	4
5	1.051	0.9515	0.1960	0.2060	5.101	4.853	1.980	9.610	5
6	1.062	0.9420	0.1625	0.1725	6.152	5.795	2.471	14.32	6
7	1.072	0.9327	0.1386	0.1486	7.214	6.728	2.960	19.92	7
8	1.083	0.9235	0.1207	0.1307	8.286	7.652	3.448	26.38	8
9	1.094	0.9143	0.1067	0.1167	9.369	8.566	3.934	33.70	9
10	1.105	0.9053	0.0956	0.1056	10.46	9.471	4.418	41.84	10
11	1.116	0.8963	0.0865	0.0965	11.57	10.37	4.901	50.81	11
12	1.127	0.8874	0.0788	0.0888	12.68	11.26	5.381	60.57	12
13	1.138	0.8787	0.0724	0.0824	13.81	12.13	5.861	71.11	13
14	1.149	0.8700	0.0669	0.0769	14.95	13.00	6.338	82.42	14
15	1.161	0.8613	0.0621	0.0721	16.10	13.87	6.814	94.48	15
16	1.173	0.8528	0.0579	0.0679	17.26	14.72	7.289	107.3	16
17	1.184	0.8444	0.0543	0.0643	18.43	15.56	7.761	120.8	17
18	1.196	0.8360	0.0510	0.0610	19.62	16.40	8.232	135.0	18
19	1.208	0.8277	0.0481	0.0581	20.81	17.23	8.702	149.9	19
20	1.220	0.8195	0.0454	0.0554	22.02	18.05	9.169	165.5	20
21	1.232	0.8114	0.0430	0.0530	23.24	18.86	9.635	181.7	21
22	1.245	0.8034	0.0409	0.0509	24.47	19.66	10.10	198.6	22
23	1.257	0.7954	0.0389	0.0489	25.72	20.46	10.56	216.1	23
24	1.270	0.7876	0.0371	0.0471	26.97	21.24	11.02	234.2	24
25	1.282	0.7798	0.0354	0.0454	28.24	22.02	11.48	252.9	25
26	1.295	0.7721	0.0339	0.0439	29.53	22.80	11.94	272.2	26
27	1.308	0.7644	0.0325	0.0424	30.82	23.56	12.40	292.1	27
28	1.321	0.7568	0.0311	0.0411	32.13	24.32	12.85	312.5	28
29	1.335	0.7493	0.0299	0.0399	33.45	25.07	13.30	333.5	29
30	1.348	0.7419	0.0288	0.0387	34.79	25.81	13.76	355.0	30
31	1.361	0.7346	0.0277	0.0377	36.13	26.54	14.21	377.0	31
32	1.375	0.7273	0.0267	0.0367	37.49	27.27	14.65	399.6	32
33	1.389	0.7201	0.0257	0.0357	38.87	27.99	15.10	422.6	33
34	1.403	0.7130	0.0248	0.0348	40.26	28.70	15.54	446.2	34
35	1.417	0.7059	0.0240	0.0340	41.66	29.41	15.99	470.2	35
36	1.431	0.6989	0.0232	0.0332	43.08	30.11	16.43	494.6	36
37	1.445	0.6920	0.0225	0.0325	44.51	30.80	16.87	519.5	37
38	1.460	0.6852	0.0218	0.0318	45.95	31.49	17.31	544.9	38
39	1.474	0.6784	0.0211	0.0311	47.41	32.16	17.74	570.7	39
40	1.489	0.6717	0.0205	0.0305	48.89	32.84	18.18	596.9	40
50	1.645	0.6080	0.0155	0.0255	64.46	39.20	22.44	879.4	50
55	1.729	0.5785	0.0137	0.0237	72.85	42.15	24.51	1,033	55
60	1.817	0.5505	0.0122	0.0222	81.67	44.96	26.53	1,193	60
65	1.909	0.5237	0.0110	0.0210	90.94	47.63	28.52	1,358	65
70	2.007	0.4983	0.0099	0.0199	100.7	50.17	30.47	1,529	70
75	2.109	0.4741	0.0090	0.0190	110.9	52.59	32.38	1,703	75
80	2.217	0.4511	0.0082	0.0182	121.7	54.89	34.25	1,880	80
85	2.330	0.4292	0.0075	0.0175	133.0	57.08	36.08	2,059	85
90	2.449	0.4084	0.0069	0.0169	144.9	59.16	37.87	2,241	90
95	2.574	0.3886	0.0064	0.0164	157.4	61.14	39.63	2,423	95
100	2.705	0.3697	0.0059	0.0159	170.5	63.03	41.34	2,606	100
200	7.316	0.1367	0.0016	0.0116	631.6	86.33	68.33	5,899	200

Table C-5 1¼% CFIs

n	F/P	P/F	A/F	A/P	F/A	P/A	A/G	P/G	n
1	1.013	0.9876	1.000	1.013	1.000	0.9880	0.0000	0.0000	1
2	1.025	0.9754	0.4969	0.5094	2.013	1.963	0.4970	0.9750	2
3	1.038	0.9634	0.3292	0.3417	3.038	2.926	0.9920	2.902	3
4	1.051	0.9515	0.2454	0.2579	4.076	3.878	1.484	5.757	4
5	1.064	0.9397	0.1951	0.2076	5.127	4.818	1.975	9.516	5
6	1.077	0.9281	0.1615	0.1740	6.191	5.746	2.464	14.16	6
7	1.091	0.9167	0.1376	0.1501	7.268	6.662	2.950	19.66	7
8	1.105	0.9053	0.1196	0.1321	8.359	7.568	3.435	25.99	8
9	1.118	0.8941	0.1057	0.1182	9.464	8.462	3.917	33.15	9
10	1.132	0.8831	0.0945	0.1070	10.58	9.345	4.397	41.09	10
11	1.147	0.8722	0.0854	0.0979	11.72	10.22	4.876	49.82	11
12	1.161	0.8614	0.0778	0.0903	12.86	11.08	5.352	59.29	12
13	1.175	0.8508	0.0713	0.0838	14.02	11.93	5.826	69.50	13
14	1.190	0.8403	0.0658	0.0783	15.20	12.77	6.298	80.42	14
15	1.205	0.8299	0.0610	0.0735	16.39	13.60	6.768	92.04	15
16	1.220	0.8196	0.0568	0.0694	17.59	14.42	7.236	104.3	16
17	1.235	0.8095	0.0532	0.0657	18.81	15.23	7.702	117.3	17
18	1.251	0.7995	0.0499	0.0624	20.05	16.03	8.166	130.9	18
19	1.266	0.7896	0.0470	0.0595	21.30	16.82	8.627	145.1	19
20	1.282	0.7799	0.0443	0.0568	22.57	17.60	9.087	159.9	20
21	1.298	0.7702	0.0419	0.0544	23.85	18.37	9.545	175.3	21
22	1.315	0.7607	0.0398	0.0523	25.15	19.13	10.00	191.3	22
23	1.331	0.7513	0.0378	0.0503	26.46	19.88	10.45	207.8	23
24	1.348	0.7420	0.0360	0.0485	27.79	20.62	10.91	224.9	24
25	1.365	0.7329	0.0343	0.0468	29.14	21.36	11.36	242.5	25
26	1.382	0.7238	0.0328	0.0453	30.50	22.08	11.80	260.6	26
27	1.399	0.7149	0.0314	0.0439	31.89	22.79	12.25	279.2	27
28	1.416	0.7060	0.0300	0.0426	33.28	23.50	12.69	298.2	28
29	1.434	0.6973	0.0288	0.0413	34.70	24.20	13.13	317.7	29
30	1.452	0.6887	0.0277	0.0402	36.14	24.89	13.57	337.7	30
31	1.470	0.6802	0.0266	0.0391	37.59	25.57	14.01	358.1	31
32	1.489	0.6718	0.0256	0.0381	39.06	26.24	14.44	378.9	32
33	1.507	0.6635	0.0247	0.0372	40.55	26.90	14.88	400.2	33
34	1.526	0.6553	0.0238	0.0363	42.05	27.56	15.31	421.8	34
35	1.545	0.6472	0.0230	0.0355	43.58	28.20	15.74	443.8	35
36	1.564	0.6392	0.0222	0.0347	45.12	28.84	16.16	466.2	36
37	1.584	0.6313	0.0214	0.0339	46.69	29.47	16.59	488.9	37
38	1.604	0.6235	0.0207	0.0332	48.27	30.10	17.01	512.0	38
39	1.624	0.6158	0.0201	0.0326	49.88	30.71	17.43	535.4	39
40	1.644	0.6082	0.0194	0.0319	51.50	31.32	17.85	559.1	40
50	1.862	0.5371	0.0145	0.0270	68.90	37.00	21.93	811.4	50
55	1.981	0.5047	0.0128	0.0253	78.45	39.59	23.89	945.9	55
60	2.108	0.4743	0.0113	0.0238	88.60	42.02	25.81	1,084	60
65	2.244	0.4457	0.0101	0.0226	99.41	44.31	27.67	1,226	65
70	2.388	0.4188	0.0090	0.0215	110.9	46.46	29.49	1,370	70
75	2.541	0.3936	0.0081	0.0206	123.2	48.47	31.26	1,515	75
80	2.704	0.3699	0.0073	0.0199	136.2	50.37	32.98	1,661	80
85	2.877	0.3476	0.0067	0.0192	150.0	52.15	34.65	1,807	85
90	3.062	0.3266	0.0061	0.0186	164.8	53.83	36.28	1,953	90
95	3.258	0.3070	0.0055	0.0181	180.5	55.40	37.86	2,098	95
100	3.467	0.2885	0.0051	0.0176	197.2	56.88	39.40	2,241	100
200	12.02	0.0832	0.0011	0.0136	880.8	73.29	61.79	4,528	200

Table C-6 1½% CFIs

n	F/P	P/F	A/F	A/P	F/A	P/A	A/G	P/G	n
1	1.015	0.9852	1.000	1.015	1.000	0.9850	0.0000	0.0000	1
2	1.030	0.9707	0.4963	0.5113	2.015	1.956	0.4960	0.9710	2
3	1.046	0.9563	0.3284	0.3434	3.045	2.912	0.9900	2.883	3
4	1.061	0.9422	0.2444	0.2594	4.091	3.854	1.481	5.710	4
5	1.077	0.9283	0.1941	0.2091	5.152	4.783	1.970	9.423	5
6	1.093	0.9145	0.1605	0.1755	6.230	5.697	2.457	14.00	6
7	1.110	0.9010	0.1366	0.1516	7.323	6.598	2.940	19.40	7
8	1.126	0.8877	0.1186	0.1336	8.433	7.486	3.422	25.62	8
9	1.143	0.8746	0.1046	0.1196	9.559	8.361	3.901	32.61	9
10	1.161	0.8617	0.0934	0.1084	10.70	9.222	4.377	40.37	10
11	1.178	0.8489	0.0843	0.0993	11.86	10.07	4.851	48.86	11
12	1.196	0.8364	0.0767	0.0917	13.04	10.91	5.323	58.06	12
13	1.214	0.8240	0.0702	0.0852	14.24	11.73	5.792	67.95	13
14	1.232	0.8118	0.0647	0.0797	15.45	12.54	6.258	78.50	14
15	1.250	0.7999	0.0599	0.0749	16.68	13.34	6.722	89.70	15
16	1.269	0.7880	0.0558	0.0708	17.93	14.13	7.184	101.5	16
17	1.288	0.7764	0.0521	0.0671	19.20	14.91	7.643	113.9	17
18	1.307	0.7649	0.0488	0.0638	20.49	15.67	8.100	126.9	18
19	1.327	0.7536	0.0459	0.0609	21.80	16.43	8.554	140.5	19
20	1.347	0.7425	0.0432	0.0582	23.12	17.17	9.006	154.6	20
21	1.367	0.7315	0.0409	0.0559	24.47	17.90	9.455	169.2	21
22	1.388	0.7207	0.0387	0.0537	25.84	18.62	9.902	184.4	22
23	1.408	0.7100	0.0367	0.0517	27.23	19.33	10.35	200.0	23
24	1.430	0.6995	0.0349	0.0499	28.63	20.03	10.79	216.1	24
25	1.451	0.6892	0.0333	0.0483	30.06	20.72	11.23	232.6	25
26	1.473	0.6790	0.0317	0.0467	31.51	21.40	11.67	249.6	26
27	1.495	0.6690	0.0303	0.0453	32.99	22.07	12.10	267.0	27
28	1.517	0.6591	0.0290	0.0440	34.48	22.73	12.53	284.8	28
29	1.540	0.6494	0.0278	0.0428	36.00	23.38	12.96	303.0	29
30	1.563	0.6398	0.0266	0.0416	37.54	24.02	13.39	321.5	30
31	1.587	0.6303	0.0256	0.0406	39.10	24.65	13.81	340.4	31
32	1.610	0.6210	0.0246	0.0396	40.69	25.27	14.24	359.7	32
33	1.634	0.6118	0.0236	0.0386	42.30	25.88	14.66	379.3	33
34	1.659	0.6028	0.0228	0.0378	43.93	26.48	15.07	399.2	34
35	1.684	0.5939	0.0219	0.0369	45.59	27.08	15.49	419.4	35
36	1.709	0.5851	0.0212	0.0362	47.28	27.66	15.90	439.8	36
37	1.735	0.5764	0.0204	0.0354	48.99	28.24	16.31	460.6	37
38	1.761	0.5679	0.0197	0.0347	50.72	28.81	16.72	481.6	38
39	1.787	0.5595	0.0191	0.0341	52.48	29.37	17.13	502.9	39
40	1.814	0.5513	0.0184	0.0334	54.27	29.92	17.53	524.4	40
50	2.105	0.4750	0.0136	0.0286	73.68	35.00	21.43	750.0	50
55	2.268	0.4409	0.0118	0.0268	84.53	37.27	23.29	868.0	55
60	2.443	0.4093	0.0104	0.0254	96.22	39.38	25.09	988.2	60
65	2.632	0.3799	0.0092	0.0242	108.8	41.34	26.84	1,109	65
70	2.835	0.3527	0.0082	0.0232	122.4	43.16	28.53	1,231	70
75	3.055	0.3274	0.0073	0.0223	137.0	44.84	30.16	1,353	75
80	3.291	0.3039	0.0066	0.0215	152.7	46.41	31.74	1,473	80
85	3.545	0.2821	0.0059	0.0209	169.7	47.86	33.27	1,592	85
90	3.819	0.2619	0.0053	0.0203	187.9	49.21	34.74	1,710	90
95	4.114	0.2431	0.0048	0.0198	207.6	50.46	36.16	1,825	95
100	4.432	0.2256	0.0044	0.0194	228.8	51.63	37.53	1,937	100
200	19.64	0.0509	0.0008	0.0158	1,243	63.27	55.94	3,539	200

Table C-7 1¾% CFIs

n	F/P	P/F	A/F	A/P	F/A	P/A	A/G	P/G	n
1	1.018	0.9828	1.000	1.018	1.000	0.9830	0.0000	0.0000	1
2	1.035	0.9659	0.4957	0.5132	2.018	1.949	0.4960	0.9660	2
3	1.053	0.9493	0.3276	0.3451	3.053	2.898	0.9880	2.864	3
4	1.072	0.9330	0.2435	0.2610	4.106	3.831	1.478	5.663	4
5	1.091	0.9169	0.1931	0.2106	5.178	4.748	1.965	9.331	5
6	1.110	0.9011	0.1595	0.1770	6.269	5.649	2.449	13.84	6
7	1.129	0.8856	0.1355	0.1530	7.378	6.535	2.931	19.15	7
8	1.149	0.8704	0.1175	0.1350	8.508	7.405	3.409	25.24	8
9	1.169	0.8554	0.1036	0.1211	9.656	8.260	3.884	32.09	9
10	1.189	0.8407	0.0924	0.1099	10.83	9.101	4.357	39.65	10
11	1.210	0.8263	0.0832	0.1007	12.02	9.927	4.827	47.92	11
12	1.231	0.8121	0.0756	0.0931	13.23	10.74	5.293	56.85	12
13	1.253	0.7981	0.0692	0.0867	14.46	11.54	5.757	66.43	13
14	1.275	0.7844	0.0637	0.0812	15.71	12.32	6.218	76.62	14
15	1.297	0.7709	0.0589	0.0764	16.98	13.09	6.677	87.42	15
16	1.320	0.7576	0.0547	0.0722	18.28	13.85	7.132	98.78	16
17	1.343	0.7446	0.0510	0.0685	19.60	14.60	7.584	110.7	17
18	1.367	0.7318	0.0477	0.0652	20.95	15.33	8.034	123.1	18
19	1.390	0.7192	0.0448	0.0623	22.31	16.05	8.480	136.1	19
20	1.415	0.7068	0.0422	0.0597	23.70	16.75	8.924	149.5	20
21	1.440	0.6947	0.0398	0.0573	25.12	17.45	9.365	163.4	21
22	1.465	0.6827	0.0377	0.0552	26.56	18.13	9.803	177.7	22
23	1.490	0.6710	0.0357	0.0532	28.02	18.80	10.24	192.5	23
24	1.516	0.6594	0.0339	0.0514	29.51	19.46	10.67	207.7	24
25	1.543	0.6481	0.0322	0.0497	31.03	20.11	11.10	223.2	25
26	1.570	0.6370	0.0307	0.0482	32.57	20.75	11.53	239.1	26
27	1.597	0.6260	0.0293	0.0468	34.14	21.37	11.95	255.4	27
28	1.625	0.6152	0.0280	0.0455	35.74	21.99	12.37	272.0	28
29	1.654	0.6047	0.0268	0.0443	37.36	22.59	12.79	289.0	29
30	1.683	0.5943	0.0256	0.0431	39.02	23.19	13.21	306.2	30
31	1.712	0.5840	0.0246	0.0421	40.70	23.77	13.62	323.7	31
32	1.742	0.5740	0.0236	0.0411	42.41	24.34	14.03	341.5	32
33	1.773	0.5641	0.0227	0.0401	44.15	24.91	14.44	359.6	33
34	1.804	0.5544	0.0218	0.0393	45.93	25.46	14.84	377.9	34
35	1.835	0.5449	0.0210	0.0385	47.73	26.01	15.24	396.4	35
36	1.867	0.5355	0.0202	0.0377	49.57	26.54	15.64	415.1	36
37	1.900	0.5263	0.0194	0.0369	51.43	27.07	16.04	434.1	37
38	1.933	0.5172	0.0188	0.0362	53.33	27.59	16.43	453.2	38
39	1.967	0.5083	0.0181	0.0356	55.27	28.10	16.82	472.5	39
40	2.002	0.4996	0.0175	0.0350	57.23	28.59	17.21	492.0	40
50	2.381	0.4200	0.0127	0.0302	78.90	33.14	20.93	693.7	50
55	2.597	0.3851	0.0110	0.0285	91.23	35.14	22.69	797.3	55
60	2.832	0.3531	0.0096	0.0271	104.7	36.96	24.39	901.5	60
65	3.088	0.3238	0.0084	0.0259	119.3	38.64	26.02	1,005	65
70	3.368	0.2969	0.0074	0.0249	135.3	40.18	27.59	1,108	70
75	3.674	0.2722	0.0066	0.0240	152.8	41.59	29.09	1,210	75
80	4.006	0.2496	0.0058	0.0233	171.8	42.88	30.53	1,309	80
85	4.369	0.2289	0.0052	0.0227	192.5	44.07	31.92	1,406	85
90	4.765	0.2099	0.0047	0.0221	215.2	45.15	33.24	1,501	90
95	5.197	0.1924	0.0042	0.0217	239.8	46.15	34.51	1,593	95
100	5.668	0.1764	0.0038	0.0212	266.8	47.06	35.72	1,681	100
200	32.13	0.0311	0.0006	0.0181	1,779	55.36	50.72	2,808	200

Table C-8 2% CFIs

n	F/P	P/F	A/F	A/P	F/A	P/A	A/G	P/G	n
1	1.020	0.9804	1.000	1.020	1.000	0.9800	0.0000	0.0000	1
2	1.040	0.9612	0.4950	0.5150	2.020	1.942	0.4950	0.9610	2
3	1.061	0.9423	0.3268	0.3468	3.060	2.884	0.9870	2.846	3
4	1.082	0.9238	0.2426	0.2626	4.122	3.808	1.475	5.617	4
5	1.104	0.9057	0.1922	0.2122	5.204	4.713	1.960	9.240	5
6	1.126	0.8880	0.1585	0.1785	6.308	5.601	2.442	13.68	6
7	1.149	0.8706	0.1345	0.1545	7.434	6.472	2.921	18.90	7
8	1.172	0.8535	0.1165	0.1365	8.583	7.325	3.396	24.88	8
9	1.195	0.8368	0.1025	0.1225	9.755	8.162	3.868	31.57	9
10	1.219	0.8203	0.0913	0.1113	10.95	8.983	4.337	38.96	10
11	1.243	0.8043	0.0822	0.1022	12.17	9.787	4.802	47.00	11
12	1.268	0.7885	0.0746	0.0946	13.41	10.58	5.264	55.67	12
13	1.294	0.7730	0.0681	0.0881	14.68	11.35	5.723	64.95	13
14	1.319	0.7579	0.0626	0.0826	15.97	12.11	6.179	74.80	14
15	1.346	0.7430	0.0578	0.0778	17.29	12.85	6.631	85.20	15
16	1.373	0.7284	0.0537	0.0737	18.64	13.58	7.080	96.13	16
17	1.400	0.7142	0.0500	0.0700	20.01	14.29	7.526	107.6	17
18	1.428	0.7002	0.0467	0.0667	21.41	14.99	7.968	119.5	18
19	1.457	0.6864	0.0438	0.0638	22.84	15.68	8.407	131.8	19
20	1.486	0.6730	0.0412	0.0612	24.30	16.35	8.843	144.6	20
21	1.516	0.6598	0.0388	0.0588	25.78	17.01	9.276	157.8	21
22	1.546	0.6468	0.0366	0.0566	27.30	17.66	9.705	171.4	22
23	1.577	0.6342	0.0347	0.0547	28.85	18.29	10.13	185.3	23
24	1.608	0.6217	0.0329	0.0529	30.42	18.91	10.56	199.6	24
25	1.641	0.6095	0.0312	0.0512	32.03	19.52	10.97	214.3	25
26	1.673	0.5976	0.0297	0.0497	33.67	20.12	11.39	229.2	26
27	1.707	0.5859	0.0283	0.0483	35.34	20.71	11.80	244.4	27
28	1.741	0.5744	0.0270	0.0470	37.05	21.28	12.21	259.9	28
29	1.776	0.5631	0.0258	0.0458	38.79	21.84	12.62	275.7	29
30	1.811	0.5521	0.0247	0.0446	40.57	22.40	13.03	291.7	30
31	1.848	0.5413	0.0236	0.0436	42.38	22.94	13.43	308.0	31
32	1.885	0.5306	0.0226	0.0426	44.23	23.47	13.82	324.4	32
33	1.922	0.5202	0.0217	0.0417	46.11	23.99	14.22	341.1	33
34	1.961	0.5100	0.0208	0.0408	48.03	24.50	14.61	357.9	34
35	2.000	0.5000	0.0200	0.0400	49.99	25.00	15.00	374.9	35
36	2.040	0.4902	0.0192	0.0392	51.99	25.49	15.38	392.0	36
37	2.081	0.4806	0.0185	0.0385	54.03	25.97	15.76	409.3	37
38	2.122	0.4712	0.0178	0.0378	56.12	26.44	16.14	426.8	38
39	2.165	0.4620	0.0172	0.0372	58.24	26.90	16.52	444.3	39
40	2.208	0.4529	0.0166	0.0366	60.40	27.36	16.89	462.0	40
50	2.692	0.3715	0.0118	0.0318	84.58	31.42	20.44	642.4	50
55	2.972	0.3365	0.0101	0.0301	98.59	33.18	22.11	733.4	55
60	3.281	0.3048	0.0088	0.0288	114.1	34.76	23.70	823.7	60
65	3.623	0.2761	0.0076	0.0276	131.1	36.20	25.22	912.7	65
70	4.000	0.2500	0.0067	0.0267	150.0	37.50	26.66	1,000	70
75	4.416	0.2265	0.0059	0.0259	170.8	38.68	28.04	1,085	75
80	4.875	0.2051	0.0052	0.0252	193.8	39.75	29.36	1,167	80
85	5.383	0.1858	0.0046	0.0246	219.1	40.71	30.61	1,246	85
90	5.943	0.1683	0.0041	0.0240	247.2	41.59	31.79	1,322	90
95	6.562	0.1524	0.0036	0.0236	278.1	42.38	32.92	1,395	95
100	7.245	0.1380	0.0032	0.0232	312.2	43.10	33.99	1,465	100
200	52.49	0.0191	0.0004	0.0204	2,574	49.05	46.12	2,262	200

Table C-9 2½% CFIs

n	F/P	P/F	A/F	A/P	F/A	P/A	A/G	P/G	n
1	1.025	0.9756	1.000	1.025	1.000	0.9760	0.0000	0.0000	1
2	1.051	0.9518	0.4938	0.5188	2.025	1.927	0.4940	0.9520	2
3	1.077	0.9286	0.3251	0.3501	3.076	2.856	0.9840	2.809	3
4	1.104	0.9060	0.2408	0.2658	4.153	3.762	1.469	5.527	4
5	1.131	0.8839	0.1902	0.2152	5.256	4.646	1.951	9.062	5
6	1.160	0.8623	0.1565	0.1815	6.388	5.508	2.428	13.37	6
7	1.189	0.8413	0.1325	0.1575	7.547	6.349	2.901	18.42	7
8	1.218	0.8207	0.1145	0.1395	8.736	7.170	3.370	24.17	8
9	1.249	0.8007	0.1005	0.1255	9.955	7.971	3.836	30.57	9
10	1.280	0.7812	0.0893	0.1143	11.20	8.752	4.296	37.60	10
11	1.312	0.7621	0.0801	0.1051	12.48	9.514	4.753	45.23	11
12	1.345	0.7436	0.0725	0.0975	13.80	10.26	5.206	53.40	12
13	1.379	0.7254	0.0660	0.0910	15.14	10.98	5.655	62.11	13
14	1.413	0.7077	0.0605	0.0855	16.52	11.69	6.100	71.31	14
15	1.448	0.6905	0.0558	0.0808	17.93	12.38	6.540	80.98	15
16	1.485	0.6736	0.0516	0.0766	19.38	13.06	6.977	91.08	16
17	1.522	0.6572	0.0479	0.0729	20.87	13.71	7.409	101.6	17
18	1.560	0.6412	0.0447	0.0697	22.39	14.35	7.838	112.5	18
19	1.599	0.6255	0.0418	0.0668	23.95	14.98	8.262	123.8	19
20	1.639	0.6103	0.0391	0.0641	25.55	15.59	8.682	135.4	20
21	1.680	0.5954	0.0368	0.0618	27.18	16.19	9.099	147.3	21
22	1.722	0.5809	0.0346	0.0596	28.86	16.77	9.511	159.5	22
23	1.765	0.5667	0.0327	0.0577	30.58	17.33	9.919	171.9	23
24	1.809	0.5529	0.0309	0.0559	32.35	17.89	10.32	184.6	24
25	1.854	0.5394	0.0293	0.0543	34.16	18.42	10.72	197.6	25
26	1.900	0.5262	0.0278	0.0528	36.01	18.95	11.12	210.7	26
27	1.948	0.5134	0.0264	0.0514	37.91	19.46	11.51	224.1	27
28	1.996	0.5009	0.0251	0.0501	39.86	19.97	11.90	237.6	28
29	2.046	0.4887	0.0239	0.0489	41.86	20.45	12.29	251.3	29
30	2.098	0.4767	0.0228	0.0478	43.90	20.93	12.67	265.1	30
31	2.150	0.4651	0.0217	0.0467	46.00	21.40	13.04	279.1	31
32	2.204	0.4538	0.0208	0.0458	48.15	21.85	13.42	293.1	32
33	2.259	0.4427	0.0199	0.0449	50.35	22.29	13.79	307.3	33
34	2.315	0.4319	0.0190	0.0440	52.61	22.72	14.15	321.6	34
35	2.373	0.4214	0.0182	0.0432	54.93	23.15	14.51	335.9	35
36	2.433	0.4111	0.0175	0.0425	57.30	23.56	14.87	350.3	36
37	2.493	0.4011	0.0167	0.0417	59.73	23.96	15.22	364.7	37
38	2.556	0.3913	0.0161	0.0411	62.23	24.35	15.57	379.2	38
39	2.620	0.3817	0.0154	0.0404	64.78	24.73	15.92	393.7	39
40	2.685	0.3724	0.0148	0.0398	67.40	25.10	16.26	408.2	40
50	3.437	0.2909	0.0103	0.0353	97.48	28.36	19.48	552.6	50
55	3.889	0.2572	0.0087	0.0337	115.6	29.71	20.96	622.8	55
60	4.400	0.2273	0.0074	0.0324	136.0	30.91	22.35	690.9	60
65	4.978	0.2009	0.0063	0.0313	159.1	31.97	23.66	756.3	65
70	5.632	0.1776	0.0054	0.0304	185.3	32.90	24.89	818.8	70
75	6.372	0.1569	0.0047	0.0297	214.9	33.72	26.04	878.1	75
80	7.210	0.1387	0.0040	0.0290	248.4	34.45	27.12	934.2	80
85	8.157	0.1226	0.0035	0.0285	286.3	35.10	28.12	987.0	85
90	9.229	0.1084	0.0030	0.0280	329.2	35.67	29.06	1,037	90
95	10.44	0.0958	0.0027	0.0276	377.7	36.17	29.94	1,083	95
100	11.81	0.0847	0.0023	0.0273	432.5	36.61	30.75	1,126	100
200	139.56	0.0072	0.0002	0.0252	5,543	39.71	38.56	1,531	200

Table C-10 3% CFIs

n	F/P	P/F	A/F	A/P	F/A	P/A	A/G	P/G	n
1	1.030	0.9709	1.000	1.030	1.000	0.9710	0.0000	0.0000	1
2	1.061	0.9426	0.4926	0.5226	2.030	1.913	0.4930	0.9430	2
3	1.093	0.9151	0.3235	0.3535	3.091	2.829	0.9800	2.773	3
4	1.126	0.8885	0.2390	0.2690	4.184	3.717	1.463	5.438	4
5	1.159	0.8626	0.1884	0.2184	5.309	4.580	1.941	8.889	5
6	1.194	0.8375	0.1546	0.1846	6.468	5.417	2.414	13.08	6
7	1.230	0.8131	0.1305	0.1605	7.662	6.230	2.882	17.96	7
8	1.267	0.7894	0.1125	0.1425	8.892	7.020	3.345	23.48	8
9	1.305	0.7664	0.0984	0.1284	10.16	7.786	3.803	29.61	9
10	1.344	0.7441	0.0872	0.1172	11.46	8.530	4.256	36.31	10
11	1.384	0.7224	0.0781	0.1081	12.81	9.253	4.705	43.53	11
12	1.426	0.7014	0.0705	0.1005	14.19	9.954	5.148	51.25	12
13	1.469	0.6810	0.0640	0.0940	15.62	10.64	5.587	59.42	13
14	1.513	0.6611	0.0585	0.0885	17.09	11.30	6.021	68.01	14
15	1.558	0.6419	0.0538	0.0838	18.60	11.94	6.450	77.00	15
16	1.605	0.6232	0.0496	0.0796	20.16	12.56	6.874	86.35	16
17	1.653	0.6050	0.0460	0.0760	21.76	13.17	7.294	96.03	17
18	1.702	0.5874	0.0427	0.0727	23.41	13.75	7.708	106.0	18
19	1.754	0.5703	0.0398	0.0698	25.12	14.32	8.118	116.3	19
20	1.806	0.5537	0.0372	0.0672	26.87	14.88	8.523	126.8	20
21	1.860	0.5375	0.0349	0.0649	28.68	15.42	8.923	137.6	21
22	1.916	0.5219	0.0327	0.0627	30.54	15.94	9.319	148.5	22
23	1.974	0.5067	0.0308	0.0608	32.45	16.44	9.709	159.7	23
24	2.033	0.4919	0.0291	0.0590	34.43	16.94	10.10	171.0	24
25	2.094	0.4776	0.0274	0.0574	36.46	17.41	10.48	182.4	25
26	2.157	0.4637	0.0259	0.0559	38.55	17.88	10.85	194.0	26
27	2.221	0.4502	0.0246	0.0546	40.71	18.33	11.23	205.7	27
28	2.288	0.4371	0.0233	0.0533	42.93	18.76	11.59	217.5	28
29	2.357	0.4244	0.0221	0.0521	45.22	19.19	11.96	229.4	29
30	2.427	0.4120	0.0210	0.0510	47.58	19.60	12.31	241.4	30
31	2.500	0.4000	0.0200	0.0500	50.00	20.00	12.67	253.4	31
32	2.575	0.3883	0.0191	0.0490	52.50	20.39	13.02	265.4	32
33	2.652	0.3770	0.0182	0.0482	55.08	20.77	13.36	277.5	33
34	2.732	0.3660	0.0173	0.0473	57.73	21.13	13.70	289.5	34
35	2.814	0.3554	0.0165	0.0465	60.46	21.49	14.04	301.6	35
36	2.898	0.3450	0.0158	0.0458	63.28	21.83	14.37	313.7	36
37	2.985	0.3350	0.0151	0.0451	66.17	22.17	14.70	325.8	37
38	3.075	0.3252	0.0145	0.0445	69.16	22.49	15.02	337.8	38
39	3.167	0.3158	0.0138	0.0438	72.23	22.81	15.34	349.8	39
40	3.262	0.3066	0.0133	0.0433	75.40	23.12	15.65	361.8	40
50	4.384	0.2281	0.0089	0.0389	112.80	25.73	18.56	477.5	50
55	5.082	0.1968	0.0074	0.0373	136.1	26.77	19.86	531.7	55
60	5.892	0.1697	0.0061	0.0361	163.1	27.68	21.07	583.1	60
65	6.830	0.1464	0.0052	0.0351	194.3	28.45	22.18	631.2	65
70	7.918	0.1263	0.0043	0.0343	230.6	29.12	23.22	676.1	70
75	9.179	0.1090	0.0037	0.0337	272.6	29.70	24.16	717.7	75
80	10.64	0.0940	0.0031	0.0331	321.4	30.20	25.04	756.1	80
85	12.34	0.0811	0.0027	0.0326	377.9	30.63	25.84	791.4	85
90	14.30	0.0699	0.0023	0.0323	443.3	31.00	26.57	823.6	90
95	16.58	0.0603	0.0019	0.0319	519.3	31.32	27.24	853.1	95
100	19.22	0.0520	0.0017	0.0316	607.3	31.60	27.84	879.9	100
200	369.36	0.0027	0.0001	0.0301	12,279	33.24	32.79	1,090	200

Table C-11 3.5% 3 ½ % CFI's 3 ½ %

n	F/P	P/F	A/F	A/P	F/A	P/A	A/G	P/G	n
1	1.035	0.9662	1.000	1.035	1.000	0.9660	0.0000	0.0000	1
2	1.071	0.9335	0.4914	0.5264	2.035	1.900	0.4910	0.9340	2
3	1.109	0.9019	0.3219	0.3569	3.106	2.802	0.9770	2.737	3
4	1.148	0.8714	0.2373	0.2723	4.215	3.673	1.457	5.352	4
5	1.188	0.8420	0.1865	0.2215	5.362	4.515	1.931	8.720	5
6	1.229	0.8135	0.1527	0.1877	6.550	5.329	2.400	12.79	6
7	1.272	0.7860	0.1285	0.1635	7.779	6.115	2.863	17.50	7
8	1.317	0.7594	0.1105	0.1455	9.052	6.874	3.320	22.82	8
9	1.363	0.7337	0.0964	0.1314	10.37	7.608	3.771	28.69	9
10	1.411	0.7089	0.0852	0.1202	11.73	8.317	4.217	35.07	10
11	1.460	0.6849	0.0761	0.1111	13.14	9.002	4.657	41.92	11
12	1.511	0.6618	0.0685	0.1035	14.60	9.663	5.091	49.20	12
13	1.564	0.6394	0.0621	0.0971	16.11	10.30	5.520	56.87	13
14	1.619	0.6178	0.0566	0.0916	17.68	10.92	5.943	64.90	14
15	1.675	0.5969	0.0518	0.0868	19.30	11.52	6.361	73.26	15
16	1.734	0.5767	0.0477	0.0827	20.97	12.09	6.773	81.91	16
17	1.795	0.5572	0.0440	0.0790	22.71	12.65	7.179	90.82	17
18	1.857	0.5384	0.0408	0.0758	24.50	13.19	7.580	100.0	18
19	1.923	0.5202	0.0379	0.0729	26.36	13.71	7.975	109.3	19
20	1.990	0.5026	0.0354	0.0704	28.28	14.21	8.365	118.9	20
21	2.059	0.4856	0.0330	0.0680	30.27	14.70	8.749	128.6	21
22	2.132	0.4692	0.0309	0.0659	32.33	15.17	9.128	138.5	22
23	2.206	0.4533	0.0290	0.0640	34.46	15.62	9.502	148.4	23
24	2.283	0.4380	0.0273	0.0623	36.67	16.06	9.870	158.5	24
25	2.363	0.4231	0.0257	0.0607	38.95	16.48	10.23	168.7	25
26	2.446	0.4088	0.0242	0.0592	41.31	16.89	10.59	178.9	26
27	2.532	0.3950	0.0229	0.0579	43.76	17.29	10.94	189.1	27
28	2.620	0.3817	0.0216	0.0566	46.29	17.67	11.29	199.4	28
29	2.712	0.3688	0.0205	0.0554	48.91	18.04	11.63	209.8	29
30	2.807	0.3563	0.0194	0.0544	51.62	18.39	11.97	220.1	30
31	2.905	0.3442	0.0184	0.0534	54.43	18.74	12.30	230.4	31
32	3.007	0.3326	0.0174	0.0524	57.34	19.07	12.63	240.7	32
33	3.112	0.3213	0.0166	0.0516	60.34	19.39	12.95	251.0	33
34	3.221	0.3105	0.0158	0.0508	63.45	19.70	13.26	261.3	34
35	3.334	0.3000	0.0150	0.0500	66.67	20.00	13.57	271.5	35
36	3.450	0.2898	0.0143	0.0493	70.01	20.29	13.88	281.6	36
37	3.571	0.2800	0.0136	0.0486	73.46	20.57	14.18	291.7	37
38	3.696	0.2706	0.0130	0.0480	77.03	20.84	14.48	301.7	38
39	3.825	0.2614	0.0124	0.0474	80.73	21.10	14.77	311.6	39
40	3.959	0.2526	0.0118	0.0468	84.55	21.36	15.06	321.5	40
50	5.585	0.1791	0.0076	0.0426	131.0	23.46	17.67	414.4	50
55	6.633	0.1508	0.0062	0.0412	160.9	24.26	18.81	456.4	55
60	7.878	0.1269	0.0051	0.0401	196.5	24.95	19.85	495.1	60
65	9.357	0.1069	0.0042	0.0392	238.8	25.52	20.79	530.6	65
70	11.11	0.0900	0.0035	0.0385	288.9	26.00	21.65	562.9	70
75	13.20	0.0758	0.0029	0.0379	348.5	26.41	22.42	592.1	75
80	15.68	0.0638	0.0024	0.0374	419.3	26.75	23.12	618.4	80
85	18.62	0.0537	0.0020	0.0370	503.4	27.04	23.75	642.0	85
90	22.11	0.0452	0.0017	0.0367	603.2	27.28	24.31	663.1	90
95	26.26	0.0381	0.0014	0.0364	721.8	27.48	24.81	681.9	95
100	31.19	0.0321	0.0012	0.0362	862.6	27.66	25.26	698.6	100
200	972.9	0.0010	0.0000	0.0350	27,769	28.54	28.37	809.6	200

Table C-12 4% CFIs

n	F/P	P/F	A/F	A/P	F/A	P/A	A/G	P/G	n
1	1.040	0.9615	1.000	1.040	1.000	0.9620	0.0000	0.0000	1
2	1.082	0.9246	0.4902	0.5302	2.040	1.886	0.4900	0.9250	2
3	1.125	0.8890	0.3203	0.3603	3.122	2.775	0.9740	2.703	3
4	1.170	0.8548	0.2355	0.2755	4.246	3.630	1.451	5.267	4
5	1.217	0.8219	0.1846	0.2246	5.416	4.452	1.922	8.555	5
6	1.265	0.7903	0.1508	0.1908	6.633	5.242	2.386	12.51	6
7	1.316	0.7599	0.1266	0.1666	7.898	6.002	2.843	17.07	7
8	1.369	0.7307	0.1085	0.1485	9.214	6.733	3.294	22.18	8
9	1.423	0.7026	0.0945	0.1345	10.58	7.435	3.739	27.80	9
10	1.480	0.6756	0.0833	0.1233	12.01	8.111	4.177	33.88	10
11	1.539	0.6496	0.0741	0.1141	13.49	8.760	4.609	40.38	11
12	1.601	0.6246	0.0666	0.1066	15.03	9.385	5.034	47.25	12
13	1.665	0.6006	0.0601	0.1001	16.63	9.986	5.453	54.46	13
14	1.732	0.5775	0.0547	0.0947	18.29	10.56	5.866	61.96	14
15	1.801	0.5553	0.0499	0.0899	20.02	11.12	6.272	69.74	15
16	1.873	0.5339	0.0458	0.0858	21.83	11.65	6.672	77.74	16
17	1.948	0.5134	0.0422	0.0822	23.70	12.17	7.066	85.96	17
18	2.026	0.4936	0.0390	0.0790	25.65	12.66	7.453	94.35	18
19	2.107	0.4746	0.0361	0.0761	27.67	13.13	7.834	102.9	19
20	2.191	0.4564	0.0336	0.0736	29.78	13.59	8.209	111.6	20
21	2.279	0.4388	0.0313	0.0713	31.97	14.03	8.578	120.3	21
22	2.370	0.4220	0.0292	0.0692	34.25	14.45	8.941	129.2	22
23	2.465	0.4057	0.0273	0.0673	36.62	14.86	9.297	138.1	23
24	2.563	0.3901	0.0256	0.0656	39.08	15.25	9.648	147.1	24
25	2.666	0.3751	0.0240	0.0640	41.65	15.62	9.993	156.1	25
26	2.772	0.3607	0.0226	0.0626	44.31	15.98	10.33	165.1	26
27	2.883	0.3468	0.0212	0.0612	47.08	16.33	10.66	174.1	27
28	2.999	0.3335	0.0200	0.0600	49.97	16.66	10.99	183.1	28
29	3.119	0.3207	0.0189	0.0589	52.97	16.98	11.31	192.1	29
30	3.243	0.3083	0.0178	0.0578	56.09	17.29	11.63	201.1	30
31	3.373	0.2965	0.0169	0.0569	59.33	17.59	11.94	210.0	31
32	3.508	0.2851	0.0160	0.0559	62.70	17.87	12.24	218.8	32
33	3.648	0.2741	0.0151	0.0551	66.21	18.15	12.54	227.6	33
34	3.794	0.2636	0.0143	0.0543	69.86	18.41	12.83	236.3	34
35	3.946	0.2534	0.0136	0.0536	73.65	18.67	13.12	244.9	35
36	4.104	0.2437	0.0129	0.0529	77.60	18.91	13.40	253.4	36
37	4.268	0.2343	0.0122	0.0522	81.70	19.14	13.68	261.8	37
38	4.439	0.2253	0.0116	0.0516	85.97	19.37	13.95	270.2	38
39	4.616	0.2166	0.0111	0.0511	90.41	19.58	14.22	278.4	39
40	4.801	0.2083	0.0105	0.0505	95.03	19.79	14.48	286.5	40
50	7.107	0.1407	0.0066	0.0466	152.7	21.48	16.81	361.2	50
55	8.646	0.1157	0.0052	0.0452	191.2	22.11	17.81	393.7	55
60	10.52	0.0951	0.0042	0.0442	238.0	22.62	18.70	423.0	60
65	12.80	0.0781	0.0034	0.0434	295.0	23.05	19.49	449.2	65
70	15.57	0.0642	0.0028	0.0427	364.3	23.40	20.20	472.5	70
75	18.95	0.0528	0.0022	0.0422	448.6	23.68	20.82	493.0	75
80	23.05	0.0434	0.0018	0.0418	551.2	23.92	21.37	511.1	80
85	28.04	0.0357	0.0015	0.0415	676.1	24.11	21.86	526.9	85
90	34.12	0.0293	0.0012	0.0412	828.0	24.27	22.28	540.7	90
95	41.51	0.0241	0.0010	0.0410	1,013	24.40	22.66	552.7	95
100	50.51	0.0198	0.0008	0.0408	1,238	24.51	22.98	563.1	100
200	2551	0.0004	0.0000	0.0400	63,744	24.99	24.92	622.8	200

Table C-13 4½% CFIs

n	F/P	P/F	A/F	A/P	F/A	P/A	A/G	P/G	n
1	1.045	0.9569	1.000	1.045	1.000	0.9570	0.0000	0.0000	1
2	1.092	0.9157	0.4890	0.5340	2.045	1.873	0.4890	0.9160	2
3	1.141	0.8763	0.3188	0.3638	3.137	2.749	0.9710	2.668	3
4	1.193	0.8386	0.2337	0.2787	4.278	3.588	1.445	5.184	4
5	1.246	0.8025	0.1828	0.2278	5.471	4.390	1.912	8.394	5
6	1.302	0.7679	0.1489	0.1939	6.717	5.158	2.372	12.23	6
7	1.361	0.7348	0.1247	0.1697	8.019	5.893	2.824	16.64	7
8	1.422	0.7032	0.1066	0.1516	9.380	6.596	3.269	21.57	8
9	1.486	0.6729	0.0926	0.1376	10.80	7.269	3.707	26.95	9
10	1.553	0.6439	0.0814	0.1264	12.29	7.913	4.138	32.74	10
11	1.623	0.6162	0.0722	0.1172	13.84	8.529	4.562	38.91	11
12	1.696	0.5897	0.0647	0.1097	15.46	9.119	4.978	45.39	12
13	1.772	0.5643	0.0583	0.1033	17.16	9.683	5.387	52.16	13
14	1.852	0.5400	0.0528	0.0978	18.93	10.22	5.789	59.18	14
15	1.935	0.5167	0.0481	0.0931	20.78	10.74	6.184	66.42	15
16	2.022	0.4945	0.0440	0.0890	22.72	11.23	6.572	73.83	16
17	2.113	0.4732	0.0404	0.0854	24.74	11.71	6.953	81.40	17
18	2.208	0.4528	0.0372	0.0822	26.86	12.16	7.327	89.10	18
19	2.308	0.4333	0.0344	0.0794	29.06	12.59	7.695	96.90	19
20	2.412	0.4146	0.0319	0.0769	31.37	13.01	8.055	104.8	20
21	2.520	0.3968	0.0296	0.0746	33.78	13.41	8.409	112.7	21
22	2.634	0.3797	0.0275	0.0725	36.30	13.78	8.755	120.7	22
23	2.752	0.3634	0.0257	0.0707	38.94	14.15	9.096	128.7	23
24	2.876	0.3477	0.0240	0.0690	41.69	14.50	9.429	136.7	24
25	3.005	0.3327	0.0224	0.0674	44.57	14.83	9.756	144.7	25
26	3.141	0.3184	0.0210	0.0660	47.57	15.15	10.08	152.6	26
27	3.282	0.3047	0.0197	0.0647	50.71	15.45	10.39	160.5	27
28	3.430	0.2916	0.0185	0.0635	53.99	15.74	10.70	168.4	28
29	3.584	0.2790	0.0174	0.0624	57.42	16.02	11.00	176.2	29
30	3.745	0.2670	0.0164	0.0614	61.01	16.29	11.30	184.0	30
31	3.914	0.2555	0.0154	0.0604	64.75	16.54	11.58	191.6	31
32	4.090	0.2445	0.0146	0.0596	68.67	16.79	11.87	199.2	32
33	4.274	0.2340	0.0137	0.0587	72.76	17.02	12.14	206.7	33
34	4.466	0.2239	0.0130	0.0580	77.03	17.25	12.41	214.1	34
35	4.667	0.2143	0.0123	0.0573	81.50	17.46	12.68	221.4	35
36	4.877	0.2050	0.0116	0.0566	86.16	17.67	12.94	228.6	36
37	5.097	0.1962	0.0110	0.0560	91.04	17.86	13.19	235.6	37
38	5.326	0.1878	0.0104	0.0554	96.14	18.05	13.44	242.6	38
39	5.566	0.1797	0.0099	0.0549	101.5	18.23	13.68	249.4	39
40	5.816	0.1719	0.0093	0.0543	107.0	18.40	13.92	256.1	40
50	9.033	0.1107	0.0056	0.0506	178.5	19.76	16.00	316.1	50
55	11.26	0.0888	0.0044	0.0494	227.9	20.25	16.86	341.4	55
60	14.03	0.0713	0.0035	0.0485	289.5	20.64	17.62	363.6	60
65	17.48	0.0572	0.0027	0.0477	366.2	20.95	18.28	382.9	65
70	21.78	0.0459	0.0022	0.0472	461.9	21.20	18.85	399.8	70
75	27.15	0.0368	0.0017	0.0467	581.0	21.40	19.35	414.2	75
80	33.83	0.0296	0.0014	0.0464	729.6	21.57	19.79	426.7	80
85	42.16	0.0237	0.0011	0.0461	914.6	21.70	20.16	437.3	85
90	52.54	0.0190	0.0009	0.0459	1,145	21.80	20.48	446.4	90
95	65.47	0.0153	0.0007	0.0457	1,433	21.88	20.75	454.0	95
100	81.59	0.0123	0.0006	0.0456	1,791	21.95	20.98	460.5	100
200	6657	0.0002	0.0000	0.0450	147,904	22.22	22.19	493.1	200

Table C-14 5% CFIs

n	F/P	P/F	A/F	A/P	F/A	P/A	A/G	P/G	n
1	1.050	0.9524	1.000	1.050	1.000	0.9520	0.0000	0.0000	1
2	1.103	0.9070	0.4878	0.5378	2.050	1.859	0.4880	0.9070	2
3	1.158	0.8638	0.3172	0.3672	3.153	2.723	0.9670	2.635	3
4	1.216	0.8227	0.2320	0.2820	4.310	3.546	1.439	5.103	4
5	1.276	0.7835	0.1810	0.2310	5.526	4.329	1.903	8.237	5
6	1.340	0.7462	0.1470	0.1970	6.802	5.076	2.358	11.97	6
7	1.407	0.7107	0.1228	0.1728	8.142	5.786	2.805	16.23	7
8	1.477	0.6768	0.1047	0.1547	9.549	6.463	3.245	20.97	8
9	1.551	0.6446	0.0907	0.1407	11.03	7.108	3.676	26.13	9
10	1.629	0.6139	0.0795	0.1295	12.58	7.722	4.099	31.65	10
11	1.710	0.5847	0.0704	0.1204	14.21	8.306	4.514	37.50	11
12	1.796	0.5568	0.0628	0.1128	15.92	8.863	4.922	43.62	12
13	1.886	0.5303	0.0565	0.1065	17.71	9.394	5.322	49.99	13
14	1.980	0.5051	0.0510	0.1010	19.60	9.899	5.713	56.55	14
15	2.079	0.4810	0.0463	0.0963	21.58	10.38	6.097	63.29	15
16	2.183	0.4581	0.0423	0.0923	23.66	10.84	6.474	70.16	16
17	2.292	0.4363	0.0387	0.0887	25.84	11.27	6.842	77.14	17
18	2.407	0.4155	0.0355	0.0855	28.13	11.69	7.203	84.20	18
19	2.527	0.3957	0.0327	0.0827	30.54	12.09	7.557	91.33	19
20	2.653	0.3769	0.0302	0.0802	33.07	12.46	7.903	98.49	20
21	2.786	0.3589	0.0280	0.0780	35.72	12.82	8.242	105.7	21
22	2.925	0.3418	0.0260	0.0760	38.51	13.16	8.573	112.8	22
23	3.072	0.3256	0.0241	0.0741	41.43	13.49	8.897	120.0	23
24	3.225	0.3101	0.0225	0.0725	44.50	13.80	9.214	127.1	24
25	3.386	0.2953	0.0210	0.0710	47.73	14.09	9.524	134.2	25
26	3.556	0.2812	0.0196	0.0696	51.11	14.38	9.827	141.3	26
27	3.733	0.2679	0.0183	0.0683	54.67	14.64	10.12	148.2	27
28	3.920	0.2551	0.0171	0.0671	58.40	14.90	10.41	155.1	28
29	4.116	0.2430	0.0161	0.0660	62.32	15.14	10.69	161.9	29
30	4.322	0.2314	0.0151	0.0651	66.44	15.37	10.97	168.6	30
31	4.538	0.2204	0.0141	0.0641	70.76	15.59	11.24	175.2	31
32	4.765	0.2099	0.0133	0.0633	75.30	15.80	11.50	181.7	32
33	5.003	0.1999	0.0125	0.0625	80.06	16.00	11.76	188.1	33
34	5.253	0.1904	0.0118	0.0618	85.07	16.19	12.01	194.4	34
35	5.516	0.1813	0.0111	0.0611	90.32	16.37	12.25	200.6	35
36	5.792	0.1727	0.0104	0.0604	95.84	16.55	12.49	206.6	36
37	6.081	0.1644	0.0098	0.0598	101.6	16.71	12.72	212.5	37
38	6.385	0.1566	0.0093	0.0593	107.7	16.87	12.94	218.3	38
39	6.705	0.1492	0.0088	0.0588	114.1	17.02	13.16	224.0	39
40	7.040	0.1421	0.0083	0.0583	120.8	17.16	13.38	229.5	40
50	11.47	0.0872	0.0048	0.0548	209.3	18.26	15.22	277.9	50
55	14.64	0.0683	0.0037	0.0537	272.7	18.63	15.97	297.5	55
60	18.68	0.0535	0.0028	0.0528	353.6	18.93	16.61	314.3	60
65	23.84	0.0420	0.0022	0.0522	456.8	19.16	17.15	328.7	65
70	30.43	0.0329	0.0017	0.0517	588.5	19.34	17.62	340.8	70
75	38.83	0.0258	0.0013	0.0513	756.7	19.49	18.02	351.1	75
80	49.56	0.0202	0.0010	0.0510	971.2	19.60	18.35	359.6	80
85	63.25	0.0158	0.0008	0.0508	1,245	19.68	18.64	366.8	85
90	80.73	0.0124	0.0006	0.0506	1,595	19.75	18.87	372.7	90
95	103.0	0.0097	0.0005	0.0505	2,041	19.81	19.07	377.7	95
100	131.5	0.0076	0.0004	0.0504	2,610	19.85	19.23	381.7	100
200	17293	0.0001	0.0000	0.0500	345,832	20.00	19.99	399.7	200

Table C-15 6% CFIs

n	F/P	P/F	A/F	A/P	F/A	P/A	A/G	P/G	n
1	1.060	0.9434	1.000	1.060	1.000	0.9430	0.0000	0.0000	1
2	1.124	0.8900	0.4854	0.5454	2.060	1.833	0.4850	0.8900	2
3	1.191	0.8396	0.3141	0.3741	3.184	2.673	0.9610	2.569	3
4	1.262	0.7921	0.2286	0.2886	4.375	3.465	1.427	4.946	4
5	1.338	0.7473	0.1774	0.2374	5.637	4.212	1.884	7.935	5
6	1.419	0.7050	0.1434	0.2034	6.975	4.917	2.330	11.46	6
7	1.504	0.6651	0.1191	0.1791	8.394	5.582	2.768	15.45	7
8	1.594	0.6274	0.1010	0.1610	9.897	6.210	3.195	19.84	8
9	1.689	0.5919	0.0870	0.1470	11.49	6.802	3.613	24.58	9
10	1.791	0.5584	0.0759	0.1359	13.18	7.360	4.022	29.60	10
11	1.898	0.5268	0.0668	0.1268	14.97	7.887	4.421	34.87	11
12	2.012	0.4970	0.0593	0.1193	16.87	8.384	4.811	40.34	12
13	2.133	0.4688	0.0530	0.1130	18.88	8.853	5.192	45.96	13
14	2.261	0.4423	0.0476	0.1076	21.02	9.295	5.564	51.71	14
15	2.397	0.4173	0.0430	0.1030	23.28	9.712	5.926	57.56	15
16	2.540	0.3936	0.0390	0.0990	25.67	10.11	6.279	63.46	16
17	2.693	0.3714	0.0354	0.0954	28.21	10.48	6.624	69.40	17
18	2.854	0.3503	0.0324	0.0924	30.91	10.83	6.960	75.36	18
19	3.026	0.3305	0.0296	0.0896	33.76	11.16	7.287	81.31	19
20	3.207	0.3118	0.0272	0.0872	36.79	11.47	7.605	87.23	20
21	3.400	0.2942	0.0250	0.0850	39.99	11.76	7.915	93.11	21
22	3.604	0.2775	0.0230	0.0830	43.39	12.04	8.217	98.94	22
23	3.820	0.2618	0.0213	0.0813	47.00	12.30	8.510	104.7	23
24	4.049	0.2470	0.0197	0.0797	50.82	12.55	8.795	110.4	24
25	4.292	0.2330	0.0182	0.0782	54.87	12.78	9.072	116.0	25
26	4.549	0.2198	0.0169	0.0769	59.16	13.00	9.341	121.5	26
27	4.822	0.2074	0.0157	0.0757	63.71	13.21	9.603	126.9	27
28	5.112	0.1956	0.0146	0.0746	68.53	13.41	9.857	132.1	28
29	5.418	0.1846	0.0136	0.0736	73.64	13.59	10.10	137.3	29
30	5.743	0.1741	0.0127	0.0726	79.06	13.77	10.34	142.4	30
31	6.088	0.1643	0.0118	0.0718	84.80	13.93	10.57	147.3	31
32	6.453	0.1550	0.0110	0.0710	90.89	14.08	10.80	152.1	32
33	6.841	0.1462	0.0103	0.0703	97.34	14.23	11.02	156.8	33
34	7.251	0.1379	0.0096	0.0696	104.2	14.37	11.23	161.3	34
35	7.686	0.1301	0.0090	0.0690	111.4	14.50	11.43	165.7	35
36	8.147	0.1227	0.0084	0.0684	119.1	14.62	11.63	170.0	36
37	8.636	0.1158	0.0079	0.0679	127.3	14.74	11.82	174.2	37
38	9.154	0.1092	0.0074	0.0674	135.9	14.85	12.01	178.2	38
39	9.704	0.1031	0.0069	0.0669	145.1	14.95	12.19	182.2	39
40	10.29	0.0972	0.0065	0.0665	154.8	15.05	12.36	186.0	40
50	18.42	0.0543	0.0034	0.0634	290.3	15.76	13.80	217.5	50
55	24.65	0.0406	0.0025	0.0625	394.2	15.99	14.34	229.3	55
60	32.99	0.0303	0.0019	0.0619	533.1	16.16	14.79	239.0	60
65	44.15	0.0227	0.0014	0.0614	719.1	16.29	15.16	246.9	65
70	59.08	0.0169	0.0010	0.0610	967.9	16.39	15.46	253.3	70
75	79.06	0.0127	0.0008	0.0608	1,301	16.46	15.71	258.5	75
80	105.8	0.0095	0.0006	0.0606	1,747	16.51	15.90	262.5	80
85	141.6	0.0071	0.0004	0.0604	2,343	16.55	16.06	265.8	85
90	189.5	0.0053	0.0003	0.0603	3,141	16.58	16.19	268.4	90
95	253.5	0.0039	0.0002	0.0602	4,209	16.60	16.29	270.4	95
100	339.3	0.0030	0.0002	0.0602	5,638	16.62	16.37	272.0	100
200	115126	0.0000	0.0000	0.0600	1,918,748	16.67	16.67	277.7	200

Table C-16 7% CFIs

n	F/P	P/F	A/F	A/P	F/A	P/A	A/G	P/G	n
1	1.070	0.9346	1.000	1.070	1.000	0.9350	0.0000	0.0000	1
2	1.145	0.8734	0.4831	0.5531	2.070	1.808	0.4830	0.8730	2
3	1.225	0.8163	0.3111	0.3811	3.215	2.624	0.9550	2.506	3
4	1.311	0.7629	0.2252	0.2952	4.440	3.387	1.416	4.795	4
5	1.403	0.7130	0.1739	0.2439	5.751	4.100	1.865	7.647	5
6	1.501	0.6663	0.1398	0.2098	7.153	4.767	2.303	10.98	6
7	1.606	0.6227	0.1156	0.1856	8.654	5.389	2.730	14.72	7
8	1.718	0.5820	0.0975	0.1675	10.26	5.971	3.147	18.79	8
9	1.838	0.5439	0.0835	0.1535	11.98	6.515	3.552	23.14	9
10	1.967	0.5083	0.0724	0.1424	13.82	7.024	3.946	27.72	10
11	2.105	0.4751	0.0634	0.1334	15.78	7.499	4.330	32.47	11
12	2.252	0.4440	0.0559	0.1259	17.89	7.943	4.703	37.35	12
13	2.410	0.4150	0.0497	0.1197	20.14	8.358	5.065	42.33	13
14	2.579	0.3878	0.0443	0.1143	22.55	8.745	5.417	47.37	14
15	2.759	0.3624	0.0398	0.1098	25.13	9.108	5.758	52.45	15
16	2.952	0.3387	0.0359	0.1059	27.89	9.447	6.090	57.53	16
17	3.159	0.3166	0.0324	0.1024	30.84	9.763	6.411	62.59	17
18	3.380	0.2959	0.0294	0.0994	34.00	10.06	6.722	67.62	18
19	3.617	0.2765	0.0268	0.0968	37.38	10.34	7.024	72.60	19
20	3.870	0.2584	0.0244	0.0944	41.00	10.59	7.316	77.51	20
21	4.141	0.2415	0.0223	0.0923	44.87	10.84	7.599	82.34	21
22	4.430	0.2257	0.0204	0.0904	49.01	11.06	7.872	87.08	22
23	4.741	0.2109	0.0187	0.0887	53.44	11.27	8.137	91.72	23
24	5.072	0.1971	0.0172	0.0872	58.18	11.47	8.392	96.26	24
25	5.427	0.1842	0.0158	0.0858	63.25	11.65	8.639	100.7	25
26	5.807	0.1722	0.0146	0.0846	68.68	11.83	8.877	105.0	26
27	6.214	0.1609	0.0134	0.0834	74.48	11.99	9.107	109.2	27
28	6.649	0.1504	0.0124	0.0824	80.70	12.14	9.329	113.2	28
29	7.114	0.1406	0.0115	0.0814	87.35	12.28	9.543	117.2	29
30	7.612	0.1314	0.0106	0.0806	94.46	12.41	9.749	121.0	30
31	8.145	0.1228	0.0098	0.0798	102.1	12.53	9.947	124.7	31
32	8.715	0.1147	0.0091	0.0791	110.2	12.65	10.14	128.2	32
33	9.325	0.1072	0.0084	0.0784	118.9	12.75	10.32	131.6	33
34	9.978	0.1002	0.0078	0.0778	128.3	12.85	10.50	135.0	34
35	10.68	0.0937	0.0072	0.0772	138.2	12.95	10.67	138.1	35
36	11.42	0.0875	0.0067	0.0767	148.9	13.04	10.83	141.2	36
37	12.22	0.0818	0.0062	0.0762	160.3	13.12	10.99	144.1	37
38	13.08	0.0765	0.0058	0.0758	172.6	13.19	11.14	147.0	38
39	14.00	0.0715	0.0054	0.0754	185.6	13.27	11.29	149.7	39
40	14.97	0.0668	0.0050	0.0750	199.6	13.33	11.42	152.3	40
50	29.46	0.0340	0.0025	0.0725	406.5	13.80	12.53	172.9	50
55	41.32	0.0242	0.0017	0.0717	575.9	13.94	12.92	180.1	55
60	57.95	0.0173	0.0012	0.0712	813.5	14.04	13.23	185.8	60
65	81.27	0.0123	0.0009	0.0709	1,147	14.11	13.48	190.1	65
70	114.0	0.0088	0.0006	0.0706	1,614	14.16	13.67	193.5	70
75	159.9	0.0063	0.0004	0.0704	2,270	14.20	13.81	196.1	75
80	224.2	0.0045	0.0003	0.0703	3,189	14.22	13.93	198.1	80
85	314.5	0.0032	0.0002	0.0702	4,479	14.24	14.02	199.6	85
90	441.1	0.0023	0.0002	0.0702	6,287	14.25	14.08	200.7	90
95	618.7	0.0016	0.0001	0.0701	8,824	14.26	14.13	201.6	95
100	867.7	0.0012	0.0001	0.0701	12,382	14.27	14.17	202.2	100
200	752932	0.0000	0.0000	0.0700	10,756,152	14.29	14.29	204.1	200

Table C-17 8% CFIs

n	F/P	P/F	A/F	A/P	F/A	P/A	A/G	P/G	n
1	1.080	0.9259	1.000	1.080	1.000	0.9260	0.0000	0.0000	1
2	1.166	0.8573	0.4808	0.5608	2.080	1.783	0.4810	0.8570	2
3	1.260	0.7938	0.3080	0.3880	3.246	2.577	0.9490	2.445	3
4	1.360	0.7350	0.2219	0.3019	4.506	3.312	1.404	4.650	4
5	1.469	0.6806	0.1705	0.2505	5.867	3.993	1.846	7.372	5
6	1.587	0.6302	0.1363	0.2163	7.336	4.623	2.276	10.52	6
7	1.714	0.5835	0.1121	0.1921	8.923	5.206	2.694	14.02	7
8	1.851	0.5403	0.0940	0.1740	10.64	5.747	3.099	17.81	8
9	1.999	0.5002	0.0801	0.1601	12.49	6.247	3.491	21.81	9
10	2.159	0.4632	0.0690	0.1490	14.49	6.710	3.871	25.98	10
11	2.332	0.4289	0.0601	0.1401	16.65	7.139	4.240	30.27	11
12	2.518	0.3971	0.0527	0.1327	18.98	7.536	4.596	34.63	12
13	2.720	0.3677	0.0465	0.1265	21.50	7.904	4.940	39.05	13
14	2.937	0.3405	0.0413	0.1213	24.22	8.244	5.273	43.47	14
15	3.172	0.3152	0.0368	0.1168	27.15	8.559	5.594	47.89	15
16	3.426	0.2919	0.0330	0.1130	30.32	8.851	5.905	52.26	16
17	3.700	0.2703	0.0296	0.1096	33.75	9.122	6.204	56.59	17
18	3.996	0.2502	0.0267	0.1067	37.45	9.372	6.492	60.84	18
19	4.316	0.2317	0.0241	0.1041	41.45	9.604	6.770	65.01	19
20	4.661	0.2145	0.0219	0.1019	45.76	9.818	7.037	69.09	20
21	5.034	0.1987	0.0198	0.0998	50.42	10.02	7.294	73.06	21
22	5.437	0.1839	0.0180	0.0980	55.46	10.20	7.541	76.93	22
23	5.871	0.1703	0.0164	0.0964	60.89	10.37	7.779	80.67	23
24	6.341	0.1577	0.0150	0.0950	66.77	10.53	8.007	84.30	24
25	6.848	0.1460	0.0137	0.0937	73.11	10.68	8.225	87.80	25
26	7.396	0.1352	0.0125	0.0925	79.95	10.81	8.435	91.18	26
27	7.988	0.1252	0.0115	0.0914	87.35	10.94	8.636	94.44	27
28	8.627	0.1159	0.0105	0.0905	95.34	11.05	8.829	97.57	28
29	9.317	0.1073	0.0096	0.0896	104.0	11.16	9.013	100.6	29
30	10.06	0.0994	0.0088	0.0888	113.3	11.26	9.190	103.5	30
31	10.87	0.0920	0.0081	0.0881	123.3	11.35	9.358	106.2	31
32	11.74	0.0852	0.0075	0.0875	134.2	11.44	9.520	108.9	32
33	12.68	0.0789	0.0069	0.0869	146.0	11.51	9.674	111.4	33
34	13.69	0.0731	0.0063	0.0863	158.6	11.59	9.821	113.8	34
35	14.79	0.0676	0.0058	0.0858	172.3	11.66	9.961	116.1	35
36	15.97	0.0626	0.0053	0.0853	187.1	11.72	10.10	118.3	36
37	17.25	0.0580	0.0049	0.0849	203.1	11.78	10.22	120.4	37
38	18.63	0.0537	0.0045	0.0845	220.3	11.83	10.34	122.4	38
39	20.12	0.0497	0.0042	0.0842	238.9	11.88	10.46	124.2	39
40	21.73	0.0460	0.0039	0.0839	259.1	11.93	10.57	126.0	40
50	46.90	0.0213	0.0017	0.0817	573.8	12.23	11.41	139.6	50
55	68.91	0.0145	0.0012	0.0812	848.9	12.32	11.69	144.0	55
60	101.3	0.0099	0.0008	0.0808	1,253	12.38	11.90	147.3	60
65	148.8	0.0067	0.0005	0.0805	1,847	12.42	12.06	149.7	65
70	218.6	0.0046	0.0004	0.0804	2,720	12.44	12.18	151.5	70
75	321.2	0.0031	0.0003	0.0802	4,003	12.46	12.27	152.8	75
80	472.0	0.0021	0.0002	0.0802	5,887	12.47	12.33	153.8	80
85	693.5	0.0014	0.0001	0.0801	8,656	12.48	12.38	154.5	85
90	1019	0.0010	0.0001	0.0801	12,724	12.49	12.41	155.0	90
95	1497	0.0007	0.0001	0.0801	18,702	12.49	12.44	155.4	95
100	2200	0.0005	0.0000	0.0800	27,485	12.49	12.46	155.6	100
200	4838950	0.0000	0.0000	0.0800	60,486,857	12.50	12.50	156.2	200

Table C-18 9% CFIs

n	F/P	P/F	A/F	A/P	F/A	P/A	A/G	P/G	n
1	1.090	0.9174	1.000	1.090	1.000	0.9170	0.0000	0.0000	1
2	1.188	0.8417	0.4785	0.5685	2.090	1.759	0.4780	0.8420	2
3	1.295	0.7722	0.3051	0.3951	3.278	2.531	0.9430	2.386	3
4	1.412	0.7084	0.2187	0.3087	4.573	3.240	1.393	4.511	4
5	1.539	0.6499	0.1671	0.2571	5.985	3.890	1.828	7.111	5
6	1.677	0.5963	0.1329	0.2229	7.523	4.486	2.250	10.09	6
7	1.828	0.5470	0.1087	0.1987	9.200	5.033	2.657	13.38	7
8	1.993	0.5019	0.0907	0.1807	11.03	5.535	3.051	16.89	8
9	2.172	0.4604	0.0768	0.1668	13.02	5.995	3.431	20.57	9
10	2.367	0.4224	0.0658	0.1558	15.19	6.418	3.798	24.37	10
11	2.580	0.3875	0.0569	0.1469	17.56	6.805	4.151	28.25	11
12	2.813	0.3555	0.0497	0.1397	20.14	7.161	4.491	32.16	12
13	3.066	0.3262	0.0436	0.1336	22.95	7.487	4.818	36.07	13
14	3.342	0.2992	0.0384	0.1284	26.02	7.786	5.133	39.96	14
15	3.642	0.2745	0.0341	0.1241	29.36	8.061	5.435	43.81	15
16	3.970	0.2519	0.0303	0.1203	33.00	8.313	5.724	47.59	16
17	4.328	0.2311	0.0270	0.1170	36.97	8.544	6.002	51.28	17
18	4.717	0.2120	0.0242	0.1142	41.30	8.756	6.269	54.89	18
19	5.142	0.1945	0.0217	0.1117	46.02	8.950	6.524	58.39	19
20	5.604	0.1784	0.0195	0.1095	51.16	9.129	6.767	61.78	20
21	6.109	0.1637	0.0176	0.1076	56.77	9.292	7.001	65.05	21
22	6.659	0.1502	0.0159	0.1059	62.87	9.442	7.223	68.21	22
23	7.258	0.1378	0.0144	0.1044	69.53	9.580	7.436	71.24	23
24	7.911	0.1264	0.0130	0.1030	76.79	9.707	7.638	74.14	24
25	8.623	0.1160	0.0118	0.1018	84.70	9.823	7.832	76.93	25
26	9.399	0.1064	0.0107	0.1007	93.32	9.929	8.016	79.59	26
27	10.25	0.0976	0.0097	0.0997	102.7	10.03	8.191	82.12	27
28	11.17	0.0896	0.0089	0.0989	113.0	10.12	8.357	84.54	28
29	12.17	0.0822	0.0081	0.0981	124.1	10.20	8.515	86.84	29
30	13.27	0.0754	0.0073	0.0973	136.3	10.27	8.666	89.03	30
31	14.46	0.0692	0.0067	0.0967	149.6	10.34	8.808	91.10	31
32	15.76	0.0634	0.0061	0.0961	164.0	10.41	8.944	93.07	32
33	17.18	0.0582	0.0056	0.0956	179.8	10.46	9.072	94.93	33
34	18.73	0.0534	0.0051	0.0951	197.0	10.52	9.193	96.69	34
35	20.41	0.0490	0.0046	0.0946	215.7	10.57	9.308	98.36	35
36	22.25	0.0449	0.0042	0.0942	236.1	10.61	9.417	99.93	36
37	24.25	0.0412	0.0039	0.0939	258.4	10.65	9.520	101.4	37
38	26.44	0.0378	0.0035	0.0935	282.6	10.69	9.617	102.8	38
39	28.82	0.0347	0.0032	0.0932	309.1	10.73	9.709	104.1	39
40	31.41	0.0318	0.0030	0.0930	337.9	10.76	9.796	105.4	40
50	74.36	0.0135	0.0012	0.0912	815.1	10.96	10.43	114.3	50
55	114.4	0.0087	0.0008	0.0908	1,260	11.01	10.63	117.0	55
60	176.0	0.0057	0.0005	0.0905	1,945	11.05	10.77	119.0	60
65	270.8	0.0037	0.0003	0.0903	2,998	11.07	10.87	120.3	65
70	416.7	0.0024	0.0002	0.0902	4,619	11.08	10.94	121.3	70
75	641.2	0.0016	0.0001	0.0901	7,113	11.09	10.99	122.0	75
80	986.6	0.0010	0.0001	0.0901	10,951	11.10	11.03	122.4	80
85	1518	0.0007	0.0001	0.0901	16,855	11.10	11.06	122.8	85
90	2336	0.0004	0.0000	0.0900	25,939	11.11	11.07	123.0	90
95	3593	0.0003	0.0000	0.0900	39,917	11.11	11.09	123.1	95
100	5529	0.0002	0.0000	0.0900	61,423	11.11	11.09	123.2	100
200	30570292	0.0000	0.0000	0.0900	339,669,901	11.11	11.11	123.5	200

Table C-19 10% CFIs

n	F/P	P/F	A/F	A/P	F/A	P/A	A/G	P/G	n
1	1.100	0.9091	1.000	1.100	1.000	0.9090	0.0000	0.0000	1
2	1.210	0.8264	0.4762	0.5762	2.100	1.736	0.4760	0.8260	2
3	1.331	0.7513	0.3021	0.4021	3.310	2.487	0.9370	2.329	3
4	1.464	0.6830	0.2155	0.3155	4.641	3.170	1.381	4.378	4
5	1.611	0.6209	0.1638	0.2638	6.105	3.791	1.810	6.862	5
6	1.772	0.5645	0.1296	0.2296	7.716	4.355	2.224	9.684	6
7	1.949	0.5132	0.1054	0.2054	9.487	4.868	2.622	12.76	7
8	2.144	0.4665	0.0874	0.1874	11.44	5.335	3.004	16.03	8
9	2.358	0.4241	0.0736	0.1736	13.58	5.759	3.372	19.42	9
10	2.594	0.3855	0.0627	0.1627	15.94	6.145	3.725	22.89	10
11	2.853	0.3505	0.0540	0.1540	18.53	6.495	4.064	26.40	11
12	3.138	0.3186	0.0468	0.1468	21.38	6.814	4.388	29.90	12
13	3.452	0.2897	0.0408	0.1408	24.52	7.103	4.699	33.38	13
14	3.797	0.2633	0.0357	0.1357	27.98	7.367	4.996	36.80	14
15	4.177	0.2394	0.0315	0.1315	31.77	7.606	5.279	40.15	15
16	4.595	0.2176	0.0278	0.1278	35.95	7.824	5.549	43.42	16
17	5.054	0.1978	0.0247	0.1247	40.55	8.022	5.807	46.58	17
18	5.560	0.1799	0.0219	0.1219	45.60	8.201	6.053	49.64	18
19	6.116	0.1635	0.0195	0.1195	51.16	8.365	6.286	52.58	19
20	6.727	0.1486	0.0175	0.1175	57.28	8.514	6.508	55.41	20
21	7.400	0.1351	0.0156	0.1156	64.00	8.649	6.719	58.11	21
22	8.140	0.1228	0.0140	0.1140	71.40	8.772	6.919	60.69	22
23	8.954	0.1117	0.0126	0.1126	79.54	8.883	7.108	63.15	23
24	9.850	0.1015	0.0113	0.1113	88.50	8.985	7.288	65.48	24
25	10.84	0.0923	0.0102	0.1102	98.35	9.077	7.458	67.70	25
26	11.92	0.0839	0.0092	0.1092	109.2	9.161	7.619	69.79	26
27	13.11	0.0763	0.0083	0.1083	121.1	9.237	7.770	71.78	27
28	14.42	0.0693	0.0075	0.1075	134.2	9.307	7.914	73.65	28
29	15.86	0.0630	0.0067	0.1067	148.6	9.370	8.049	75.42	29
30	17.45	0.0573	0.0061	0.1061	164.5	9.427	8.176	77.08	30
31	19.19	0.0521	0.0055	0.1055	181.9	9.479	8.296	78.64	31
32	21.11	0.0474	0.0050	0.1050	201.1	9.526	8.409	80.11	32
33	23.23	0.0431	0.0045	0.1045	222.3	9.569	8.515	81.49	33
34	25.55	0.0391	0.0041	0.1041	245.5	9.609	8.615	82.78	34
35	28.10	0.0356	0.0037	0.1037	271.0	9.644	8.709	83.99	35
36	30.91	0.0323	0.0033	0.1033	299.1	9.677	8.796	85.12	36
37	34.00	0.0294	0.0030	0.1030	330.0	9.706	8.879	86.18	37
38	37.40	0.0267	0.0028	0.1027	364.0	9.733	8.956	87.17	38
39	41.15	0.0243	0.0025	0.1025	401.4	9.757	9.029	88.09	39
40	45.26	0.0221	0.0023	0.1023	442.6	9.779	9.096	88.95	40
50	117.4	0.0085	0.0009	0.1009	1,164	9.915	9.570	94.89	50
55	189.1	0.0053	0.0005	0.1005	1,881	9.947	9.708	96.56	55
60	304.5	0.0033	0.0003	0.1003	3,035	9.967	9.802	97.70	60
65	490.4	0.0020	0.0002	0.1002	4,894	9.980	9.867	98.47	65
70	789.7	0.0013	0.0001	0.1001	7,887	9.987	9.911	98.99	70
75	1272	0.0008	0.0001	0.1001	12,709	9.992	9.941	99.33	75
80	2048	0.0005	0.0001	0.1000	20,474	9.995	9.961	99.56	80
85	3299	0.0003	0.0000	0.1000	32,980	9.997	9.974	99.71	85
90	5313	0.0002	0.0000	0.1000	53,120	9.998	9.983	99.81	90
95	8557	0.0001	0.0000	0.1000	85,557	9.999	9.989	99.88	95
100	13781	0.0001	0.0000	0.1000	137,796	10.00	9.993	99.92	100

Table C-20 12% CFIs

n	F/P	P/F	A/F	A/P	F/A	P/A	A/G	P/G	n
1	1.120	0.8929	1.000	1.120	1.000	0.8930	0.0000	0.0000	1
2	1.254	0.7972	0.4717	0.5917	2.120	1.690	0.4720	0.7970	2
3	1.405	0.7118	0.2963	0.4163	3.374	2.402	0.9250	2.221	3
4	1.574	0.6355	0.2092	0.3292	4.779	3.037	1.359	4.127	4
5	1.762	0.5674	0.1574	0.2774	6.353	3.605	1.775	6.397	5
6	1.974	0.5066	0.1232	0.2432	8.115	4.111	2.172	8.930	6
7	2.211	0.4523	0.0991	0.2191	10.09	4.564	2.551	11.64	7
8	2.476	0.4039	0.0813	0.2013	12.30	4.968	2.913	14.47	8
9	2.773	0.3606	0.0677	0.1877	14.78	5.328	3.257	17.36	9
10	3.106	0.3220	0.0570	0.1770	17.55	5.650	3.585	20.25	10
11	3.479	0.2875	0.0484	0.1684	20.66	5.938	3.895	23.13	11
12	3.896	0.2567	0.0414	0.1614	24.13	6.194	4.190	25.95	12
13	4.363	0.2292	0.0357	0.1557	28.03	6.424	4.468	28.70	13
14	4.887	0.2046	0.0309	0.1509	32.39	6.628	4.732	31.36	14
15	5.474	0.1827	0.0268	0.1468	37.28	6.811	4.980	33.92	15
16	6.130	0.1631	0.0234	0.1434	42.75	6.974	5.215	36.37	16
17	6.866	0.1456	0.0205	0.1405	48.88	7.120	5.435	38.70	17
18	7.690	0.1300	0.0179	0.1379	55.75	7.250	5.643	40.91	18
19	8.613	0.1161	0.0158	0.1358	63.44	7.366	5.838	43.00	19
20	9.646	0.1037	0.0139	0.1339	72.05	7.469	6.020	44.97	20
21	10.80	0.0926	0.0122	0.1322	81.70	7.562	6.191	46.82	21
22	12.10	0.0826	0.0108	0.1308	92.50	7.645	6.351	48.55	22
23	13.55	0.0738	0.0096	0.1296	104.6	7.718	6.501	50.18	23
24	15.18	0.0659	0.0085	0.1285	118.2	7.784	6.641	51.69	24
25	17.00	0.0588	0.0075	0.1275	133.3	7.843	6.771	53.11	25
26	19.04	0.0525	0.0067	0.1267	150.3	7.896	6.892	54.42	26
27	21.33	0.0469	0.0059	0.1259	169.4	7.943	7.005	55.64	27
28	23.88	0.0419	0.0052	0.1252	190.7	7.984	7.110	56.77	28
29	26.75	0.0374	0.0047	0.1247	214.6	8.022	7.207	57.81	29
30	29.96	0.0334	0.0041	0.1241	241.3	8.055	7.297	58.78	30
31	33.56	0.0298	0.0037	0.1237	271.3	8.085	7.381	59.68	31
32	37.58	0.0266	0.0033	0.1233	304.8	8.112	7.459	60.50	32
33	42.09	0.0238	0.0029	0.1229	342.4	8.135	7.530	61.26	33
34	47.14	0.0212	0.0026	0.1226	384.5	8.157	7.596	61.96	34
35	52.80	0.0189	0.0023	0.1223	431.7	8.176	7.658	62.61	35
36	59.14	0.0169	0.0021	0.1221	484.5	8.192	7.714	63.20	36
37	66.23	0.0151	0.0018	0.1218	543.6	8.208	7.766	63.74	37
38	74.18	0.0135	0.0016	0.1216	609.8	8.221	7.814	64.24	38
39	83.08	0.0120	0.0015	0.1215	684.0	8.233	7.858	64.70	39
40	93.05	0.0108	0.0013	0.1213	767.1	8.244	7.899	65.12	40
50	289.0	0.0035	0.0004	0.1204	2,400	8.304	8.160	67.76	50
55	509.3	0.0020	0.0002	0.1202	4,236	8.317	8.225	68.41	55
60	897.6	0.0011	0.0001	0.1201	7,472	8.324	8.266	68.81	60
65	1582	0.0006	0.0001	0.1201	13,174	8.328	8.292	69.06	65
70	2788	0.0004	0.0000	0.1200	23,223	8.330	8.308	69.21	70
75	4913	0.0002	0.0000	0.1200	40,934	8.332	8.318	69.30	75
80	8658	0.0001	0.0000	0.1200	72,146	8.332	8.324	69.36	80
85	15259	0.0001	0.0000	0.1200	127,152	8.333	8.328	69.39	85
90	26892	0.0000	0.0000	0.1200	224,091	8.333	8.330	69.41	90
95	47393	0.0000	0.0000	0.1200	394,931	8.333	8.331	69.43	95
100	83522	0.0000	0.0000	0.1200	696,011	8.333	8.332	69.43	100

Table C-21 15% CFIs

n	F/P	P/F	A/F	A/P	F/A	P/A	A/G	P/G	n
1	1.150	0.8696	1.000	1.150	1.000	0.8700	0.0000	0.0000	1
2	1.323	0.7561	0.4651	0.6151	2.150	1.626	0.4650	0.7560	2
3	1.521	0.6575	0.2880	0.4380	3.473	2.283	0.9070	2.071	3
4	1.749	0.5718	0.2003	0.3503	4.993	2.855	1.326	3.786	4
5	2.011	0.4972	0.1483	0.2983	6.742	3.352	1.723	5.775	5
6	2.313	0.4323	0.1142	0.2642	8.754	3.784	2.097	7.937	6
7	2.660	0.3759	0.0904	0.2404	11.07	4.160	2.450	10.19	7
8	3.059	0.3269	0.0729	0.2229	13.73	4.487	2.781	12.48	8
9	3.518	0.2843	0.0596	0.2096	16.79	4.772	3.092	14.76	9
10	4.046	0.2472	0.0493	0.1993	20.30	5.019	3.383	16.98	10
11	4.652	0.2149	0.0411	0.1911	24.35	5.234	3.655	19.13	11
12	5.350	0.1869	0.0345	0.1845	29.00	5.421	3.908	21.19	12
13	6.153	0.1625	0.0291	0.1791	34.35	5.583	4.144	23.14	13
14	7.076	0.1413	0.0247	0.1747	40.51	5.724	4.362	24.97	14
15	8.137	0.1229	0.0210	0.1710	47.58	5.847	4.565	26.69	15
16	9.358	0.1069	0.0179	0.1679	55.72	5.954	4.752	28.30	16
17	10.76	0.0929	0.0154	0.1654	65.08	6.047	4.925	29.78	17
18	12.38	0.0808	0.0132	0.1632	75.84	6.128	5.084	31.16	18
19	14.23	0.0703	0.0113	0.1613	88.21	6.198	5.231	32.42	19
20	16.37	0.0611	0.0098	0.1598	102.4	6.259	5.365	33.58	20
21	18.82	0.0531	0.0084	0.1584	118.8	6.312	5.488	34.65	21
22	21.65	0.0462	0.0073	0.1573	137.6	6.359	5.601	35.62	22
23	24.89	0.0402	0.0063	0.1563	159.3	6.399	5.704	36.50	23
24	28.63	0.0349	0.0054	0.1554	184.2	6.434	5.798	37.30	24
25	32.92	0.0304	0.0047	0.1547	212.8	6.464	5.883	38.03	25
26	37.86	0.0264	0.0041	0.1541	245.7	6.491	5.961	38.69	26
27	43.54	0.0230	0.0035	0.1535	283.6	6.514	6.032	39.29	27
28	50.07	0.0200	0.0031	0.1531	327.1	6.534	6.096	39.83	28
29	57.58	0.0174	0.0027	0.1527	377.2	6.551	6.154	40.32	29
30	66.21	0.0151	0.0023	0.1523	434.7	6.566	6.207	40.75	30
31	76.14	0.0131	0.0020	0.1520	501.0	6.579	6.254	41.15	31
32	87.57	0.0114	0.0017	0.1517	577.1	6.591	6.297	41.50	32
33	100.7	0.0099	0.0015	0.1515	664.7	6.600	6.336	41.82	33
34	115.8	0.0086	0.0013	0.1513	765.4	6.609	6.371	42.10	34
35	133.2	0.0075	0.0011	0.1511	881.2	6.617	6.402	42.36	35
36	153.2	0.0065	0.0010	0.1510	1,014	6.623	6.430	42.59	36
37	176.1	0.0057	0.0009	0.1509	1,167	6.629	6.455	42.79	37
38	202.5	0.0049	0.0007	0.1507	1,344	6.634	6.478	42.97	38
39	232.9	0.0043	0.0007	0.1506	1,546	6.638	6.499	43.14	39
40	267.9	0.0037	0.0006	0.1506	1,779	6.642	6.517	43.28	40
50	1084	0.0009	0.0001	0.1501	7,218	6.661	6.620	44.10	50
55	2180	0.0005	0.0001	0.1501	14,524	6.664	6.641	44.26	55
60	4384	0.0002	0.0000	0.1500	29,220	6.665	6.653	44.34	60
65	8818	0.0001	0.0000	0.1500	58,779	6.666	6.659	44.39	65
70	17736	0.0001	0.0000	0.1500	118,231	6.666	6.663	44.42	70
75	35673	0.0000	0.0000	0.1500	237,812	6.666	6.665	44.43	75
80	71751	0.0000	0.0000	0.1500	478,333	6.667	6.666	44.44	80
85	144317	0.0000	0.0000	0.1500	962,104	6.667	6.666	44.44	85
90	290272	0.0000	0.0000	0.1500	1,935,142	6.667	6.666	44.44	90
95	583841	0.0000	0.0000	0.1500	3,892,269	6.667	6.667	44.44	95
100	1174313	0.0000	0.0000	0.1500	7,828,750	6.667	6.667	44.44	100

Table C-22 20% CFIs

n	F/P	P/F	A/F	A/P	F/A	P/A	A/G	P/G	n
1	1.200	0.8333	1.000	1.200	1.000	0.8330	0.0000	0.0000	1
2	1.440	0.6944	0.4545	0.6545	2.200	1.528	0.4550	0.6940	2
3	1.728	0.5787	0.2747	0.4747	3.640	2.106	0.8790	1.852	3
4	2.074	0.4823	0.1863	0.3863	5.368	2.589	1.274	3.299	4
5	2.488	0.4019	0.1344	0.3344	7.442	2.991	1.641	4.906	5
6	2.986	0.3349	0.1007	0.3007	9.930	3.326	1.979	6.581	6
7	3.583	0.2791	0.0774	0.2774	12.92	3.605	2.290	8.255	7
8	4.300	0.2326	0.0606	0.2606	16.50	3.837	2.576	9.883	8
9	5.160	0.1938	0.0481	0.2481	20.80	4.031	2.836	11.43	9
10	6.192	0.1615	0.0385	0.2385	25.96	4.192	3.074	12.89	10
11	7.430	0.1346	0.0311	0.2311	32.15	4.327	3.289	14.23	11
12	8.916	0.1122	0.0253	0.2253	39.58	4.439	3.484	15.47	12
13	10.70	0.0935	0.0206	0.2206	48.50	4.533	3.660	16.59	13
14	12.84	0.0779	0.0169	0.2169	59.20	4.611	3.817	17.60	14
15	15.41	0.0649	0.0139	0.2139	72.04	4.675	3.959	18.51	15
16	18.49	0.0541	0.0114	0.2114	87.44	4.730	4.085	19.32	16
17	22.19	0.0451	0.0094	0.2094	105.9	4.775	4.198	20.04	17
18	26.62	0.0376	0.0078	0.2078	128.1	4.812	4.298	20.68	18
19	31.95	0.0313	0.0065	0.2065	154.7	4.843	4.386	21.24	19
20	38.34	0.0261	0.0054	0.2054	186.7	4.870	4.464	21.74	20
21	46.01	0.0217	0.0044	0.2044	225.0	4.891	4.533	22.17	21
22	55.21	0.0181	0.0037	0.2037	271.0	4.909	4.594	22.56	22
23	66.25	0.0151	0.0031	0.2031	326.2	4.925	4.647	22.89	23
24	79.50	0.0126	0.0026	0.2025	392.5	4.937	4.694	23.18	24
25	95.40	0.0105	0.0021	0.2021	472.0	4.948	4.735	23.43	25
26	114.5	0.0087	0.0018	0.2018	567.4	4.956	4.771	23.65	26
27	137.4	0.0073	0.0015	0.2015	681.9	4.964	4.802	23.84	27
28	164.8	0.0061	0.0012	0.2012	819.2	4.970	4.829	24.00	28
29	197.8	0.0051	0.0010	0.2010	984.1	4.975	4.853	24.14	29
30	237.4	0.0042	0.0009	0.2008	1,182	4.979	4.873	24.26	30
31	284.9	0.0035	0.0007	0.2007	1,419	4.982	4.891	24.37	31
32	341.8	0.0029	0.0006	0.2006	1,704	4.985	4.906	24.46	32
33	410.2	0.0024	0.0005	0.2005	2,046	4.988	4.919	24.54	33
34	492.2	0.0020	0.0004	0.2004	2,456	4.990	4.931	24.60	34
35	590.7	0.0017	0.0003	0.2003	2,948	4.992	4.941	24.66	35
36	708.8	0.0014	0.0003	0.2003	3,539	4.993	4.949	24.71	36
37	850.6	0.0012	0.0002	0.2002	4,248	4.994	4.956	24.75	37
38	1021	0.0010	0.0002	0.2002	5,098	4.995	4.963	24.79	38
39	1225	0.0008	0.0002	0.2002	6,119	4.996	4.968	24.82	39
40	1470	0.0007	0.0001	0.2001	7,344	4.997	4.973	24.85	40
50	9100	0.0001	0.0000	0.2000	45,497	4.999	4.995	24.97	50
55	22645	0.0000	0.0000	0.2000	113,219	5.000	4.998	24.99	55
60	56348	0.0000	0.0000	0.2000	281,733	5.000	4.999	24.99	60
65	140211	0.0000	0.0000	0.2000	701,048	5.000	5.000	25.00	65
70	348889	0.0000	0.0000	0.2000	1,744,440	5.000	5.000	25.00	70
75	868147	0.0000	0.0000	0.2000	4,340,732	5.000	5.000	25.00	75
80	2160228	0.0000	0.0000	0.2000	10,801,137	5.000	5.000	25.00	80
85	5375340	0.0000	0.0000	0.2000	26,876,693	5.000	5.000	25.00	85
90	13375565	0.0000	0.0000	0.2000	66,877,821	5.000	5.000	25.00	90
95	33282687	0.0000	0.0000	0.2000	166,413,428	5.000	5.000	25.00	95
100	82817975	0.0000	0.0000	0.2000	414,089,868	5.000	5.000	25.00	100

Table C-23 25% CFIs

n	F/P	P/F	A/F	A/P	F/A	P/A	A/G	P/G	n
1	1.250	0.8000	1.000	1.250	1.000	0.8000	0.0000	0.0000	1
2	1.563	0.6400	0.4444	0.6944	2.250	1.440	0.4440	0.6400	2
3	1.953	0.5120	0.2623	0.5123	3.813	1.952	0.8520	1.664	3
4	2.441	0.4096	0.1734	0.4234	5.766	2.362	1.225	2.893	4
5	3.052	0.3277	0.1218	0.3718	8.207	2.689	1.563	4.204	5
6	3.815	0.2621	0.0888	0.3388	11.26	2.951	1.868	5.514	6
7	4.768	0.2097	0.0663	0.3163	15.07	3.161	2.142	6.773	7
8	5.960	0.1678	0.0504	0.3004	19.84	3.329	2.387	7.947	8
9	7.451	0.1342	0.0388	0.2888	25.80	3.463	2.605	9.021	9
10	9.313	0.1074	0.0301	0.2801	33.25	3.571	2.797	9.987	10
11	11.64	0.0859	0.0235	0.2735	42.57	3.656	2.966	10.85	11
12	14.55	0.0687	0.0184	0.2684	54.21	3.725	3.115	11.60	12
13	18.19	0.0550	0.0145	0.2645	68.76	3.780	3.244	12.26	13
14	22.74	0.0440	0.0115	0.2615	86.95	3.824	3.356	12.83	14
15	28.42	0.0352	0.0091	0.2591	109.7	3.859	3.453	13.33	15
16	35.53	0.0281	0.0072	0.2572	138.1	3.887	3.537	13.75	16
17	44.41	0.0225	0.0058	0.2558	173.6	3.910	3.608	14.11	17
18	55.51	0.0180	0.0046	0.2546	218.0	3.928	3.670	14.42	18
19	69.39	0.0144	0.0037	0.2537	273.6	3.942	3.722	14.67	19
20	86.74	0.0115	0.0029	0.2529	342.9	3.954	3.767	14.89	20
21	108.4	0.0092	0.0023	0.2523	429.7	3.963	3.805	15.08	21
22	135.5	0.0074	0.0019	0.2519	538.1	3.970	3.836	15.23	22
23	169.4	0.0059	0.0015	0.2515	673.6	3.976	3.863	15.36	23
24	211.8	0.0047	0.0012	0.2512	843.0	3.981	3.886	15.47	24
25	264.7	0.0038	0.0010	0.2509	1,055	3.985	3.905	15.56	25
26	330.9	0.0030	0.0008	0.2508	1,319	3.988	3.921	15.64	26
27	413.6	0.0024	0.0006	0.2506	1,650	3.990	3.935	15.70	27
28	517.0	0.0019	0.0005	0.2505	2,064	3.992	3.946	15.75	28
29	646.2	0.0016	0.0004	0.2504	2,581	3.994	3.955	15.80	29
30	807.8	0.0012	0.0003	0.2503	3,227	3.995	3.963	15.83	30
31	1010	0.0010	0.0003	0.2502	4,035	3.996	3.969	15.86	31
32	1262	0.0008	0.0002	0.2502	5,045	3.997	3.975	15.89	32
33	1578	0.0006	0.0002	0.2502	6,307	3.997	3.979	15.91	33
34	1972	0.0005	0.0001	0.2501	7,885	3.998	3.983	15.92	34
35	2465	0.0004	0.0001	0.2501	9,857	3.998	3.986	15.94	35
36	3081	0.0003	0.0001	0.2501	12,322	3.999	3.988	15.95	36
37	3852	0.0003	0.0001	0.2501	15,403	3.999	3.990	15.96	37
38	4815	0.0002	0.0001	0.2501	19,255	3.999	3.992	15.97	38
39	6019	0.0002	0.0000	0.2500	24,070	3.999	3.994	15.97	39
40	7523	0.0001	0.0000	0.2500	30,089	3.999	3.995	15.98	40
50	70065	0.0000	0.0000	0.2500	280,256	4.000	3.999	16.00	50

Table C-24 30% CFIs

n	F/P	P/F	A/F	A/P	F/A	P/A	A/G	P/G	n
1	1.300	0.7692	1.000	1.300	1.000	0.7690	0.0000	0.0000	1
2	1.690	0.5917	0.4348	0.7348	2.300	1.361	0.4350	0.5920	2
3	2.197	0.4552	0.2506	0.5506	3.990	1.816	0.8270	1.502	3
4	2.856	0.3501	0.1616	0.4616	6.187	2.166	1.178	2.552	4
5	3.713	0.2693	0.1106	0.4106	9.043	2.436	1.490	3.630	5
6	4.827	0.2072	0.0784	0.3784	12.76	2.643	1.765	4.666	6
7	6.275	0.1594	0.0569	0.3569	17.58	2.802	2.006	5.622	7

Table C-24 (*Cont.*)

n	F/P	P/F	A/F	A/P	F/A	P/A	A/G	P/G	n
8	8.157	0.1226	0.0419	0.3419	23.86	2.925	2.216	6.480	8
9	10.60	0.0943	0.0312	0.3312	32.02	3.019	2.396	7.234	9
10	13.79	0.0725	0.0235	0.3235	42.62	3.092	2.551	7.887	10
11	17.92	0.0558	0.0177	0.3177	56.41	3.147	2.683	8.445	11
12	23.30	0.0429	0.0135	0.3135	74.33	3.190	2.795	8.917	12
13	30.29	0.0330	0.0102	0.3102	97.63	3.223	2.889	9.314	13
14	39.37	0.0254	0.0078	0.3078	127.9	3.249	2.969	9.644	14
15	51.19	0.0195	0.0060	0.3060	167.3	3.268	3.034	9.917	15
16	66.54	0.0150	0.0046	0.3046	218.5	3.283	3.089	10.14	16
17	86.50	0.0116	0.0035	0.3035	285.0	3.295	3.135	10.33	17
18	112.5	0.0089	0.0027	0.3027	371.5	3.304	3.172	10.48	18
19	146.2	0.0068	0.0021	0.3021	484.0	3.311	3.202	10.60	19
20	190.1	0.0053	0.0016	0.3016	630.2	3.316	3.228	10.70	20
21	247.1	0.0040	0.0012	0.3012	820.2	3.320	3.248	10.78	21
22	321.2	0.0031	0.0009	0.3009	1,067	3.323	3.265	10.85	22
23	417.5	0.0024	0.0007	0.3007	1,388	3.325	3.278	10.90	23
24	542.8	0.0018	0.0006	0.3006	1,806	3.327	3.289	10.94	24
25	705.6	0.0014	0.0004	0.3004	2,349	3.329	3.298	10.98	25
26	917.3	0.0011	0.0003	0.3003	3,054	3.330	3.305	11.01	26
27	1193	0.0008	0.0003	0.3003	3,972	3.331	3.311	11.03	27
28	1550	0.0007	0.0002	0.3002	5,164	3.331	3.315	11.04	28
29	2015	0.0005	0.0002	0.3001	6,715	3.332	3.319	11.06	29
30	2620	0.0004	0.0001	0.3001	8,730	3.332	3.322	11.07	30
31	3406	0.0003	0.0001	0.3001	11,350	3.332	3.324	11.08	31
32	4428	0.0002	0.0001	0.3001	14,756	3.333	3.326	11.09	32
33	5756	0.0002	0.0001	0.3001	19,184	3.333	3.328	11.09	33
34	7483	0.0001	0.0000	0.3000	24,940	3.333	3.329	11.09	34
35	9728	0.0001	0.0000	0.3000	32,423	3.333	3.330	11.10	35
36	12646	0.0001	0.0000	0.3000	42,151	3.333	3.330	11.10	36
37	16440	0.0001	0.0000	0.3000	54,797	3.333	3.331	11.10	37
38	21372	0.0001	0.0000	0.3000	71,237	3.333	3.332	11.11	38
39	27784	0.0000	0.0000	0.3000	92,609	3.333	3.332	11.11	39
40	36119	0.0000	0.0000	0.3000	120,393	3.333	3.332	11.11	40
50	497929	0.0000	0.0000	0.3000	1,659,761	3.333	3.333	11.11	50

Table C-25 40% CFIs

n	F/P	P/F	A/F	A/P	F/A	P/A	A/G	P/G	n
1	1.400	0.7143	1.000	1.400	1.000	0.7140	0.0000	0.0000	1
2	1.960	0.5102	0.4167	0.8167	2.400	1.224	0.4170	0.5100	2
3	2.744	0.3644	0.2294	0.6294	4.360	1.589	0.7800	1.239	3
4	3.842	0.2603	0.1408	0.5408	7.104	1.849	1.092	2.020	4
5	5.378	0.1859	0.0914	0.4914	10.95	2.035	1.358	2.764	5
6	7.530	0.1328	0.0613	0.4613	16.32	2.168	1.581	3.428	6
7	10.54	0.0949	0.0419	0.4419	23.85	2.263	1.766	3.997	7
8	14.76	0.0678	0.0291	0.4291	34.40	2.331	1.919	4.471	8
9	20.66	0.0484	0.0203	0.4203	49.15	2.379	2.042	4.858	9
10	28.93	0.0346	0.0143	0.4143	69.81	2.414	2.142	5.170	10
11	40.50	0.0247	0.0101	0.4101	98.74	2.438	2.221	5.417	11
12	56.69	0.0176	0.0072	0.4072	139.2	2.456	2.285	5.611	12
13	79.37	0.0126	0.0051	0.4051	195.9	2.469	2.334	5.762	13
14	111.1	0.0090	0.0036	0.4036	275.3	2.478	2.373	5.879	14
15	155.6	0.0064	0.0026	0.4026	386.4	2.484	2.403	5.969	15

n	F/P	P/F	A/F	A/P	F/A	P/A	A/G	P/G	n
16	217.8	0.0046	0.0018	0.4018	542.0	2.489	2.426	6.04	16
17	304.9	0.0033	0.0013	0.4013	759.8	2.492	2.444	6.09	17
18	426.9	0.0023	0.0009	0.4009	1,065	2.494	2.458	6.13	18
19	597.6	0.0017	0.0007	0.4007	1,492	2.496	2.468	6.16	19
20	836.7	0.0012	0.0005	0.4005	2,089	2.497	2.476	6.18	20
21	1171	0.0009	0.0003	0.4003	2,926	2.498	2.482	6.20	21
22	1640	0.0006	0.0002	0.4002	4,097	2.498	2.487	6.21	22
23	2296	0.0004	0.0002	0.4002	5,737	2.499	2.490	6.22	23
24	3214	0.0003	0.0001	0.4001	8,033	2.499	2.493	6.23	24
25	4500	0.0002	0.0001	0.4001	11,247	2.499	2.494	6.24	25
26	6300	0.0002	0.0001	0.4001	15,747	2.500	2.496	6.24	26
27	8820	0.0001	0.0001	0.4000	22,047	2.500	2.497	6.24	27
28	12348	0.0001	0.0000	0.4000	30,867	2.500	2.498	6.24	28
29	17287	0.0001	0.0000	0.4000	43,214	2.500	2.498	6.25	29
30	24201	0.0000	0.0000	0.4000	60,501	2.500	2.499	6.25	30
31	33882	0.0000	0.0000	0.4000	84,703	2.500	2.499	6.25	31
32	47435	0.0000	0.0000	0.4000	118,585	2.500	2.499	6.25	32
33	66409	0.0000	0.0000	0.4000	166,019	2.500	2.500	6.25	33
34	92972	0.0000	0.0000	0.4000	232,428	2.500	2.500	6.25	34
35	130161	0.0000	0.0000	0.4000	325,400	2.500	2.500	6.25	35
36	182226	0.0000	0.0000	0.4000	455,561	2.500	2.500	6.25	36
37	255116	0.0000	0.0000	0.4000	637,787	2.500	2.500	6.25	37
38	357162	0.0000	0.0000	0.4000	892,903	2.500	2.500	6.25	38
39	500027	0.0000	0.0000	0.4000	1,250,065	2.500	2.500	6.25	39
40	700038	0.0000	0.0000	0.4000	1,750,092	2.500	2.500	6.25	40
50	20248916	0.0000	0.0000	0.4000	50,622,288	2.500	2.500	6.25	50

Table C-26 50% CFIs

n	F/P	P/F	A/F	A/P	F/A	P/A	A/G	P/G	n
1	1.500	0.6667	1.000	1.500	1.000	0.6670	0.0000	0.0000	1
2	2.250	0.4444	0.4000	0.9000	2.500	1.111	0.4000	0.4440	2
3	3.375	0.2963	0.2105	0.7105	4.750	1.407	0.7370	1.037	3
4	5.063	0.1975	0.1231	0.6231	8.125	1.605	1.015	1.630	4
5	7.594	0.1317	0.0758	0.5758	13.19	1.737	1.242	2.156	5
6	11.39	0.0878	0.0481	0.5481	20.78	1.824	1.423	2.595	6
7	17.09	0.0585	0.0311	0.5311	32.17	1.883	1.565	2.947	7
8	25.63	0.0390	0.0203	0.5203	49.26	1.922	1.675	3.220	8
9	38.44	0.0260	0.0134	0.5134	74.89	1.948	1.760	3.428	9
10	57.67	0.0173	0.0088	0.5088	113.3	1.965	1.824	3.584	10
11	86.50	0.0116	0.0058	0.5058	171.0	1.977	1.871	3.699	11
12	129.7	0.0077	0.0039	0.5039	257.5	1.985	1.907	3.784	12
13	194.6	0.0051	0.0026	0.5026	387.2	1.990	1.933	3.846	13
14	291.9	0.0034	0.0017	0.5017	581.9	1.993	1.952	3.890	14
15	437.9	0.0023	0.0011	0.5011	873.8	1.995	1.966	3.922	15
16	656.8	0.0015	0.0008	0.5008	1,312	1.997	1.976	3.945	16
17	985.3	0.0010	0.0005	0.5005	1,969	1.998	1.983	3.961	17
18	1478	0.0007	0.0003	0.5003	2,954	1.999	1.988	3.973	18
19	2217	0.0005	0.0002	0.5002	4,432	1.999	1.991	3.981	19
20	3325	0.0003	0.0002	0.5002	6,649	1.999	1.994	3.987	20
21	4988	0.0002	0.0001	0.5001	9,974	2.000	1.996	3.991	21
22	7482	0.0001	0.0001	0.5001	14,962	2.000	1.997	3.994	22
23	11223	0.0001	0.0000	0.5000	22,443	2.000	1.998	3.996	23
24	16834	0.0001	0.0000	0.5000	33,666	2.000	1.999	3.997	24

Table C-24 (Cont.)

n	F/P	P/F	A/F	A/P	F/A	P/A	A/G	P/G	n
25	25251	0.0000	0.0000	0.5000	50,500	2.000	1.999	3.998	25
26	37877	0.0000	0.0000	0.5000	75,752	2.000	1.999	3.999	26
27	56815	0.0000	0.0000	0.5000	113,628	2.000	2.000	3.999	27
28	85223	0.0000	0.0000	0.5000	170,443	2.000	2.000	3.999	28
29	127834	0.0000	0.0000	0.5000	255,666	2.000	2.000	4.000	29
30	191751	0.0000	0.0000	0.5000	383,500	2.000	2.000	4.000	30
31	287627	0.0000	0.0000	0.5000	575,251	2.000	2.000	4.000	31
32	431440	0.0000	0.0000	0.5000	862,878	2.000	2.000	4.000	32
33	647160	0.0000	0.0000	0.5000	1,294,318	2.000	2.000	4.000	33
34	970740	0.0000	0.0000	0.5000	1,941,477	2.000	2.000	4.000	34
35	1456110	0.0000	0.0000	0.5000	2,912,217	2.000	2.000	4.000	35
36	2184164	0.0000	0.0000	0.5000	4,368,327	2.000	2.000	4.000	36
37	3276247	0.0000	0.0000	0.5000	6,552,491	2.000	2.000	4.000	37
38	4914370	0.0000	0.0000	0.5000	9,828,738	2.000	2.000	4.000	38
39	7371555	0.0000	0.0000	0.5000	14,743,108	2.000	2.000	4.000	39
40	11057332	0.0000	0.0000	0.5000	22,114,663	2.000	2.000	4.000	40

Table C-27 Continuous Compounding

Continuous Compounding					
R = yearly nominal rate			**n = number of years**		
Compound Amount			**Present Worth**		
Factor e^rn			**Factor e^−rn**		
r n	**Find F Given P F/P**		**Find P Given F P/F**	**r n**	**Find F Given P F/P**	**Find P Given F P/F**
0.01	1.0101		0.9900			
0.02	1.0202		0.9802	0.52	1.6820	0.5945
0.04	1.0408		0.9608	0.54	1.7160	0.5827
0.06	1.0618		0.9418	0.56	1.7507	0.5712
0.08	1.0833		0.9231	0.58	1.7860	0.5599
0.10	1.1052		0.9048	0.60	1.8221	0.5488
0.12	1.1275		0.8869	0.62	1.8589	0.5379
0.14	1.1503		0.8694	0.64	1.8965	0.5273
0.16	1.1735		0.8521	0.66	1.9348	0.5169
0.18	1.1972		0.8353	0.68	1.9739	0.5066
0.20	1.2214		0.8187	0.70	2.0138	0.4966
0.22	1.2461		0.8025	0.72	2.0544	0.4868
0.24	1.2712		0.7866	0.74	2.0959	0.4771
0.26	1.2969		0.7711	0.76	2.1383	0.4677
0.28	1.3231		0.7558	0.78	2.1815	0.4584
0.30	1.3499		0.7408	0.80	2.2255	0.4493
0.32	1.3771		0.7261	0.82	2.2705	0.4404
0.34	1.4049		0.7118	0.84	2.3164	0.4317
0.36	1.4333		0.6977	0.86	2.3632	0.4232
0.38	1.4623		0.6839	0.88	2.4109	0.4148
0.40	1.4918		0.6703	0.90	2.4596	0.4066
0.42	1.5220		0.6570	0.92	2.5093	0.3985
0.44	1.5527		0.6440	0.94	2.5600	0.3906
0.46	1.5841		0.6313	0.96	2.6117	0.3829
0.48	1.6161		0.6188	0.98	2.6645	0.3753
0.50	1.6487		0.6065	1.00	2.7183	0.3679

Note: 10% compounded for five years continuously = 1.6487 vs. F/P(10%, 5 year) = 1.611.

Index